Environmental Impacts
of Mining

Environmental Impacts
of Mining

Monitoring, Restoration and Control

Mritunjoy Sengupta

CRC Press
Taylor & Francis Group
Boca Raton London New York

CRC Press is an imprint of the
Taylor & Francis Group, an **informa** business

First edition published 2021
by CRC Press
6000 Broken Sound Parkway NW, Suite 300, Boca Raton, FL 33487-2742

and by CRC Press
2 Park Square, Milton Park, Abingdon, Oxon, OX14 4RN

© 2021 Taylor & Francis Group, LLC

CRC Press is an imprint of Taylor & Francis Group, LLC

The right of Mritunjoy Sengupta to be identified as author of this work has been asserted by him in accordance with sections 77 and 78 of the Copyright, Designs and Patents Act 1988.

Reasonable efforts have been made to publish reliable data and information, but the author and publisher cannot assume responsibility for the validity of all materials or the consequences of their use. The authors and publishers have attempted to trace the copyright holders of all material reproduced in this publication and apologize to copyright holders if permission to publish in this form has not been obtained. If any copyright material has not been acknowledged please write and let us know so we may rectify in any future reprint.

Library of Congress Cataloging-in-Publication Data

Names: Sengupta, M., author.
Title: Environmental impacts of mining : monitoring, restoration, and
control / Mritunjoy Sengupta.
Description: Second edition. | Boca Raton, FL : CRC Press, 2021. | Includes
bibliographical references and index.
Identifiers: LCCN 2020049314 (print) | LCCN 2020049315 (ebook) | ISBN
9780367861001 (hbk) | ISBN 9781003164012 (ebk)
Subjects: LCSH: Mineral industries--Environmental aspects.
Classification: LCC TD195.M5 S46 2021 (print) | LCC TD195.M5 (ebook) |
DDC 622.028/6--dc23
LC record available at https://lccn.loc.gov/2020049314
LC ebook record available at https://lccn.loc.gov/2020049315

ISBN: 9780367861001 (hbk)
ISBN: 9780367757892 (pbk)
ISBN: 9781003164012 (ebk)

Typeset in Times
by Deanta Global Publishing Services, Chennai, India

Dedicated to my wife Nupur and my son Shyam Sundar

Contents

Preface

This book has been written to serve as an introductory text to the environmental impacts of mining and their control. It endeavours to fulfil the needs of students and professional engineers concerned with the environmental problems created by mining operations, a subject that has not received comprehensive treatment in other texts. The need for such a contribution is understood by the widespread scattering of literature in government reports, journals and conference proceedings.

Mritunjoy Sengupta
Simi Valley, California

Author Biography

Dr. Mritunjoy Sengupta received his PhD from the Colorado School of Mines. He also holds an E.M. in Mining Engineering and a Master's in Industrial Engineering from the School of Engineering and Applied Science, Columbia University, New York. He also received a B.S. degree in Mining Engineering from the Indian School of Mines. He is a registered professional engineer in Colorado, Idaho, New Mexico and Alaska. His biography is included in *Who's Who in America, Who's Who in Science and Engineering*, and *Who's Who in the World*. Dr. Sengupta was awarded a medal "Among the Outstanding Intellectuals of the 21st Century" from Cambridge, England. He holds a gold medal from the Mining, Metallurgical and Geological Institute of India.

Dr. Sengupta has 10 years of professional experience in the U.S. mining industry. His former employers include AMAX Inc., Continental Oil Company, Texas Gulf Inc., Morrison-Knudsen Company, United Nuclear Corporation and Hawley Coal Mining Company. Dr. Sengupta has served as a scientist with the Central Mining Research Station of the Government of India and as a United Nations expert to the Government of India.

Acknowledgements

The author is grateful to the Society of Mining, Metallurgy, and Exploration and to the Southern Illinois University Press for their kind permission to reproduce materials from their publications.

1 Mining and the Environment

1.1 INTRODUCTION

Mining operations have been seen by environmentalists and conservationists alike as causing problems. Undoubtedly, the operations of metal and coal producers have caused varying degrees of environmental damage in mining areas, which are often located in remote regions. In the urban, suburban, and rural settings of agricultural communities, the operators of rock quarries, gravel pits, and certain industrial mines have been considered the more visible and significant offenders. Much of the concern has been focused on the concurrent and subsequent physical and aesthetic effects that their operations have had on the land—as a basic resource. Mining is only a temporary occupier of the land surface and, hence, is of a transient nature. Although active mines at any particular time are not as widespread as other land uses, they dramatically change the landscape and tend to leave evidence of their past use. Thus, the results of abandonment or closure become most conspicuous to the public. There have been continuous confrontations between citizen groups, governmental agencies, and members of the mining industry. The degree of conflict and its nature usually depended on the current land use and the estimated consequences of proposed disturbances. The conflicts centred on the following issues:

- Destruction of the landscape
- Degradation of the visual environment
- Disturbance of watercourses
- Destruction of agricultural and forest lands
- Damage to recreational lands
- Noise pollution
- Dust
- Truck traffic
- Sedimentation and erosion
- Land subsidence
- Vibration from blasting and air blasts

Environmental conscience has developed dramatically and led to widespread public opinion that governments at all levels should be able to control the depletion of natural resources and excessive environmental damage.

1.2 UNIQUENESS OF MINING

Mineral deposits have fixed locations, so mining activities, unlike renewable resource activities (such as fishing, agriculture, and forestry), are not subject to rational selection or advanced planning. Due to unique physical conditions associated with their location, there is no choice about the characteristics of their ecological setting, the biological and chemical characteristics, mineral composition, or grade of ore in question. All of these factors influence the ultimate design, layout, and size of the operation, as well as the basic environmental problems and the potential longer range of regional impacts. The nature of the ecological setting also determines other land uses or activities that would be affected by the proposed mining.

Mines have a finite life. Because of the non-renewable nature of mineral deposits, mining is only a temporary land user. However, in some situations, ore reserves are so great that mining activities appear to be permanent fixtures in the life of the regional inhabitants.

Mines are usually located in a setting of relatively unspoiled nature. The contrast between the mine itself, its dump, mill, and newly constructed tower and the wooded valley or otherwise unscarred mountainside is always there for all to see. A generation ago, this isolated outpost of industry, winning wealth from untapped nature, was looked on as a symbol of man's ingenuity and as a proud demonstration of progress towards an ever-expanding better future. Now it is looked on by some, perhaps by an increasing number, as a forerunner of the destruction of the environment which supports us and of which we are a part.

In many instances, the original mine is the very reason for the existence of a town. Other means of economic support are generated. Often, because of the isolated nature of these mining towns in forested and lake regions, alternative activities grow in forestry, recreation, and tourism. The latter two usually rely on increased access by a travelling public whose desire for a relatively clear, unspoiled environment contributes to changing viewpoints and increased opposition to mining activities.

The major difference between alternative land uses and mining in isolated forest, barren tundra, or alpine areas is that most alternative land uses are related to renewable resources perceived as less damaging than mining. Mining activities are also related to coal, sand, gravel, stone, and potash deposits underlying prime agricultural lands or bordering expanding urban centres. In these instances, the problem of land allocation is further compounded by political, social, economic, and environmental considerations.

How can the mineral deposits be extracted in the short term without permanently altering the land values for post-mining uses? Alternative land uses with a measurable economic value can present considerable competition for mineral-rich lands or can lead to conflicts. In some cases, no matter how economically valuable a mineral deposit is, the socio-political and environmental factors in opposition are so strong that no new mine development will take place.

Another important aspect of mine development is the time lag. After the initial discovery and evaluation of a potential mine, years of development and construction pass before a mine begins production. Time can have an important effect on the ultimate impact of mining operations on the environment. Unforeseen physical and chemical changes to the environment can emerge at any stage of the life of the mine or even long after closure, despite precautions.

In the recent past, the legislation of new environmental controls and resource management procedures has resulted in a number of significant changes in the traditional approach to both mining and resource development. These developments include the following:

- Environmental impact assessment and public inquiries
- Conditions for "permit" approval
- Resource management and land-use planning
- Land reclamation and rehabilitation

These control measures have created a lengthy and complicated development process in the life of a future mine.

With a growing practice for careful and formalized proposals for all major resource developments, the use of environmental impact assessment procedures as a guide to the development and management of a new project has spread widely. Throughout the development and operation stage of a mine, two parallel processes function: (1) the engineering design, layout, and technological requirements normally associated with mining and (2) an environmental programme that meets all existing regulations and standards for air, water, and land quality. A new mine development process requires the collection of environmental baseline data, an environmental impact assessment, and, sometimes, a socio-economic impact assessment. A series of formal submissions, public hearings, and applications may follow. In most instances, additional constraints designed to reduce the environmental impacts are imposed.

The concern is growing within the industry that the increasing complexity of regulations, policies, and guidelines designed to protect and conserve our resources has led to considerable delay in the development of mines. A fundamental question has been how to balance the requirements of society for continued economic growth with its desire to preserve the environmental quality of the land resource base. This illustrates the intimate relationship between the exploitation and use of mineral resources and the consequences for the environment.

The environment is integrated, and its components are linked by dynamic processes. We cannot use or affect any part without affecting some other parts. No matter how beneficial for our own desired purposes the principal intended results of our activities may be, our actions are bound to cause effects in addition to the principal effects we have in mind. Such additional or unintended effects may not be to our advantage, and then we have the problem of controlling the environmental consequences of our own environmental control.

1.3 ENVIRONMENTAL EFFECTS OF SURFACE COAL MINING

Potentially, many adverse environmental impacts result from area surface mining of coal if no mitigating measures (reclamation practices) are used.[1] Such measures are used with varying degrees of effectiveness throughout the United States.

A summary of types of environmental impacts, causal factors, and mitigating measures is shown in Table 1.1. As shown in that table, air quality, on and off the mine site, may be affected in several ways. Fugitive dust from a variety of sources, such as coal haul roads, unvegetated spoil surfaces, topsoil stockpiles, and coal stockpiles, is a potential problem in areas of low rainfall, high winds, and erodible soils. Dust levels are not dependent upon the mining method to any great extent. Watering of haul roads and stockpiles is a commonly used dust-suppression measure.

Overburden blasting may adversely affect the environment in several ways. Two of these, noise and air shock, are loosely classified as air impacts. Blasting noise and air shock can be troublesome if people live within a radius of several miles (km) of the blasting site. The magnitude of the problem is dependent primarily upon the depth and type of overburden being blasted, the powder factor, the amount of explosive detonated at a given instant, the population density in the vicinity of the blasting site, and the times of day during which blasting takes place. The common practice is to use millisecond delays between rows of blast holes in a given blasting pattern in order to reduce the amount of explosive charge detonated at any given instant. Reduction of the powder factor, that is, use of less explosive per cubic yard of overburden, and restriction of blasting to daylight hours are additional mitigating measures which are sometimes used.

Ground vibration due to overburden blasting is another potential problem where people live near the mine site. The practices used to reduce blasting noise, and air shock also reduces ground vibration levels.

The quality and quantity of surface water and groundwater, both on and off the mine site, can be adversely affected if effective reclamation practices are not used. Sedimentation of surface waters may result from several factors in combination. Inside cast-area mining, the out slope side of the box-cut spoil pile drains externally. Since, immediately after placement of the box-cut spoils the out slope is steep and unvegetated, erosion and transport of sediment from the out slope to external drainage ways may occur, particularly if rainfall is frequent and intense and spoil surface materials are erodible. Procedures that may be used to prevent these kinds of problems include reduction of the out slope angle by grading, construction of terraces or contour ditches on the out slope, mulching and revegetation of the out slope, and placement of sediment basins in drainage ways below the box-cut spoil pile.

Where the side cast method of spoil placement is used, erosion and sedimentation from spoil areas in the second and subsequent cuts during mining generally are not a problem. This is because,

TABLE 1.1
Summary of Possible Environmental Impacts, Causal Factors, and Possible Mitigating Measures: Area Mining

Environment	Environmental Impact Category	Specific Environmental Impact	Uncontrollable Causal Factors	Controllable Causal Factors	Possible Mitigating Measures
Air	Air quality	Fugitive dual	Precipitation (lack of) Wind Soil types	Coal haul road surfaces Haulage road surface areas Surface material (soil) Vegetative density on mined areas	Dust control (watering) on coal haul roads Revegetation of mined areas
	Noise levels	Blasting noise	Overburden characteristics	Spacing and of overburden blast holes Blasting sequence	Delaying shots: hole-to-hole or row-to-row Decking (delaying shots vertically within holes) Use of less explosive charge Use of blasting machines
	Air pressure	Air shock	Same as above	Same as above	Same as above Reduction of spoil grades
Surface water	Physical quality	Erosion and sedimentation	Precipitation Natural topography Natural drainage patterns Natural vegetative density	Length and slope of spoil surfaces Water intercepted by open cut Type of "soil" on mined areas Vegetative density on mined areas	Reduction of length of unbroken spoil slopes Diversion of surface drainage around active mining areas Settlement of suspended solid prior to discharge to natural drainage ways Restoration of approximate original drainage patterns Revegetation of mined areas
	Chemical quality	Acid or mineralized surface water	Precipitation Overburden Geochemistry Overburden stratigraphy	Inversion of overburden materials on spoil piles Water intercepted by open cut Change in permeability (spoil) Length and slope of spoil surface	Identification and selective placement of undesirable overburden materials Drainage diversion Spreading of topsoil on spoil surfaces Grading of spoils soon after placement Chemical water treatment
Groundwater	Quantity	Drawdown	Natural height of water table Rates and directions of natural groundwater flow	Interception of groundwater by open cut	Reduce length of open cut Deepen wells on properties adjoining the mine site

(Continued)

TABLE 1.1 (CONTINUED)
Summary of Possible Environmental Impacts, Causal Factors, and Possible Mitigating Measures: Area Mining

Environment	Environmental Impact Category	Specific Environmental Impact	Uncontrollable Causal Factors	Controllable Causal Factors	Possible Mitigating Measures
		Altered flow rates	Overburden characteristics Aquifer characteristics	Replacement of coal seam aquifer by spoil material Differences in percolation rates for overburden and spoil Open final cut (can be beneficial)	Selective placement of overburden materials Spreading of topsoil on spoil surfaces Backfilling of final cut
	Chemical quality	Acid or mineralized groundwater	Precipitation Nature of aquifer Overburden geochemistry	Same as above	Same as above
Land-use potential	Topography	Major changes in topography (more rugged)	Natural topography Spoil swell factor Coal seam thickness Natural repose angle of spoil	Method of spoil placement Removal of coal (thick coal only)	Spoil grading (during or after mining) Backfilling of final cut
	Drainage	Disruption of natural drainage	Natural drainage patterns Overburden geochemistry Precipitation	Method of spoil placement	Selective placement of overburden materials Spoil grading
	Vegetation	Removal of native vegetation	Native vegetation	Overburden removal	Revegetation
Surface texture		Rocks on spoil surface	Overburden characteristics	Method of spoil placement	Selective placement of overburden materials Spreading of topsoil on spoil surfaces
	Appearance	Changed appearance	Natural topography Natural vegetation	Overburden removal Method of spoil placement	Spoil grading Revegetation Backfilling of final cut

Source: From Cook, F. Evaluation of Current Surface Coal Mining Overburden Handling Techniques and Reclamation Practices, U.S. Department of Commerce, National Technical Information Service, PB-264–111 (1976), pp. 60–123.

before extensive grading of the spoil piles, most drainage in the areas disturbed by mining is internal. That is, surface runoff from spoil piles does not enter external drainage systems unless the water is pumped from the pit into those systems. At large-area mining operations where 10–50 cuts may be made, the out slopes of box-cut spoils, which drain externally, constitute a relatively small percentage of the total mining disturbance.

As mining progresses, side cast spoils are graded to make the topography unsuitable for the planned postmining land use. In practice, this generally means the restoration of approximate original contours. First, in the standard procedure used for grading of side cast spoils, dozers are used to construct roads or flat working areas along the ridgelines of the spoil piles to be graded. Next, working from these roads, spoil is pushed sideways or in a herringbone pattern by the dozers into the vee's between adjacent spoil piles.

As large areas are graded, erosion and transport of sediment to external drainage systems become a potential problem. The magnitude of the problem is dependent upon the lengths and steepness of slopes on graded areas, the frequency and intensity of rainfall, the erodibility of spoil surface materials, and the types and density of vegetative cover on reclaimed areas. The erosion and sediment control procedures that may be used in these areas are identical to those described above in discussing box-cut spoils. These include reduction through grading, terracing, contour ditching, mulching, revegetation, and use of sediment basins.

Diversion of surface water around the active pit and interception of surface water and groundwater by the pit are further potential sources of sedimentation. Drainage diversion systems, which are used primarily for production reasons to keep the pits from flooding, are of two basic types. The first type is diversion or rerouting of a major drainage way, such as a perennial stream crossed by the pit. Standard practice in such cases is to divert the stream around the mining area by constructing a new stream channel. The tendency is to straighten and channelize the stream, thereby increasing the hydraulic gradient and possibly causing head cutting in alluvial material. In order to prevent head cutting and the associated sedimentation, the gradients of diverted streams must be carefully controlled.

A more common kind of drainage diversion used at area mines involves the construction of ditches, usually near the perimeter of the area to be mined, to divert surface runoff around the pit and generally around reclaimed areas as well. These ditches discharge directly onto undisturbed land or into sediment basins. If the hydraulic gradient in the ditches is too great or the ditches are not adequately vegetated, sediment loads in surface water carried by the ditches may increase over natural levels. Additionally, unless some means for dissipation of water energy at the outlet of the ditch is provided, the erosion of the land surface near the outlet and sedimentation may occur. After all mining and reclamation have been completed, the drainage diversion ditches may be removed, or they may be left in place permanently. The choice, among other factors, depends upon prevailing reclamation regulations.

Some water will enter the pit as direct rainfall, surface runoff from the highwall side of the pit, groundwater seepage from the highwall, surface runoff down the inclines, and seepage from the spoil. Water that collects in the pit will usually become heavily sedimented, even if sediment loads in the pit were low prior to the interception of water. One reason is the transit of coal haulage trucks through water that has collected in the pit.

Unless the pit is self-draining, water will be pumped from the pit. If the quality of the pit water is acceptable, the water may be discharged directly onto undisturbed or reclaimed land. More typically, however, pit water is discharged into a diversion ditch, the outlet of which is at or near a sediment basin or basins. Alternately, the pit discharge may be piped all the way to the sediment basin from the pit via plastic tubing.

The chemical quality of surface waters may also be adversely affected if preventive measures are not used. There are two possible causes of chemical pollution of surface water. One, just discussed, is the interception of surface and groundwater by the pit. Where the coal, overburden, spoil, or incline surfaces are high in pyrite or marcasite—as is generally the case when the coal being mined is medium to high in sulphur—or where there are soluble minerals or trace elements in the overburden, spoil, coal, or pit floor, then the collection of water in the pit may lead to chemical changes in that water. These changes may result from oxidation of pyritic materials (thereby increasing the acidity of the water) or from the dissolution of soluble salts in the spoil, overburden, or coal (thereby increasing levels of dissolved solids in the water). The oxidation process requires that both air and water come in contact with pyritic materials, whereas air is not required for dissolution of soluble salts.

Several measures are used to prevent or minimize oxidation and dissolution. The standard ones are to minimize the amount of water that enters the pit, using the procedures described above, and where acid water is a problem, to chemically treat the water in the pit or in an out-of-pit treatment facility prior to its discharge to offsite drainage ways. An additional procedure, discussed below, is to identify and selectively place toxic or otherwise undesirable materials during placement of spoil.

Runoff from graded or ungraded spoil surfaces may be chemically altered if the surface spoil materials are pyritic or high in soluble minerals or trace elements. This can be a problem where undesirable overburden materials are located stratigraphically close to and above the coal seam or seams being mined and the side cast method of spoil placement is used.

A characteristic of this method, in the absence of special handling procedures, is that overburden materials are inverted on the spoil pile. That is, the overburden material stratigraphically just above the coal seam being mined is placed on the top of the spoil piles. If this material is undesirable, as is often the case, the chemical quality of surface runoff and water that percolate into the spoil may be altered.

The problem can be particularly acute where two or more coal seams are mined in one pit using the side cast method of spoil placement. In the absence of special spoil handling procedures, the interburden separating two coal seams (which usually has undesirable chemical or physical characteristics) is placed on the tops of spoil piles. A procedure frequently used to alleviate this problem, with varying degrees of effectiveness, is to bury the undesirable materials during the overburden removal and spoil placement process.

An additional technique used in part to bury undesirable materials, but primarily to aid revegetation of graded lands is to salvage top soiling material prior to overburden removal and then to spread that material over spoil surfaces soon after grading of the spoil piles. Experience indicates that covering acid-producing spoil with as little as 2 ft (0.6 m) of clean fill is generally sufficient to greatly inhibit acid production. Erosion on top soiled spoils must, of course, be controlled, or the top soiling material may be washed away, exposing the underlying toxic material. As mentioned above, revegetation is a major erosion-control practice.

Whether or not topsoil is used where undesirable spoil materials are a potential problem, it is important to grade and revegetate spoil soon after placement. Several factors are affected when these procedures may be used. The most important of these are the proximity of grading activities to the active pit, as well as the width and length of the pit itself.

If mining takes place below the water table, groundwater will be intercepted by the open cut, pumped out, or lost by evaporation, and the water table will be lowered in the mining and adjacent areas. This could result in loss of head or dewatering of wells within a radius of several miles (km) of the pit. Where this situation occurs, responsible coal companies generally remedy it by deepening wells or hauling water to affected residents for as long as groundwater levels are affected by mining. Generally, after mining, the water table will rise close to its original level.

After mining and reclamation have been completed, groundwater quantity can still be affected, although not necessarily adversely. If the mine is located in a groundwater recharge area that is, an area in which surface waters percolate into the groundwater system, the recharge characteristics may be affected because the coal seam has been removed or because the spoil has different permeability characteristics than the overburden. Suppose, for example, that the coal seam itself was an aquifer. Sometimes, water enters or recharges a coal seam aquifer where the seam outcrops. If so, removal of the crop line coal and replacement of coal with spoil may increase or decrease recharge rates. Alternatively, the aquifer may be recharged from above by percolation of surface water or by groundwater from higher-lying aquifers. Replacement of overburden with spoil that has different permeability characteristics from the original overburden might also affect groundwater recharge rates.

Rates of groundwater flow (gallons per minute or litres per minute) may be altered if spoil materials have different physical characteristics from the original overburden or coal strata. If spoil placed below the water table is less permeable than the original aquifer, the groundwater flow rate after completion of mining and reclamation may be lower than that before mining. This possibility is of greatest concern where the coal seam (or seams) being mined are themselves aquifers, which may occur when the coal is thick. This is of concern because the coal seam is replaced stratigraphically by spoil whose characteristics may not be well known.

Oxidation of acid-producing materials is not likely because of the relative lack of airflow in the spoil profile. However, groundwater flow through spoil materials may result in chemical changes in that groundwater due to dissolution of soluble minerals or trace elements. A way to prevent this kind of occurrence is to place the "best" spoil materials below the water table during the spoil placement phase of mining.

It should be noted that the effects of mining on groundwater, as well as on land use and other factors, may be beneficial rather than adverse. For example, groundwater recharge rate, flow rate, or quality may be improved by mining and reclamation. The purpose here, however, is not to present a carefully balanced cost-benefit analysis of surface coal mining, but rather to identify the known problems and the issues.

In addition to potentially affecting air and water, surface mining also affects land or, more specifically, land-use potential. After mining and reclamation, the land-use potential is usually dependent upon topography, drainage, vegetation, surface texture, and appearance.

As described above, inside cast-area mining situations, topography suitable for the planned postmining land use is restored by dozer grading of side cast spoils. During the grading process, drainage ways can be reconstructed so that there will be positive drainage from reclaimed areas. If adequate drainage patterns are not restored, the use potential of the reclaimed land may be diminished because of the formation of bogs (among other adverse effects).

Revegetation of graded areas is a further means of improving land productivity. Burial of undesirable materials during spoil placement, reduction of slopes through grading, replacement of topsoil, fertilization or liming of spoil, seeding, mulching, and irrigation are all practices which may be used to aid in revegetating mined lands. The success of revegetation efforts depends on precipitation patterns, physical and chemical characteristics of surface and near-surface spoil materials, reclaimed topography, and vegetative species mixes.

In arid or semi-arid areas, supplemental irrigation may be required to establish vegetation. Acid-producing spoil materials, if present, must be covered to prevent oxidation and acid formation. Alternatively, or in addition, lime may be used to neutralize the acid. If surface spoil materials are clayey in texture and therefore relatively impermeable, permeability may need to be improved, either by chemical treatment or covering of clayey spoils with suitable top soiling material. Soil fertility may be improved by adding fertilizers. Mulch may be used to control erosion and retain soil moisture.

The texture of surface spoil or soil materials also affects use potential. In the case of timberland, rangeland, and possibly pasture, occasional rocks on land surfaces do not necessarily diminish the use potential. However, in agricultural and some pasture areas, rocks on the surface may degrade the use potential. If overburden, and thus spoil, is unconsolidated, or consolidated spoil materials weather rapidly, special rock-prevention techniques may not be needed. In other cases, rocky spoil materials must be buried during or after the spoil placement process.

The appearance of reclaimed lands affects use potential in a real but intangible way. Standard practice in area mining situations is to grade spoil piles so that reclaimed terrain blends with surrounding natural terrain. In addition, the appearance of the land is affected by types of vegetation and, to a lesser degree, by surface texture.

The general physical factors causally related to the environmental effects of surface coal mining are climate, topography, overburden, coal, hydrology, and land use. Each factor is described by a series of parameters, and each parameter may have several possible values. These parameters and possible values are shown in Table 1.2. The rationale for the choice of these parameters and possible values is described above.

Five "technological" factors or parameters are felt to have major effects on mining costs and reclamation performance. These factors are topography, mining method, type of stripping equipment, number of coal seams mined per pit, and type of stripping subsystem. The definitions of these terms are summarized in Table 1.3 and are discussed further below.

Bucket wheel excavators (BWEs) were observed in use only in tandem with side cast stripping machines. In each case, the side cast machines were the prime movers. For this reason, bucket wheel excavators were not included here as an equipment type.

Topography, although physical rather than technological, has been included because of its effect on the applicability and environmental characteristics of alternative mining methods.

Five mining methods have been considered. These are area mining, modified open-pit mining, mountaintop removal, haul back mining, and conventional contour mining. It should be noted, however, that there are no sharp-line dividing-area mining methods from steep-slope mining methods where the topography is hilly or sharply rolling. The distinction has been made here on the basis of ground slope angles, the number of stripping cuts, and the type of overburden removal and placement (stripping) equipment.

Three types of stripping equipment have been considered. Draglines and stripping shovels are of the side cast type. Open-pit equipment consists of loading shovels for overburden removal and off-highway trucks for spoil haulage and placement. Construction equipment includes dozers, end loaders, scrapers, and if loaded by end loaders off-highway haul trucks.

The number of coal seams mined per pit may be single or multiple. Multiple seam mining is the mining of two or more coal seams in one pit. Practically speaking, however, multiple seam mining usually connotes two-seam mining. The non-coal strata separating two coal seams or splits in a given seam are termed partings if the total thickness of all strata is 5 ft (1.5 m) or less. In such cases, a mining situation is classified as a single seam because removal of the parting usually presents neither mining nor reclamation problems. Where the total thickness of all separating strata exceeds 5 ft, the strata are collectively termed "interburden," and the mining situation is classified as multiple seam.

The fifth and last technological parameter is the stripping subsystem, which may be single, tandem, or dual. A single-machine stripping subsystem is one in which one machine is used to remove and place all overburden materials. Prime examples are single dragline and single stripping shovel subsystems. In such cases, construction equipment used for clearing and grubbing, topsoil removal, or construction of shallow dragline benches is considered to be auxiliary equipment only. Infrequently used single-machine subsystems include those in which a scraper or end loader is used for all overburden removal and spoil placement.

A tandem machine subsystem is one in which two or more machines are used for overburden removal, and one machine follows or works behind another. Tandem machine stripping is by definition always a multi-lift overburden removal operation, with the first machine removing the top overburden lift, and so on. Tandem machine stripping may be used in both single and multiple seam situations. In a situation in which two draglines are used in tandem to strip two coal seams in one pit, the first dragline removes all overburden. After removal of the upper coal seam, the second dragline follows and removes the interburden.

1.4 WATER POLLUTION

During underground operations, exposed material within the mine oxidizes and gives rise to acid production, which contaminates the water. The most common case is the exposure of sulphide to air

TABLE 1.2

Summary of Possible Values of Important Physical Parameters

Environmental Category	Physical Parameter	Possible Parameter Value	Remarks
Climate	Climate	Arid or semi-arid Humid	
	Wind speed	High	
		Low	
Topography	Topography	Flat to hilly	Slopes of up to 17° from the horizontal
		Steep or mountainous	Slopes in excess of 17°
Overburden	Soil	High productivity	Encompasses fertility, erodibility, texture
		Low or medium productivity	drainage, depth
	Overburden	Acid producing	Alluvium, loess, glacial drift (sand, silt,
		Non-acid producing	clay)
		Unconsolidated	Hard clays, soft shales
		Semi-consolidated	Limestone, sandstone, shale, slate
		Consolidated	
	Stratigraphy	Semi-consolidated	Hard clays, soft shales
		Consolidated	Limestone, sandstone, shale, slate
		Undesirable strata immediately above coal	
		Undesirable strata not immediately above coal	
Coal	Average sulphur content (% by weight after washing)	High	More than 2%
		Medium	1–2%
		Low	Less than 1%
		Very thick	More than 40 ft
	Average individual seam thickness (ft)	Thick	8–40 ft
		Thin	3–8 ft
		Very thin	Less than 3 ft
	Dip	Flat or slightly dipping	Up to 5
		Pitching	More than 5°
Hydrology	Surface drainage (on mine site)	Perennial (continuous)	Up to 150 ft deep
		Ephemeral (intermittent)	More than 150 ft deep
	Height of water table (in mining area)	Shallow	
		Deep	
	Groundwater use (in mining area)	Stock or domestic drinking	
Land Use	Land use	Row crops	Irrigation
			Corn, soybeans, sugar beets, other crops grown in rows
		Closely spaced crops	Small grains and other close-seeded crops not usually grown in rows and tilled
		Hay or pasture	Cattle and sheep grazing
		Range	Productive or non-productive forest land
		Forest	
		Recreation	
		Unused	

Source: From Cook, F. Evaluation of Current Surface Coal Mining Overburden Handling Techniques and Reclamation Practices, U.S. Department of Commerce, National Technical Information Service, PB-264–111 (1976), pp. 60–123.

TABLE 1.3

Definition of Technological Parameters

Parameter	Parameter Value	Definition	Remarks
Topography	Rolling terrain	Average natural ground slope angles up to and including 17° from the horizontal.	Includes flat, gradually rolling, and hilly terrain.
	Steep slopes	Average natural ground slope angles greater than 17° from the horizontal.	Includes sharply rolling and mountainous terrain.
Mining method	Area mining	Method used in rolling terrain. Three or more parallel cuts are made.	
	Modified open-pit mining	Area-type method based on the use of loading shovels for overburden removal and trucks for spoil haulage and placement. Terrace type of pit or benched highwall is used.	Used only where the coal is very thick.
	Mountaintop removal	Method used in steep-slope areas to mine high-lying coal seams by complete removal of mountaintop. Usually, there are no final highwalls after mining has been completed. Reclaimed topography is usually flat or gently rolling.	
	Haulback mining	Method generally used in steep-slope areas. One or two cuts are made along the contour. Some spoil is hauled back to a previously mined area and placed on the solid bench.	
	Conventional contour mining	Steep-slope method in which a single cut is generally made along the contour. Spoil is placed by pushing it to rest on natural ground immediately below the elevation of the coal seam being mined.	
Stripping equipment3	Sidecast stripping equipment	Dragline or stripping shovel as the predominant stripping machine.	In classifying mining situations, the category Sidecast Stripping Equipment includes the following situations: • Single dragline or stripping shovel • Draglines and stripping shovels in tandem with one another • Draglines or stripping shovels in tandem with bucket wheel excavators • Draglines or stripping shovels in tandem with construction equipment In each situation, the sidecast machine is considered to be predominant if it accounts for at least two-thirds of the total overburden removal yardage.
	Open-pit equipment	Loading shovel and rock haul truck combinations.	Front-end loader/truck combinations are considered to be construction equipment, rather than open-pit equipment.
	Construction equipment	Dozers, end loaders, scrapers, rock haul trucks.	

Source: From Cook, F. Evaluation of Current Surface Coal Mining Overburden Handling Techniques and Reclamation Practices, U.S. Department of Commerce, National Technical Information Service, PB-264–111 (1976), pp. 60–123.

and water, which promotes a chemical reaction resulting in sulphuric acid. Problems of acid mine drainage (AMD) are predominantly chemical and extremely complex and originate with the extraction of sulphide minerals. Many reactions are possible, depending upon the environmental conditions existing in the mine and the nature of other minerals that may be present. Not all mines have AMD problems. It is normally associated with sulphide-bearing metallic ores (pyrite, pyrrhotite, chalcopyrite, sphalerite, marcasite, arsenopyrite, etc.).

Acid contamination is increased by water entering the mine from external sources. In addition to seepage from natural watercourses, which constitutes the larger part of the external supply, water (usually one-third by weight) may be used to transport mill tailings underground for the purposes of backfilling excavations and for processing and servicing needs. The average composition of underground drainage waters has been estimated at 51% from natural watercourses, 14% from mine backfill (where employed), 34% from service and process sources, and the remaining 1% from other unspecified sources.

As the total average water flow into an underground mine is estimated at 1,000 L/min, pumping must be maintained to ensure continuous operation. The estimated flow is an average for the industry and is subject to wide variations between mines and overtime in any particular mine. Following pumping, the drainage water, which frequently contains significant quantities of highly toxic dissolved minerals, is impounded at the surface prior to discharge and recycling. The impoundment areas may range from small ponds accepting a few tonnes per day to large tailings dams enclosing several square kilometres and receiving slurry wastes at the rate of thousands of tonnes per day. A sequential approach to water clarification is normally employed and involves the use of a single pond, or a series, to allow the settling out of the contained solids. The effluent is then subject to additional treatment to neutralize acids and remove heavy metals and radioactive wastes.

During impoundment, the effluent is partially depleted through seepage, percolation, runoff, and evaporation. The effect on the environment of water losses is dependent, to a large degree, upon the location of the mine within the localized drainage basin and its interconnecting watercourses. The nature of the materials discharged can also affect the environment profoundly. The heavier particulate matter in suspension will fall out under gravity relatively close to the point of discharge. Dissolved metals, on the other hand, are capable of being transported over much greater distances and can affect water quality at the regional scale.

Table 1.4 illustrates the effects of mining pollution on aquatic ecosystems in Canada. Together with Table 1.5, the values listed demonstrate the adverse effects of AMD, lowered pH values that eliminate aquatic organisms, and increased levels of acidity, sulphate iron, and total solids.

As in underground mines, surface operations (both open pit and strip) are also prone to the effects of AMD. While the mean flow of water into an underground mine is approximately 1,000 L/min, for open-pit mining the mean value is 13,800 L/min, 48% derived from a natural watercourse, 9% from service and process water, and 13% from other sources. The substantial quantities of water involved in surface mining operations enhance the potential for effluents to enter the natural environment. The problem is compounded by the runoff, leaching, and percolation processes acting upon the residuals contained in the solid wastes, which are considerably in excess to those produced in underground operations.

1.5 LAND USE

The nature and characteristics of the mining industry illustrate the wide range and intensity of land-use activities that can take place during the various stages of mining. Millions of hectares may be subject to wide-ranging exploration techniques, but only a fraction is directly affected by the development and production stages. In most cases, the factors determining the area of land affected are (1) the characteristics of the mineral being produced depth to the ore, density, and type of material, (2) its ore grade and reserve, (3) its percentage recovery rate, (4) the method of mining, and (5) whether or not beneficiation and further processing take place at the mine site.

TABLE 1.4

Some Details on the Effects of Mining Pollution on Aquatic Ecosystems in Canada

Locality	Source of Pollution	Chemical Effects	Biological Effects
New Brunswick			
S. Tomogonops R. and N.W. Miramichi R.	Heath Steele	Heavy metals, increased hardness acid generation	Decline in *Salmo salar* since 1965; reduced diversity and abundance of benthos
Nepisiguit R.	Brunswick 6	Heavy metals	*Salmo salar* fishery approaching zero
Ontario			
Crowe R. Basin	Bancroft area mines	TDS, SO_4^{2-} and hardness increased $8–10 \times 4$ in Bow L.	Aquatic biota affected only in immediate vicinity of tailings decant
Serpent R. Basin	Elliot L. area mines	Acid, high TDS, SO_4^{2-}, NO_3^-, and Ca^{2+}. Most affected were Quirke L. and Pecors L. Biological effects mainly because of acidity	Reduced productivity, altered phytoplankton and zooplankton communities. Elimination of *Stizostedion vitreum* and reduction of *Salvelinus namaycush*. Reduced benthos diversity
LaCloche Mtn. lakes	Sudbury smelters	Increased acidity from aerial fallout	Extinction of fish populations
Manitouwadge L. area	Noranda and Wiilroy mines	Acidity; toxic concentration of NH_3, zinc, copper, iron; nutrient enrichment; high TDS, especially SO_4^{2-}. Mose, L. has become meromictic—O_2 deoxygenated below 20 ft.	Mose, L. total elimination of macroinvertebrates; reduced benthos in all lakes.
Manitoba	Tantalum Mining Co.	Slight increase in turbidity, TDS, and suspended solids	Decrease in abundance and diversity of benthos
Bernie L.			
Borden L., Clarke L., and Lily L.	Mannibridge	No effect	No effect
Grass R.	INCO, Thompson	Increased turbidity, hardness, TDS, copper, and chloride	Reduced benthos and absence of Ephemeroptera in a small area
Ospwagan Lakes	INCO, Thompson	Overburden into Upper Ospwagan L. Increase in most parameters in Lower Ospwagan L.	Major reduction in Amphipoda and Ephemeroptera in 1968, but recovering in 1969
Schist L.	Flin Fion Mines	Increased turbidity, siltation, CO_2, and metals. Pollution spreading into L. Athapapuskow.	Sparse benthos, entirely of chironomids. No *Stizostedion vitreum* or *Salvelinus namaycush* in affected area. Decreased angling success.
Eldon R. and Cockeram L.	Lynn Lake	Siltation; toxic amounts of heavy metals, especially cadmium	Decrease in fish and benthos populations. Fish and benthos absent from area receiving drainage.
British Columbia			
Pend d'Oreille R.	Reeves MacDonald	Increased turbidity and siltation	Benthos severely reduced; fewer aquatic plants, indicating reduced 1° productivity
Benson L.	Coast Copper	Increased turbidity; has become meromictic	No benthos; fish population probably reduced

Source: Draft Acid Rock Drainage Technical Guide, Vol. 1 (Vancouver, Canada: BiTech Publishers, 1989).

TABLE 1.5

Typical Assays of Acid Waters from Mines (All Concentrations Are mg/L Except pH)

Type of Operation	Cu–Pb–Zn (Mine and Surface Drainage)	Cu–Pb–Zn (Mine Water)	Uranium (Seepage)	Cu–Zn (Active Mine)	Base Metal (Abandoned)	Uranium (Abandoned Mine)
pH	4.0	2.0	2.0	3.0	2.6	2.0–2.8
Suspended solids	8.8	690	Nil	—	—	25
Total less solids	79	24,000	—	—	9,200	13,440
Hardness	293	2,960	—	—	1,390	
Ca	—	—	416	—	454	—
Mg	—	—	106	—	178	—
Cu	17	11	3.6	0.0	2.5	2.2
Zn	118	1,090	11.4	0.4	34	9.4
Pb	0.4	58	0.7	0.11	0.5	—
Fe (total)	79	1,830	3,200	11.7	11,300	300
Mn	21	0	5.6	0.4	8.2	3.6
SO$_4$	36	16,560	7,440	885	4,050	6,900
COD	—	245	270	—	110	—

Source: From Cook, F. Evaluation of Current Surface Coal Mining Overburden Handling Techniques and Reclamation Practices, U.S. Department of Commerce, National Technical Information Service, PB-264–111 (1976), pp. 60–123.

The widest geographic distribution of mining activity occurs at the exploration stage. Fluctuations in the supply of, and demand for, mineral products are such that very little activity may occur in a given region or indeed the whole country for extended periods, and then a sudden demand for certain minerals will cause a great expansion in exploration.

Depending on what mineral is being sought when, many areas were explored and re-explored, making it difficult to quantify all aspects of the exploration stage where they affect the land surface directly through seismic lines, trenches, adits, or drill sites. For the purposes of this study, the major effort to quantify the effects of mining on land use, particularly land disturbances, will be in the development and production stages.

Once exploration has successfully identified a favourable zone worthy of further assessment, regulations require the establishment of a legal basis for development, either through claim staking, leases, grants, or licenses. The records of these legal requirements provide a good indication of the level of activity at this stage. The land area actually used and disturbed in the exploration, development, and production stages is much smaller.

The direct effects of most mining operations within the "mine site," while not negligible, are localized and relatively small compared to other forms of human economic activity, such as agriculture, forestry, and urban settlement. That mining will have a considerable influence on the land surrounding its operations is inevitable, and this influence, referred to as the "shadow effect," is far more extensive than previously estimated. In addition, the indirect effects of mining can be quite large, especially the infrastructure developed for mining operations (rail, roads, housing power plants, water storage, and other facilities). It can significantly extend the land area directly influenced by all mine-related activities and often permits a large number of other activities that would be difficult or impossible to undertake without it.

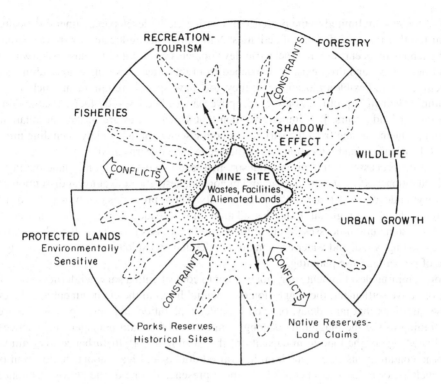

FIGURE 1.1 Mining and land use.

Indeed, the interplay of the direct uses at the "mine site" and other land uses in the "shadow" zone illustrates the areas of increased responsibility that have been assigned to mining operations. In the past decade, the responsibility of a mining company for environmental protection has been dramatically projected beyond the visible perimeters of their working operations to substantially larger neighbouring lands. In effect, changing environmental regulations have assigned an increasing degree of responsibility to the mining company to minimize the environmental impact of its operations on the use of neighbouring lands. It has resulted in a continued and increasing interaction, desired or not, with landowners and resource managers responsible for the land-use activities bordering mine sites (Figure 1.1). The range of interaction between mining and other land uses has become increasingly complex. No longer is it a straight economic or technological choice between competitive uses. Demands for resource conservation, environmental protection, and restricted use, coupled with the wide range of traditional uses agriculture, forestry, settlement, and recreation have all increased the potential constraints on development and the actual number of conflicts.

Over the past few decades, there has been a growing indication that mining developments are being given lower social approval than competing land uses. Much of this, no doubt, has been due to a past history of abandoned operations and their adverse effects on the local environment. Public attention to the allocation of land to certain uses has also been heightened by increased publicity about the scarcity of certain land resources and threats to them from an increasing array of degradation processes, both real and perceived.

Although the public has been aware for some time of the effects of mining operations, the deterioration of the land resource base attributable to other land-use activities, particularly those centred on the agriculture and forest industries, is now believed to be equally extensive, if not greater. This has only served to heighten the pressure on new industrial developments (particularly those associated with minerals and energy resources) and clouds their acceptance as a legitimate option in the highly emotional process of land allocation.

Mining options are limited to the presence of commercially developable mineral deposits. This is similar to other resource-dependent activities: National parks are located where the natural features or phenomena occur; and hydroelectric developments are limited to those waterways where sufficient head, flow, and drop exist. Approached from the same perspective, agriculture and forestry seem far more flexible in location, yet they, too, have specific requirements such as soil and temperature. Their high profile in our consumptive pattern becomes the focus of attention when competition for land occurs. It is imperative then that society be aware of the interrelationships, needs, and problems arising from the various economic sectors of our society, including mining, in order that decisions on land allocation reflect the needs and concerns of all.

Many of the processes in the mining industry that lead to a deterioration in land quality due to physical alteration, chemical degradation, and biological, aesthetic, and cultural disruption are evident to varying degrees in all areas of human activity. Those processes can be separated into two broad groups, local and regional. Land degradation, in its broadest sense, includes not only man-induced but also natural processes. Many of the natural processes involved in land degradation, for example, flooding, wind, and water erosion are accelerated by exploitation and poor management, or a lack of proper land-use planning.

Although mining developments may not come into contact with a park, wilderness zone, Indian reserve, or native settlement, the suggestion of potential harm to the environment may be enough to arouse questions that may delay, or even exclude, exploration and development. We are only now beginning to see the magnitude of long-term effects of certain management practices. Our limited knowledge of the long-term capacity of the earth to absorb disturbances, certain wastes, or airborne contaminants in any medium has raised doubts and fears about the location of new mines, smelters, or refineries. Therefore, the mere presence of one of the above-mentioned protected areas in a drainage basin or downwind from a proposed mine development is sufficient to arouse considerable opposition, demand for a public inquiry, or outright exclusion. Thus, protected or limited-use lands can play a role disproportionate to their actual size, due to their multiplicity and wide distribution. They can, and have, become important factors influencing mine developments. In the case of native groups, a new development can become the focus of attention of a much wider land-claims issue.

The mining industry has been concerned with what appears to be a trend towards increased land withdrawals from any use whatsoever or the banning of exploration and development activities from certain lands.

Proponents of unrestricted access to land by mining believe that most exploration techniques have zero environmental impact; drilling can be of slight temporary impact, less for intrinsic reasons than because of the need to construct roads or helicopter pads in remote localities. Trenching, pitting, and shaft exploration can be intrusive but, even when carried out intensively, can easily be remedied by appropriate restoration; the underlying motive seems to be that mining might follow. It is better not to know what lies in the ground.

The argument supporting unrestricted access is that effective land-use planning and management can only be made after all the resource potential has been confirmed. The industry questions the arbitrary assigning of "values" to wilderness areas, parks, etc., without any consideration of the potential mineral value. It would appear to most proponents of mining developments that banning of mineral exploration is nothing less than an admission that if mining proposals do follow, the legal and administrative machinery is inadequate to enable the full and mature assessment of the merits of mining in relation to the merits of wilderness areas.

The dilemma facing the various levels of government is how to balance the overall needs for resource development with the demand for conservation and protection. Some workable mechanism for trade-offs among the various interested parties is still necessary. This requires the very difficult task of reconciling the multiple objectives of a wide range of agencies and departments among the varying levels of government. The task is going to become increasingly difficult as pressure for more parks, wilderness areas, increased environmental protection, and native claims, coupled with the demand for mineral and energy resources continues to accelerate in the next two decades. Jurisdiction and land allocation will continue to be the focus of attention for all parties concerned, and the determining factors in any attempt to reconcile competing demands on land resources.

In this regard, a major problem facing the mining industry is its unfavourable public image. Past perceptions of mining and its environmental impacts seem to be permanently etched in the minds of the general public. It appears that the attempts of the mining industry to improve its image and the understanding of the public for its activities have progressed only marginally. It is this general perception of mining as a singular source of land degradation and pollution which has caused so much attention to be focused on it in the form of government regulation or restricted zoning. The industry continues to be confronted with its image every time it enters a public debate or inquiry. Now that it is almost mandatory that environmental impact assessment and public hearings occur prior to any new mine development, public perceptions are playing an increasingly important role in the decision-making process.

The standards for all new developments have continued to become more stringent since the quality of life, as well as the environment, has become increasingly important social and economic issues. The situation today has become considerably more complex as socio-economic and political issues merge with the more traditional conservation and environmental issues. The links between resource developments, the environment, lifestyles, and institutional values can no longer be ignored.

At some public hearings, some of the major issues raised were identified as follows:

* Waste management and its long-term storage and disposal
* Health; effects of radiation
* Land-use conflicts and their adverse economic impact
* Lack of faith in government and regulating standards
* Lack of public access to information
* Commercial interests versus human concerns
* Poor past record, in both mining and government
* Enforcement and regulation problems

It is evident that the scientific community, regulating authorities, and elected officials also suffer from a lack of trust and a greater demand for public involvement in decision making.

The views of all the various pressure or lobby groups have become highly relevant within their own frame of reference since there is no absolute way of measuring the relative importance of different environmental problems. Of overriding importance is the way in which individual decision makers and pressure groups perceive the situation in which they are confronted. The extent to which they diverge on issues may determine whether or not any decision is achieved. Differences in perception can occur on a personal basis or on up through various levels of authority and society on a local, regional, provincial, or national basis, and associations or pressure groups, although sincere, are often partisan by nature.

Mining land-use problems and conflicts often fall into one or more of the following generalized categories:

* Conflicts that arise when two or more proponents compete for the same parcel of land, for example, coal strip mining and agriculture in the western prairies of Saskatchewan or Alberta.
* Conflicts that arise when a particular land use on one parcel of land adversely affects the use of land on adjacent properties. This usually raises the question of compatibility or incompatibility of uses; for example, the impact of metallic mine wastes on the use of neighbouring water resources, fishing in particular.

- Conflicts that arise between those who wish to develop a particular mineral or energy resource and those who desire the maintenance of a pristine environment through conservation and protection, for example, the establishment of parks or wilderness preserves.

1.6   SUBSIDENCE OF MINED LAND

Subsidence has been defined as "vertical and horizontal movements of surface, sub-surface and underground points as a result of various natural and manmade activities." Because of the inherent impacts on the surface structures/features, subsidence has been a subject of deliberations on an international level since 1965 when UNESCO announced its International Hydrological Decade. The subsidence of the land surface may be due to endogenic causes (i.e., those within the earth crust itself) and exogenic causes (i.e., factors that originate from outside the planet).

The different causes that contribute to subsidence have been classified. In the case of endogenic causes, these have been classified in accordance with whether they have been caused by natural phenomena or by the activities of man Tables 1.6 and 1.7.

While landslides have been classified as cases of land subsidence, here we are concerned with subsidence of the land surface resulting from the activities of extraction of the mineral. Though vertical subsidence is a common feature in mining areas, cases are on record where there has been a lateral drift. These are the areas where the strata are highly water bearing with very high specific yield. Besides these physical ramifications, these have, as do other effects of subsidence in general, considerations of safety attached to them when the highly water-bearing strata, with high specific yield, discharge the water to the workings below.

TABLE 1.6
Causes of Land Subsidence

Endogenic	Exogenic
Volcanism	Removal of support
Folding	Weakening of support
Faulting	Increase in effective and actual loading

Source: From Cook, F. Evaluation of Current Surface Coal Mining Overburden Handling Techniques and Reclamation Practices, U.S. Department of Commerce, National Technical Information Service, PB-264–111 (1976), pp. 60–123.

TABLE 1.7
Activities Causing Subsidence

Natural	Man-made
Solution of rocks, minerals, etc.	Pumping of water and petroleum from below ground
Drainage of subsoil	
Drifting of subsoil and sliding of rocks	Underground extractions for mining and other purposes
Rodents	
Frosting and defrosting	Settlement of foundations as a result of seepage of water and otherwise
Tectonic movement	

Source: From Cook, F. Evaluation of Current Surface Coal Mining Overburden Handling Techniques and Reclamation Practices, U.S. Department of Commerce, National Technical Information Service, PB-264–111 (1976), pp. 60–123.

The subsidence can vary from a couple of millimetres, in the case of water withdrawal from the aquifers below, to above 6–7 m in the case of extraction of coal from thick seams or due to mine fires in these seams. Subsidence, which was caused by the pumping of water and petroleum, of as much as 15 m (in one case) has been reported.

Subsidence, due to shallow mining, causes direct air circulation through the goaved-out areas, which if not checked, coupled with the coal available, causes spontaneous heating and fires within the goal areas. These engulf large areas and sometimes reach the surface with devastating effects on the land surface and other features. Fires starting in one seam may travel to upper seams and across the barriers to the neighbouring mines. The subsidence in regular workings has been found to follow a regular pattern that becomes severe when other factors come into play, for example:

- Presence of faults
- Presence of dykes/sills having a breaking effect on the subsidence pattern against extension resulting from faults
- Presence of old workings in the vicinity, making things very severe and unpredictable, particularly where the interaction is with workings that were carried out before the current statute applied

The physical impacts of subsidence may vary from simple negligible lowering of the ground surface to severe damage to buildings or other surface features by wide and deep cracks. When subsidence is caused by fires, it progresses very fast within short periods of the fires starting. Other commonly observed impacts are as detailed:

- Gross changes in surface topography resulting in undesired depressions
- Disruption/disturbance in aquifers and thereby reduction in availability and contamination of water
- Retardation in the growth of vegetation due to poor availability of water
- Waterlogging in the central portion of subsided areas
- Underground fires in coal left below ground at shallow depths
- Contamination of surface atmosphere by gases produced as a result of underground fires
- Development of cracks leading to joining with surface and/or underground waterbodies, resulting in an increased make of water in underground workings
- Adverse flow condition and waterlogging in rivers, small waterlogging in rivers, small watercourses
- Abrupt changes in road gradients
- Damage to underground pipelines, cables, and drainage systems
- Tilting and damage to buildings, structures, plants, pylons, etc.
- Depression in the ground level below HFL of watercourses nearby

In overlying seams, the impact of subsidence movements due to extraction below them could be as listed below:

- Collapse of old workings standing on stocks/smaller pillars, and thereby increased subsidence on the surface
- Damage to overlying coal seams, which may sometimes be rendered unworkable
- Floor lifting, spalling, and roof falls in workings

1.7 MINING AND ITS IMPACTS, AN OVERVIEW

Environmental impacts of mining are assessed on different phases of a mining project. There are different phases of mining projects which suffer different impacts in mine development and extraction process, and post-closure phase. Each phase of a mining project is associated with a different set of environmental impacts.

1.7.1 EXPLORATION

A mining project commences with the evaluation of the extent and value of the mineral project. Information about the value of the metal content of the deposit is assessed in the exploration stage. This phase often involves surveys, field studies, and drilling test boreholes and other exploratory activities.

The exploration phase involves clearing of vegetation to facilitate surveying, to allow the operation of heavy vehicles mounted with drilling rigs. Often, the government may require a separate environmental statement on the exploration phase. The impacts of this phase may have important effects on the environment as related to the entire project.

1.7.2 DEVELOPMENT

If the exploration phase proves the existence of a large mineral deposit with sufficient grade, then the corporation may plan to develop a mine. This phase of the mining project usually has several phases and components.

Construction of access roads to pride transportation of heavy equipment to the mine site and shipping access roads to transport processed metals and ores can have significant environmental impacts, especially if access roads cut through ecologically sensitive areas or are near previously isolated communities. If a proposed mining project involves the construction of any access roads, then the environment must include a comprehensive assessment of environmental and social impacts of these roads.

If a mine site is located in a remote, undeveloped area, the project development may begin by clearing the land for construction of staging areas that would house construction employees and project equipment. Even before the mining operation starts, activities associated with the site preparation and clearing can result in significant environmental impacts. An Environmental Impact Statement (EIS) must assess the impacts of access roads, site preparation, and clearing separately.

1.7.3 ACTIVE MINING

After completion of the access roads and necessary facilities for mine development mining may commence. The activities in mining operation involve the extraction and concentration of metal ores, called beneficiation. Proposed mining projects differ considerably in the proposed method for the extraction and concentration of the metallic ore. In most of the situations, metallic ores are buried under layers of soil and rock, called overburden that must be removed or excavated to allow access to the orebody. The proposed mining methods may differ significantly.

1.7.4 BENEFICIATION

Metallic ores contain higher levels of metals, but they generate large quantities of waste rock. For example, a good grade copper deposit may contain only one-quarter of 1% copper. Similarly, the gold content of a good grade gold deposit may be only a few one-hundredth of a per cent. Therefore, the next step involves grinding (milling) of the ore and separating the relatively small quantities of the metal from the waste material. The process is called beneficiation or milling.

Milling is one of the costly parts of the beneficiation process. It results in very fine particles that allow more efficient extraction of metals. However, milling allows a complete release of contaminants when these materials become tailings. Tailings are what remains following milling of the ore.

Beneficiation includes physical/chemical separation techniques, such as gravity concentration magnetic separation, electrostatic separation, flotation, solvent extraction, electrowinning, leaching precipitation, amalgamation using mercury. Wastes from these operations produce waste dumps, tailings, heap leach materials (for gold and silver operations), and dump leach materials (for copper leaching operations).

Leaching involves the use of cyanide and is a kind of beneficiation process commonly used with gold, silver, and copper ores which result in serious environmental impacts. In the leaching operation, finely grounded ore is deposited in large piles on top of an impermeable pad, and a solution containing cyanide is sprayed on the top of the pile. The cyanide solution dissolves the desired metals, and the pregnant solution containing the desired metals is collected from the bottom of the pile using a system of pipes.

1.7.5 TAILINGS DISPOSAL

High-grade mineral ores consist of non-metallic materials, as well as toxic metals such as cadmium, lead, and arsenic. The beneficiation process generates a high volume of waste materials called tailings. Tailings often contain a high volume of toxic materials. One of the important questions in the process of awarding a permit for a mining operation is how well the toxic materials produced in the mining operation will be handled. The long-term goal of tailings disposition and management is to prevent the mobilization of toxic materials and their release to the environment in the form of tailings. The alternative procedures include the use of wet tailing impoundment facilities called tailings pond and dewatering and disposal of dry tailings as backfill and submarine tailings disposal.

Before the establishment of environmental laws, many mining companies simply dumped the tailings into nearby streams and rivers, and into the ocean.

The World Bank group state: "Riverine (e.g. rivers lakes and lagoons or shallow marine tailings disposal is not considered good international industry practice. By extension, riverine dredging which requires riverine tailings disposal is also not considered good international practice."

1.8   ENVIRONMENTAL AND SOCIAL IMPACTS

Mining produces various types of environmental effects, which includes impacts on water resources, including acid mine drainage and contaminant leaching, erosion of soils mine wastes into surface waters, impacts of tailing impounds and dump leach facilities, impacts of mine dewatering, and air pollution, impacts on air quality, incidental releases of mercury, noise and vibration, impacts on wild life, impacts on soil quality, lost access impacts on social values, impacts on livelihoods, impacts on public health, impacts on cultural and aesthetic resources, effects on climate change, etc.

1.8.1 IMPACTS ON WATER RESOURCES

The most important impact of mining projects is its effects on water quality and availability of water resources within the project area. A mining operation can also impact water resources outside its area of influence. Key issues are whether the groundwater and surface water supplies will remain fit for human consumption and whether the quality of surface waters in the project area will remain adequate to support native aquatic life and terrestrial wildlife.

1.8.2 ACID MINE DRAINAGE AND CONTAMINANT LEACHING

Potential for the generation of acid mine drainage is a critical issue for determining whether a mining project is environmentally acceptable. When mined materials such as the walls of open pits and underground mines, tailings, waste rock heap, and dump leach materials are excavated and exposed to oxygen and water, acid can form if iron sulphide minerals (especially pyrite) are abundant and neutralize materials to counteract the acid formation. The acid will, in turn, leach or dissolve metals and other contaminants from mined materials and form a solution that is acidic, high in sulphate, and metal-rich, including elevated concentrations of cadmium copper, lead, zinc, arsenic, etc.

Leaching of toxic constituents, such as arsenic, selenium, and metals, can occur even if acidic conditions are not present. Elevated levels of cyanide and nitrogen compounds (ammonia, nitrate, and nitrite) can also be found in waters at mine sites, from heap leaching and blasting.

Acid drainage and contaminant leaching are the most important source of water quality impacts related to metallic ore mining. Acid mine drainage is considered one of mining's most serious threats to water resources. A mine with acid mine drainage has the potential for long-term devastating impacts on rivers, streams, and aquatic life.

If mine waste is acid generating, their impacts on fish, animals, and plants can be severe. Many streams impacted by acid mine drainage have a pH value 4 or lower, similar to battery acid. Plants, animals, and fish are unlikely to survive in streams such as these.

1.8.3 Erosion of Soils and Mine Wastes into Surface Waters

In many operations, the potential of soil and sediment eroding into and degrading surface water quality is an important problem.

According to a study commissioned by the European Union:

Because of the large area of land disturbed by mining operations and the large quantities of earthen materials exposed at sites, erosion can be a major concern at hard rock mining sites. Consequently, erosion control must be considered from the beginning of operations through completion of reclamation. Erosion may cause significant loading of sediments (and any entrained chemical pollutions) to nearly waterbodies, especially during severe storm events and high snow melt periods.

Sediment-laden surface runoff typically originates as sheet flow and collects in rills, natural channels or gullies, or artificial convergences. The ultimate deposition of the sediment may occur in surface waters or it may be deposited within the floodplains of a stream valley. Historically, erosion and sedimentation processes have caused the build-up of thick layers of mineral fines. And sediment within regional flood plains and the alteration of aquatic habitat and the loss of storage capacity within surface waters. The main factors influencing erosion includes the volume and velocity of runoff from precipitation events, the rate of precipitation infiltration downward through the soil, the amount of vegetative cover, the slope length or the distance from the point of origin of overland flow to the point where deposition begins, and operational erosion-control structures.

Major sources of erosion/sediment loading at mining sites can include open-pit areas, heap and dump leaches, waste rock and overburden piles, tailings piles and dams, haul roads and access roads, ore stockpiles, vehicle and equipment maintenance areas, exploration areas, and reclamation areas. A further concern is that exposed materials from mining operations, (mine workings, wastes, contaminated soils, etc.) may contribute sediments with chemical pollutions, principally heavy metals. The variability in natural site conditions (e.g., geology, vegetation, topography, climate, and proximity to and characteristics of surface waters), combined with significant differences in the quantities and characteristics of exposed materials at mines, preclude any generalization of the quantities and characteristics of sediment loading.

The types of impacts associated with erosion and sedimentation are numerous, typically producing both short-term and long-term impacts. In surface waters, elevated concentrations of particulate matter in the water column can produce both chronic and acute toxic effects in fish.

Sediments deposited in layers in flood plains or terrestrial ecosystems can produce many impacts associated with surface waters, groundwater, and terrestrial ecosystems. Minerals associated with deposited sediments nay depress the pH of surface runoff thereby mobilising heavy metals that can infiltrate into the surrounding subsoil or can be carried away to nearby surface waters. The associated impacts could include substantial pH depression or metals loading to surface waters and/or persistent contaminated sediments nay also lower the pH of soils to the extent that vegetation and suitable habitat are lost.

Beyond the potential for pollutant impacts on human and aquatic life, there are potential physical impacts associated with the increased runoff velocities and volumes from new land disturbances activities. Increased velocities and volumes can lead to downstream flooding, scoring of streams channels, and structural damage to bridge footings and culvert entries. In areas where air emissions have deposited acidic particles, and the native vegetation has been destroyed, runoff has the potential to increase the rate of erosion and lead to the removal of soil from the affected area. This is particularly true where the landscape is characterised by steep and rocky slopes. Once the soil has been removed, it is difficult for the slope to be revegetated either naturally or with human assistance.

Potentially adverse effects of inadequate mine site water management and design include unacceptable high levels of suspended solids (non-filterable residue) and dissolved solids (filterable residue) in surface runoff bed and bank erosion in waterways. It is self-evident that a Sediment and Erosion Control Plan is a fundamental component of a Mine site Water Management Plan.

1.9 IMPACTS OF WET TAILING IMPOUNDMENTS, WASTE ROCK, HEAP LEACH, AND DUMP LEACH FACILITIES

The impacts on water quality can be severe. The impacts include contamination of groundwater below these facilities and surface. Toxic substances can percolate and leach through the ground and contaminate groundwater, especially if the bottom of these facilities is underlain with an impermeable liner. Tailings are a by-product of metallic ore processing, a high-volume waste that often contains harmful quantities of toxic substances which include arsenic, lead, cadmium, chromium, nickel, and cyanide if cyanide leaching is used. Most mining companies dispose of tailings by mixing them with water to form a slurry and disposing the slurry behind a dam in a large wet tailings' impoundment. As the ore is frequently extracted as the slurry, the resulting waste contains large amounts of water and forms ponds at the top of tailings that can be a threat to wildlife. Cyanide tailings in precious metals mines are particularly dangerous. Tailings ponds are either dry in arid climates or may release contaminated water, in wet climates. In both cases, specific management techniques are required to close these waste ponds and reduce environmental threats. Under heavy rainy conditions, more water may enter a tailings impoundment higher than its capacity, necessitating the discharge of tailings of impoundments effluent. Since the effluent can contain toxic substances, the release of these effluents can seriously degrade the water quality of surrounding rivers and streams if the effluent is not treated prior to discharge. Many dam failures have created some of the worst environmental impacts of the mining industry. When the tailings fail, they release large quantities of toxic waters that can kill aquatic life and poison drinking water supplies for many miles downstream of the impoundments.

1.9.1 IMPACTS OF MINE DEWATERING

When a water table is intersected, groundwater flows into the open pit for the progress of mining and the polluting water must be pumped out and discharged into another location. Pumping and discharging mine water causes a unique set of environmental impacts:

Mine water is produced when the water table is higher than the underground mine workings or the depth of an open-pit surface mine. When this occurs, the water must be pumped out of the nine. Alternatively, water may be pumped from wells surroundings the mine to create a cone of depression in the groundwater table, thereby reducing infiltration. When the mine is operational, mine water must be continually removed from the mine to facilitate the removal of the ore. However, once mining operations end, the removal and management of mine water often end, resulting in possible accumulation in the rock fractures, shafts, tunnels, and open pits, thus leading to uncontrolled releases to the environment.

Groundwater drawdown and associated impacts to surface waters and nearby wetlands can be a serious concern in some areas.

Impacts from groundwater drawdown may include reduction or elimination of surface water flows; degradation of surface water quality and beneficial uses; degradation of habitat (not only riparian zones, springs and other wetland habitats, but also upland habitats such as greasewood as groundwater levels decline below the deep root zone); reduced or eliminated production in domestic supply wells; water quality/quantity problems associated with discharge of the pumped groundwater back into surface waters downstream from the dewatered area. The impacts could last for many decades. While dewatering is occurring, discharge of the pumped water, after appropriate treatment, can often be used to mitigate adverse effects on surface waters. However, when dewatering ceases, the cones of depression may take many decades to recharge and may continue to reduce surface flows. Mitigation measures that rely on the use of pumped water to create wetlands may only last as long as dewatering occurs.

1.9.2 IMPACTS OF MINING ON AIR QUALITY

During different stages of mining operations, airborne emissions occur. This is often true during exploration, development, construction, and operational stages of mining. Mining operations mobilize large quantities of material and waste piles containing small-size particles that are easily dispersed by the wind.

The main sources of air pollution in mining operations usually include the following:

- Particulate matter transported by wind as a result of surface mining, transportation of rocks and ores
- Wind erosion, fugitive dust from tailings dumps, stockpiles, waste dumps, and haul roads
- Exhaust emissions from mobile sources such as trucks raise the particulate pollution levels
- Emissions from combustion of fuels, in stationary and mobile sources, explosions, and mineral processing

After the pollutants enter the atmosphere, they undergo physical and chemical changes before reaching a receptor. Large-scale surface mining has the potential to contribute significantly to air pollution, especially in the operation stage. All activities during various stages of ore extraction, processing, handling, and transport depend on equipment, generators, processes, and materials that generate hazardous air pollutants, such as particulate matter, heavy metals, carbon monoxide, sulphur dioxide, and nitrogen oxides. Mobile sources of air pollution include heavy vehicles and cars that operate on unravelled roads.

1.9.3 INCIDENTAL RELEASE OF MERCURY

Mercury is commonly present in gold ore. Concentrations may vary widely, even within an ore deposit. If the mercury content in a gold ore is 10 mg/kg, and 1 million tons are processed at a particular facility, 10 tons of mercury are potentially released into the environment. This is a major source of mercury pollution and should be controlled.

In some gold mining operations, gold-containing ore is crushed and, then if necessary, heated and oxidized in roasters to remove sulphur and carbonaceous materials that affect gold recovery. Mercury that is present in the ore is vapourized in roasters, and they are some of the largest sources of mercury emitted to the atmosphere.

Following roasting or autoclaving, the ore is mixed with water and reacts with cyanide leach solution. When the gold and mercury are dissolved, and solids are removed by filtration. Then, the purified solution is sent to an electrowinning process, where the gold is recovered. In this process, mercury must also be recovered and collected. If this mercury is not collected by air pollution control devices, it could be released into the atmosphere and adversely affect the environment and public health.

Volatilization of mercury from active heaps and tailings facility has been identified as another substantial source of mercury emitted into the atmosphere. This process should be assessed and controlled. In summary, mercury present in gold ore may be released into the land and air or in the gold product (as an impurity).

1.9.4 IMPACTS OF MINING ON SOIL QUALITY

Mining operations can damage and contaminate soil quality over large areas. Agricultural operations near a mining activity or a mining operation may be affected. According to a European study:

> Mining operations routinely modify the surrounding landscape by exposing previously undisturbed earthen materials. of exposed soils, extracted mineral ores tailings, and fine materials in waste rock piles can result in substantial sediment loading to surface waters and drainage ways. In addition, spills and leaks hazardous materials and the deposition of contaminated windblown dust can lead to soil contamination.
>
> Soil contamination: Human health and environmental receptors risks from soils generally fall into two categories: (1) contaminated soil resulting from windblown dust, and (2) soils contaminated from chemical spills and residues. Fugitive dust can pose significant environmental problems at some mines. The inherent toxicity of the dust depends upon the proximity of environmental receptors and types ore being mined. High levels of arsenic, lead, and radionuclides in windblown dust usually the pose the greatest risks. Soils contaminated from chemical spills and residues at mine sites may pose a direct contact risk when these materials are misused as fill materials, ornamental landscaping or soil supplements.

1.9.5 IMPACTS OF MINING PROJECTS ON WILDLIFE

Wildlife refers to plants and animals that are not domesticated. Mining affects the environment through the removal of vegetation and topsoil, displacement of fauna, the release of pollutants, and the generation of noise. Wildlife species live in communities that depend on each other. Survival of these species can depend on soil conditions, local climate, altitude, and other features of the local habitat. Mining causes direct and indirect damage to wildlife. The impacts stem primarily due to disturbing, removing, and redistributing the land surface. Some effects are short term and confined to the mine site. Some may have long-term effects.

The most important impact on wildlife is the destruction and/or displacement of species in areas of excavation and piling of mine wastes. Mobile wildlife species, like game animals, birds, and predators, leave these areas. More sedentary animals, like invertebrates, many reptiles, burrowing rodents, and small animals, may be more severely affected.

When streams, lakes, ponds, and marshes are filled or drained, fish, aquatic invertebrates, and many reptiles are severely affected. Food supplies for predators are reduced because of the disappearance of these land and water species.

Many wildlife species are highly dependent on vegetation growing in natural drainages. This vegetation provides essential food, nesting sites, and cover for an escape from predators. Any activity that destroys vegetation near ponds, reservoirs, marshes, and wetlands reduces the quality and quantity of habitat essential for waterfowl, shorebirds, and many terrestrial species.

Many animal species require special kinds of habitats that do not allow them to adjust to changes created by mining. Some of these changes reduce living spaces. The degree to which animals tolerate human competition for space varies. Some species tolerate very little disturbance. For example, where a particular critical habitat is restricted, such as lakes, ponds, or primary breeding grounds, a species could be eliminated.

Surface mining can degrade aquatic habitats with impacts felt many miles from mining sites. For example, sediment contamination of rivers and streams are common with surface mining operations. Habitat fragmentation occurs when large areas of land are broken up into smaller and smaller patches, making dispersal of native species from one patch to another difficult and cutting off migratory routes. Species that require large patches simply disappear.

1.9.6 MINING AND HUMAN DISPLACEMENT

Mining causes displacement of human habitats. According to the International Institute for Environment and Development:

> The displacement of settled communities is a significant cause of resentment and conflict associated with large-scale mineral development. Entire communities may be uprooted and forced to shift else-where, often into purpose-built settlements not necessarily of their own choosing. Besides losing their homes, communities may also lose their land, and thus their livelihoods. Community institutions and power relations may also be disrupted. Displacement communities are often settled in areas without adequate resources or are left near the mine, where they may bear the brunt of pollution and contamina-tion. Forced resettlement can be particularly disastrous for indigenous communities who have strong cultural and spiritualties to the lands of their ancestors and who may find it difficult to survive when these are broken.

As per the Institute for Environment and Environment:

> One of the most significant impact of mining activity is the migration of people into a mine area, par-ticularly in remote parts of developing countries where the mine represents the single most important economic activity. For example, at the Grasberg mine in Indonesia the local population increased from less than 1000 in 1973 to between 100,000 and 110,000 in 1999. Similarly, the population of the squat-ter settlements around Porgera in PNG, which opened in 1990, has grown from 4000 to over 18,000. This influx of newcomers can have a profound impact on the original inhabitants, and disputes may arise over land and the way benefits have been shared. These were among the factors that led to violent uprisings at Grasberg in the 1970s and the 1990s.

A sudden increase in population can also lead to pressures on land, water, and other resources as well as bringing problems of sanitization and waste disposal. Migration effects may extend far beyond the immediate vicinity of the mine. Improved infrastructure can also bring an influx of settlers. For instance, it is estimated that the 80-metre-wide, 890 kilometres-long transportation corridors built from the Atlantic Ocean to the Carajas mine in Brazil created an area of influence of 300,000 square kilometres.

1.9.7 IMPACTS ON PUBLIC HEALTH

Mining projects often underestimate the effects of lack of proper environmental management and impacts on soils, water, biodiversity, and forest resources, which are critical to the subsistence of local people. Contamination caused by mining is often transferred to other economic activities, such as agriculture and fishing. The situation becomes worse when mining activities take place in areas inhabited by populations who are historically marginalized, discriminated against, or excluded.

Advocates of mining projects must ensure that basic rights of affected people and communities are protected, and their rights are upheld. Their rights include rights to control and use of land and rights to clean water as well as rights to livelihood. Such rights are specified in national laws, based on a range of national and international instruments and agreements. All groups are equal under the law, and interests of the most vulnerable group including low-income people need to be protected.

EIAs often underestimate the potential health risks of mining projects. Hazardous substances and wastes in water, air, and soil can have serious health impacts on the public. The term "hazard-ous substances" includes all substances that can be harmful to people and the environment. Public health problems related to mining include:

- Water: Surface and groundwater contamination with metals and elements; microbiological contamination from sewage and wastes in composite and mine worker residential areas

- Air: Exposure to high concentrations of sulphur dioxide, particulate matter, heavy metals, including lead, mercury, and cadmium
- Soil: Deposition of toxic elements from air emissions

Mining operations can suddenly produce adverse effects on quality life and physical, mental, and social well-beings of local communities. Improvised mining towns and camps often threat food availability and safety. Indirect effects of mining on public health can include increased incidence of tuberculosis, asthma, chronic bronchitis, and gastrointestinal disease.

1.9.8 Impacts on Cultural and Aesthetic Resources

Mining activities can have direct and indirect impacts on cultural activities. Direct impacts can result from mining activities. Indirect impacts can result from soil erosion. Increased accessibility to current or proposed to current or proposed mining sites. Mining projects can affect sacred landscapes, historical structures, and natural landmarks. Potential impacts include:

- Complete destruction of the resources through surface disturbance or excavation
- Degradation or destruction caused by topographic or hydrological pattern changes or from soil movement through removal, erosion, sedimentation, etc.
- Unauthorized removal of artefacts or vandalism because of increased access to previously inaccessible areas
- Visual impacts due to clearing of vegetation, large excavations, dust, and the presence of large-scale equipment and vehicles

1.9.9 Climate Change Considerations

Every mining project or operation has the potential to change the global carbon budget, including an assessment of projects carbon impact. Large-scale mining projects have the potential to alter global carbon in various ways, including lost carbon dioxide uptake by forests and vegetation that is cleared. Many large-scale mining projects are proposed in heavily forested areas of tropical regions that are critical for absorbing atmospheric and maintaining carbon dioxide and carbon dioxide uptake. Some mining projects propose long-term or even permanent destruction of tropical forests. The EIS for mining projects must include a careful accounting of how any proposed disturbance of tropical forests will alter the carbon budget.

Other environmental impacts include carbon dioxide emitted by machines such as diesel-powered, heavy vehicles involved in extracting and transporting ore. Carbon dioxide emitted by the processing ore into the metal (for example, by pyrometallurgical versus hydro-metallurgical processing). Life Cycle Assessment methods can be used to estimate the life cycle emissions of greenhouse gases from copper and nickel extraction including mining.

1.10 ENVIRONMENTAL AUDITS

As with most industries, mining must now include in its plan of operation mechanisms to address a project environmental impact. New regulations have focused attention on the potential environmental impacts of industry, particularly the handling of wastes.

One of the most effective tools in managing regulatory requirements is an environmental audit. An environmental audit can be a literature review to determine regulatory requirements. The audit can also take on the form of a phase-one site assessment. This is a limited investigation to determine whether there are specific contamination problems at a site.

Phase-two or phase-three assessments are more detailed. They evaluate the feasibility of remedial action alternatives. Phase-three audits are generally performed as the result of regulatory requirements. It is a remedial investigation and feasibility study (RI-FS) if performed in the context of Superfund. Or, a phase-three audit can be a Resource Conservation and Recovery Act (RCRA) facility investigation or assessment (RCI-RCA) if performed under RCRA.

New regulations that address the storage, treatment, transport, and disposal of hazardous materials have made site assessments a major risk-reduction tool for those involved in real-estate transfers or in a business that uses hazardous materials. Most metals and many products used at mine facilities are defined as hazardous.

The concerns of purchasers arise primarily from the scope of liability under the Comprehensive Environmental Response, Compensation and Liability Act (CERCLA) or Superfund. CERCLA extends liability for clean-up costs and damages to current and former owners and operators, regardless of who caused the contamination.

Environmental risks and associated liabilities can occur due to soil, groundwater, surface water, or air quality contamination resulting from present or past land-use practices. Contamination may also occur due to past or future migration of chemicals onto a prospective site from adjacent properties. Most comprehensive insurance today will not cover any of these potential liabilities.

Lenders can also be exposed to liability even if they never actually take possession of a site. They may be considered operators if they get too involved in day-to-day decision making. They may be indirectly affected by lending money to a client whose collateral property holding is found to be contaminated or are included in a statutory super lien to cover other clean-up costs. Lenders may be indirectly affected if the borrower's ability to repay is impaired by the cost of clean-up at another site.

Those contemplating the purchase of a site, acting as a lender for a buyer, managing a real-estate based trust, or repossessing a site through foreclosure should ensure that they:

- Exercise due diligence under CERCLA-RCRA regulations (They should make all appropriate inquiries regarding potential contamination before acquiring the property.)
- Know the facility or site and what regulations apply
- Know about any potential for or the presence of contamination that might require future expenditures

A site assessment can provide this information. It gives the present and future property owners and the lender a way to reduce their financial exposure.

Mine operators must understand the requirements of all environmental laws and regulations applicable to their facility. This is necessary to ensure compliance or at least to recognize noncompliance in day-to-day operations. A complete understanding of the regulations will help to protect mine operators' interests should they be subject to a regulatory agency review and inspection.

Regulatory requirements continue to grow with the passage of state and local laws that must be reviewed on a site-by-site basis. Federal laws, however, are applicable to all mining facilities.

CERCLA establishes a mechanism of response for the immediate clean-up of hazardous waste contamination, accidental spills, or chronic contamination (abandoned, hazardous waste disposal sites). The Environmental Protection Agency (EPA) has promulgated regulations that establish the quantity of any hazardous substance that, if released, should be reported to the National Response Center.

The Superfund Amendments and Reauthorization Act (SARA)—Title III has two main components. Subtitle A establishes the framework for emergency planning by state and local governments. It creates a state emergency response commission, as well as local emergency planning committees. This section requires these local panels to work with representatives of facilities covered by the law on emergency response plans.

Subtitle B requires certain facilities to provide information to appropriate state, local, and federal officials on the type, amount, location, use, disposal, and release of chemicals. There are several reporting provisions contained in Subtitle B:

- Section 311 applies to facilities subject to the Occupational Safety and Health Act. These facilities must submit material safety data sheets or a list of the chemicals for which the facility is required to have material safety data sheets to local emergency planning committees, state emergency response commissions, and local fire departments.
- Section 312 establishes an inventory of toxic chemical emissions from facilities meeting certain criteria. Facilities subject to this reporting requirement must complete a toxic chemical release form for specified chemicals.

The RCRA hazardous waste programme regulates all aspects of the management of hazardous waste from generation to disposal. Major components of the programme include regulations for the identification of hazardous waste, notification of any hazardous waste activities, and compliance with standards for generators, transporters, and treatment-storage-disposal (TSD) facilities. Owners and operators of TSD facilities are also required to obtain a permit.

Mining waste regulations under RCRA are currently being developed.

The Clean Air Act provides the basic framework for modern air pollution control. Key elements of the act include the establishment of human health-based ambient air quality standards and technology-limited uniform national emission standards. It also provides for prevention of significant air quality deterioration. Each state has specific air regulations to implement these programmes.

The Clean Water Act addresses point and nonpoint sources of pollution. Point sources include industrial discharges. Nonpoint sources include mining and other construction activities that cause runoff into streams. Point sources are subject to five different effluent limitations administered through the National Pollutant Discharge Elimination System (NPDES). Control of nonpoint sources of pollution is left to the states. They are required to formulate plans that contain land-use regulations to control nonpoint sources. Hazardous waste and oil spills are also addressed in this act. Facility operators are liable, without fault, for the costs of cleaning up spills and for civil and criminal penalties.

The Safe Drinking Water Act (SDWA) was enacted to ensure safe drinking water supplies, protect especially valuable aquifers, and protect drinking water from contamination by underground injection of waste. Under this act, the EPA established a series of drinking water standards to protect public health. These standards apply to public water that regularly supplies water to 15 or more connections or 25 or more individuals at least 60 days a year. This definition applies to most industrial establishments that supply water to employees.

The SDWA's most direct effect on the industry is through the regulation of underground injection to protect usable aquifers from contamination. The regulations address hazardous waste disposal, the reinjection of brine from oil and gas production, and certain mining processes. The underground injection-control programme is administered through a permit process with substantive requirements depending on the type of injection taking place. The most stringent conditions are for well injection waste classified as hazardous under RCRA.

The Toxic Substances Control Act (TSCA) provides the EPA with authority to require testing of chemical substances, new and old, entering the environment and to regulate them where necessary. This authority supplements sections of existing toxic substance laws, such as Sections 112 and 307 of the Clean Water Act and Section 6 of the Occupational Safety and Health Act. These already provide regulatory control over toxic substances.

Although the heart of the TSCA is the requirement for premanufacture notification, the area of interest in an environmental audit relates to the regulation by TSCA of polychlorinated biphenyls (PCB). Section 6(a) of the TCSA directed the EPA to phase out PCB manufacture and prevent the process, distribution, or use of PCBs except in a totally enclosed manner. The EPA regulations require transformers containing PCBs to be appropriately labelled. The agency sets standards for the transportation of PCBs and recommends certain disposal techniques.

1.11 TYPES OF ENVIRONMENTAL AUDITS

There are several types of environmental audits[2] that can be tailored to suit specific needs:

Permit performance audits (compliance and monitoring). This is a review of environmental quality assurance plans, environmental permits, and agency-required operating restrictions procedures. It assesses possible or actual non-conformance (especially regarding air and water emissions and hazardous materials management). This type of audit also interprets regulatory agency permit conditions and suggests measures for ongoing permit conformance. It may also involve long-term monitoring of environmental activities.

Regulatory requirement audits. This provides a detailed evaluation of facility operations that are or may be governed by local, state, or federal environmental regulations. It identifies applicable regulations to pinpoint potential noncompliance or conflicts with such regulations. Procedures are also recommended for coming into compliance.

Environmental management practice audits. This type of audit examines existing management structure, procedures, and policies used by the client to implement environmental compliance and to communicate environmental-regulatory awareness (including health and safety) to workforce personnel. Recommendations are also provided for the remediation of deficient practices.

Technical processes-practices audits. Production practices and facility conditions are reviewed to determine whether design or process modifications should be made to accomplish specific environmental goals (minimizing hazardous waste, waste stream treatment, or technology transfer).

Risk management audits. Practices, procedures, and policies are surveyed to identify sources of risk. It suggests how risks of environmental (or health and safety) incidents, accidents, and liability exposure can be reduced or eliminated. A risk management audit may also include a formal risk assessment study or contingency planning component.

Special purpose audit. This is a one-time audit conducted in response to unusual circumstances or requirements, such as an SEC 10k report, an EPA consent decree, insurance-liability impairment determination, or emergency response plan.

Site assessment audits. This consists of a thorough examination of previous and current environmental hazards and physical conditions on or surrounding a facility site. Its purpose is to assess potential on-site problems or sources of external encroachment, contamination, or threat. This audit includes measures to remediate or reduce such problems before they affect operations. A site assessment audit is particularly useful as a planning and predevelopment decision-making tool for suspected problem sites. It is necessary before property transfer or asset sale acquisition.

BIBLIOGRAPHY

1. Cook, F., *Evaluation of Current Surface Coal Mining Overburden Handling Techniques and Reclamation Practices* (Washington, DC: U.S. Department of Commerce, National Technical Information Service, 1976), PB-264–111, pp. 60–123.
2. Philbrook, J.N., Environmental audits: Determining the need at mining facilities, *Mining Engineering*, 43(2): 207–209 (1991).

2 Surface Coal Mining with Reclamation

2.1 INTRODUCTION: ENVIRONMENTAL EFFECTS OF SURFACE COAL MINING

Surface coal mining, commonly called strip mining, produces one of the worst impacts on environment. Surface coal mining destroys landscape, disturbs ground water, raises dust cloud, creates noise, and various other environmental disturbances. Surface mining can be conducted so as to minimise these environmental effects.

2.2 DRAGLINE OPERATION

A simplified dragline operation is illustrated in Figure 2.1. The overburden is excavated, and the coal seam is uncovered along a mining panel. The spoil material is cast into the adjacent mined-out pit area. Usually, the dragline cuts a trench referred to as a keycut, adjacent to the newly formed highwall. The keycut is made from Position 1. The distance between the previous keycut position and the present keycut position is known as the digout length. Sometimes, this is also referred to as "length of block." The keycut material is deposited in the bottom of the keycut, mined-out pit. Sometimes, the keycut material can be used to turn into an extended bench. The extended bench method is described in subsequent sections.[1]

When the keycut has been completed, the dragline moves to Position 2. From this position, the dragline excavates the digout and casts the spoil into the adjacent area of the pit. Position 2 is known as the production cut. When the excavation is completed, the dragline moves to Position 3 to start a new keycut.

The operating cycle of a dragline is separated into five distinct steps:

1. The empty bucket is placed in a position ready to be filled in the cut. This position may vary from the maximum position, which is the interface between the coal seam and the overburden, to any position in the overburden above the coal.
2. The bucket is dragged towards the dragline to fill it.
3. The filled bucket is hoisted up, and the boom is swung out towards the spoil pile. When the swing has to be slowed to permit hoisting, the operation is said to be hoist critical. When hoisting to the dump point is completed before the boom reaches the dump position, the operation is considered as swing critical. While swinging out to the spoil, the drag cables are released so that the filled bucket can be lowered down to the dumping position under the boom point. If the swing is slowed to permit payout of the drag cables, the operation is said to be payout critical.
4. The bucket dumps the spoil material it is carrying.
5. The bucket is lowered, and the boom swings back to the cut. In the return part of the cycle, the empty bucket is lowered from the dumping position to the digging position by reeving in the drag cables. If the return swing is slowed to permit reeving in the drag cables, the operation is said to retrieve critical. If the return swing is retarded to allow lowering of the bucket, the operation is said to be lower critical.

Optimization of the dragline operating cycle requires minimization of the time required to position, drag, and dump; and synchronization of hoist/swing-out and retrieve/swing-in to the minimum critical state.

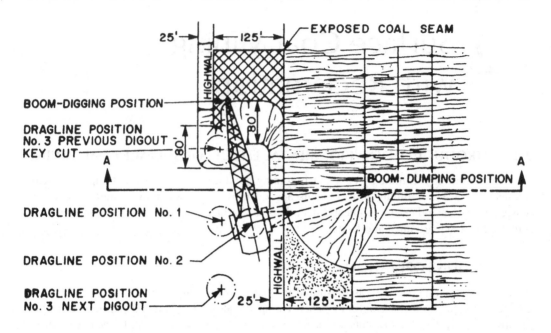

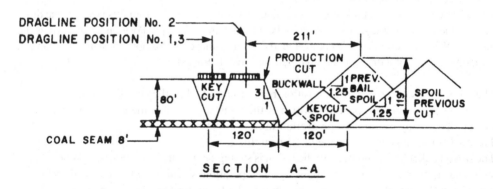

FIGURE 2.1 A typical dragline operation.

2.2.1 KEYCUTTING AND LAYERED CUTTING

When a new digout is initiated, the dragline is positioned directly over the toe of the future highwall (Position 1, Figure 2.1). In this position, a clear and safe highwall can be established on stable ground.

A keycut is made from this position over the new highwall. The keycut separates the new high-wall from the remaining overburden to be excavated. The keycut is about one to two buckets wide at the bottom. The objective of making a keycut is to establish a steep and safe highwall. A keycut may not be needed when the overburden falls truly from the highwall, and a relatively clean highwall slope can be formed by the digging actions of the dragline.

The panel width and dragline size generally dictate a two or more position digout. Only the largest dragline operating within ideal geometric parameters allows digout from one position while providing a panel wide enough for efficient coal production. In the first position, the keycut material is cast short, adjacent to the rib of the coal being stripped, so as to minimize the cycle time. This short-casted material forms a "bucket-wall" slope. The dragline then moves out towards the old

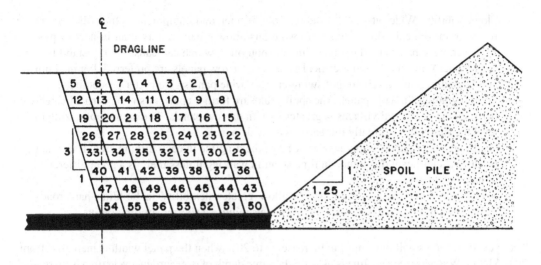

FIGURE 2.2 Layered cutting sequence. (From Fluor Engineers, Inc. Optimal Dragline Operating Techniques, U.S. National Technical Information Service, DE83-006980 (1982), pp. 10–52.)

highwall (Position 2, Figure 2.1) and completes the digout. The material is cast behind the apex of the keycut spoil pile.

In some situations where the dragline dumping radius and dumping height are compatible with overburden depth, the dragline may be able to excavate the complete digout from the position over the toe of the new highwall. The width of the digout is excavated in layers called "layered cutting." The keycut is formed as the layers are removed. Figure 2.2 illustrates the cutting sequence for layered cutting. Each layer is usually one bucket deep. The numbers in the figure refer to the digging sequence. The highwall is "dressed" prior to completion of each layer.

The layered cutting can be performed in another manner in deeper overburden by positioning the dragline in the production boiling position. A tractor dozer is needed to dress the highwall slope while the dragline is digging.

In some mines, the keycut is dug first, and then the production cut is excavated by layers. A small swing-critical dragline is used to excavate wide panels in deep overburden. An extension of the bench is necessary to provide spoiling reach. Between each layer of the production cut, the dragline would take a few steps towards the spoil pile to which it was casting (termed "layered walk"). This walking action between layers helps to minimize swing angle while digging down through the layers.

2.2.2 Panel (Pit) Width

Panel spill width is primarily a function of the dragline size. Other factors influencing optimum panel width include dragline operating rate, overburden depth, swell of spoil, and coal production requirements.

Small draglines operating on wide panels become swing critical, causing a large increase in cycle time. Medium and large draglines usually have fast swing rates but comparatively slower hoist, payout, and drag receiving rates. Therefore, they are usually hoisted, pay out, and return critical. But this does not affect cycle time as much as its being swing critical. The cycle time of medium and large draglines does not increase significantly with an increase in panel width. Due to its short reach, a small dragline cannot strip a wide panel. For working on a wide panel, a mall dragline may be required to rehandle spoil (e.g., extended bench). A large dragline has only to increase its swing angle to operate efficiently on a wide panel.

The optimum panel width is determined from consideration of the following factors:[2]

- Coal loadout: The practical minimum width for a panel is 28 m. Panel width less than 28 m greatly hampers coal haul truck manoeuvrability.

- Slope stability: Wide pits are considered safe for men and equipment as they allow more room for movement. Slope failures in wide pits allow greater safety than in narrow pits. However, wide panels tend to create higher spoil piles, which can cause spoil instability.
- Cycle time: With small, swing-critical draglines, narrow panels are preferred. For medium and large draglines which are not swing-critical, wide panels give better productivity.
- Spoil regrading: In wide panels, the spoil peaks are farther apart. Also, the vertical height between the peaks and villages is greater than in narrow panels. The amount of dozing to level the spoil piles is greatly increased with wide panels.
- Walking: Wider panels imply less walking of the dragline in each mining area. It may not produce a significant difference in time spent in walking unless there is a great difference in panel widths.
- Spoiling at entryways: Narrow panels allow greater flexibility in building haul roads through spoil, inside curves, etc. Narrow panels require a shorter spoiling radius.

The cycle time for a small dragline can increase up to 20% when the panel width is increased from 22 to 53.3 m. When a large dragline is used on the same depth of overburden, cycle time is increased by only 1.6% when panel width is increased from 75 ft (22.8 m) to 175 ft (53.3 m).

2.2.3 EXTENDED BENCH

This method is usually used to increase dragline reach. The bench width is extended with spoil, so the dragline is capable of dumping spoil off the coal. Figure 2.3 illustrates a typical extended bench with the dragline positioned on the spoil extension. The material shown in cross-hatching will require rehandling by the dragline.

Sometimes an extended bench becomes indispensable for a particular stripping situation. Reasons may be as follows:

- Overburden depth
- Highwall stability
- Necessity of a wide pit for safety during coal removal
- High swell of spoil material
- The limited reach of the dragline

The main disadvantage of the extended bench is the rehandling of the spoil, which increases the total volume of material that must be moved to uncover the coal.

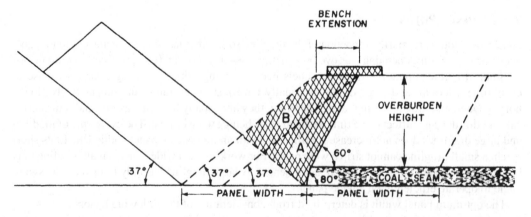

FIGURE 2.3 Sectional view of an extended bench. (From Fluor Engineers, Inc. Optimal Dragline Operating Techniques, U.S. National Technical Information Service, DE83-006980 (1982), pp. 10–52.)

2.3   DRAGLINE STRIPPING PROCEDURES

Dragline stripping procedures are divided into different systems depending upon the physical and technological parameters of the coal property. The classification of dragline systems is presented in Figure 2.4. As shown in the figure, operating procedures are dependent upon the number of coal seams mined per pit, the spoil placement method, the number of overburden and interburden lifts, and the number of dragline passes per pit. The dumping radius (range) of the dragline is a basic parameter because the classification of the overburden as shallow or deep is relative to the range of the machine. Identified in Figure 2.4 are 17 different classifications.

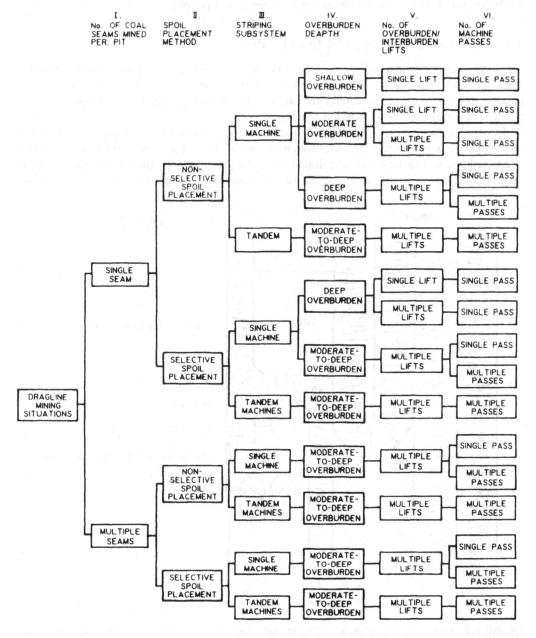

FIGURE 2.4   Classification of a dragline mining system. (From Math Tech, Inc., *Evaluation of Current Surface Coal Mining Overburden Handling Techniques* U.S. National Technical Information Service, PB-264 111 (1976).)

2.3.1 SINGLE-SEAM STRIPPING WITH NONSELECTIVE SPOIL PLACEMENT

When a single coal seam is mined and the overburden is placed non-selectively, the practice can be described as the least complex of the dragline operating systems.

2.3.2 SINGLE MACHINE SUBSYSTEMS

There are five basic procedures for mining a single coal seam using a single-dragline and nonselective overburden placement. These procedures depend upon overburden depth, the number of overburden/interburden lifts, and the number of draglines passes per pit.

2.3.3 SHALLOW OVERBURDEN, ONE LIFT, ONE PASS

The dragline operating in a shallow overburden occurs early in the life of a mine when mining takes place in a shallow overburden. Basic operating procedures are shown in Figure 2.5. The dragline operates from a position on the natural ground surface, either after a thin layer of top soil has been removed by scrapers or from a position on a bench.

The dragline is positioned directly over the desired location of the new highwall. A keycut is made. The keycut is a narrow trench, about one or two buckets wide at the bottom. It is excavated to establish a safe, new highwall. The dragline must be placed directly over the keycut in order to establish a clean, safe highwall. The spoil material from the keycut can be placed close to the old

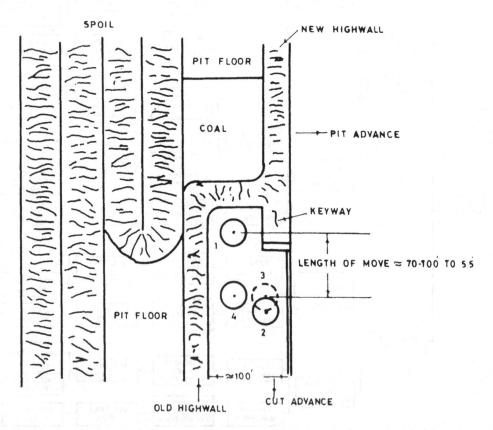

FIGURE 2.5 Plan view of dragline operating in shallow overburden. (From Math Tech, Inc. Evaluation of Current Surface Coal Mining Overburden Handling Techniques, U.S. National Technical Information Service, PB-264 111 (1976).)

highwall in the adjacent open cut, or the dragline can swing and dump over the apex of the final spoil pile. The former procedure, called logging the spoil, minimizes dragline swing angle and cycle time.

In shallow overburden, the dragline has sufficient range to excavate the overburden remaining in the keycut. Two practices are commonly observed. In the first, the spoil is dumped on a fixed apex point, resulting in the deposition of the spoil in conical piles. In the second method, the boom is swung only the required minimum distance on each cycle, and the spoil is deposited along a curvilinear ridge line. The second method is considered superior to the former as swing angles are minimized, resulting in minimum cycle times and maximum productivity, as well as more efficient use of spoil room.

After having excavated an entire block of overburden which may be from 13 to 60 m deep (depending on the size of the dragline), the machine is moved back to a new place as indicated by Position 2 in the plan view of Figure 2.5. The distance the machine is moved is called the length of the digout, the digout being the block of overburden excavated from a given machine position. The length of the digout affects both dragline efficiency and spoil grading costs. Many operators tend to make mistakes. A spoil ridge line becomes undulating because spoil is deposited in widely separated conical piles. The alternative is to shorten the digout and to develop a sharp spoil ridge line. The widely spaced spoil piles may have two major disadvantages. Some part of the spoil storage room is lost in the gaps between spoil piles as shown in Figure 2.6. During grading of the piles, dozers must be used to flatten the ridge line before the main grading activities can be performed. When the spoil ridge line is even, this procedure is fairly straightforward.

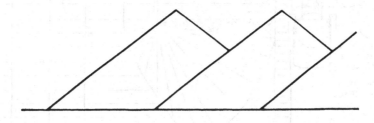

Crest-to-Crest Spacing: 90 feet.

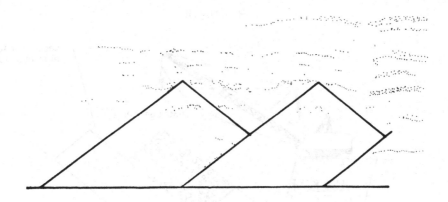

Crest-to-Crest Spacing: 120 feet

FIGURE 2.6 Comparison of spoil piles with 90- and 120-ft crest-to-crest spacing. (From Math Tech, Inc. Evaluation of Current Surface Coal Mining Overburden Handling Techniques, U.S. National Technical Information Service, PB-264 111 (1976).)

Any increased length of the digout increases the dragline cycle time. The longer the digout, the farther the bucket will be dragged towards the machine before hoisting. The solution lies in determining an optimum digout length. The dragline may move to Position 3 from Position 2 to complete the keycut, so as to prevent the drag ropes from dragging on the highwall with increasing depth of keycut. Then the dragline moves to Position 4 from Position 3 and completes stripping setup.

Figure 2.7 shows the final cut removal in the simple sidecasting method.

When the dragline reaches the end of the pit, two alternative procedures can be followed. Where the coal is not too thick and there is no need to maintain a stock of exposed coal seams in the pit,

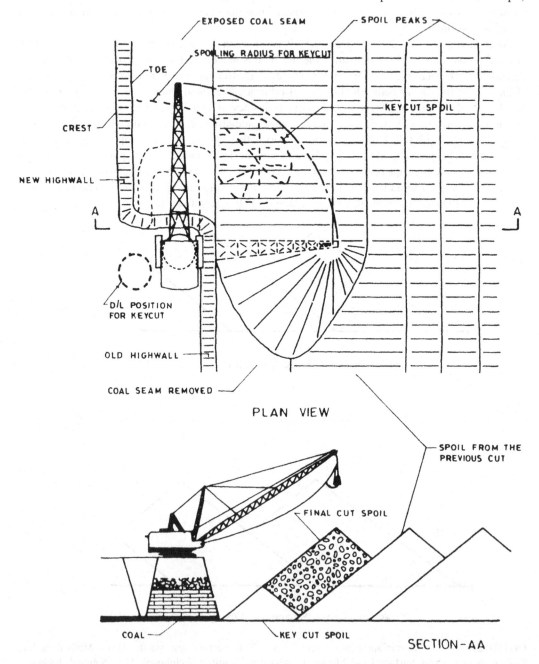

FIGURE 2.7   Sidecasting—final cut removal. (From Math Tech, Inc. Evaluation of Current Surface Coal Mining Overburden Handling Techniques, U.S. National Technical Information Service, PB-264 111 (1976).)

coal extraction closely follows the dragline. At the end of this pit, the dragline is turned around, and stripping commences back in the opposite direction. Where the coal seam is thick and the dragline stripping operation is far ahead of the coal loading, or where it is desired to maintain an inventory of exposed coal seams in the ground, the dragline is deadheaded back to the other end of the pit, and stripping begins again.

The width of the pit affects dragline productivity and spoil grading costs. The effect of pit width is less critical in shallow overburden than in deep or moderately deep overburden. Narrowing of the pit results in a decrease in dragline cycle times as swing angles are reduced. If the pit is narrowed too far, the swing angle becomes so small that the loaded bucket cannot be fully hoisted during the swing. So narrowing the pit beyond certain limits may reduce dragline productivity.

Narrowing of the pit in a specific situation reduces spoil grading volumes and costs. As the pit is narrowed, the crest-to-crest spacing of spoil piles and the depth troughs between piles are reduced. The grading cost per hectare of spoil piles with 40-m crest-to-crest spacing is roughly double that for spoil piles with 30-m spacing.

The minimum pit width is determined primarily by the room needed for coal leading and haulage equipment. At large mines, 30–35 m is the practical minimum. In very deep mines, wider pits are used for safety reasons. A rule of thumb often used is that the pit should be at least as wide as it is deep. The maximum pit width is determined by many factors, such as the dumping radius of the dragline, the overburden depth, and the volume of rehauled spoil.

As the height of the spoil pile increases, the bucket hoist time increases. Usually, the short-swing-angle sections of the overburden are dug early in the digout when the height of the spoil pile is small.

The choice of pit widths is also influenced by topography. If the overburden depths quickly increase in successive cuts, the early cuts in shallow overburden may be made fairly wide, and the widths of successive cuts can be progressively narrowed.

Summarizing, with shallow overburden, operating decisions must be made according to the following factors:

- Width of the keycut
- Spoil casting pattern—conical piles or curvilinear ridge lines
- Bucket loading method—layer loading or non-layer digout
- Length of the digout
- Pit width
- Digging and casting pattern

2.4 BOX PITS

In some cases, the first cut to begin a new mine is commonly a box cut. This produces an initial highwall. This pit gets its name from the long box-like shape of the completed cut. There are two distinct methods of making the box pit without rehandling material:

- End cut
- Side cut

For very thick overburdens relative to dragline size, methods which entail certain quantities of the rehandle are necessary:

- Spoiling on both sides of the box cut
- First producing a borrow pit on the spoil side of the box cuts

Each of these four methods is discussed below.

2.4.1   END CUT

The end-cut box pit is produced with the dragline positioned at the end of the pit, as shown in Figure 2.8. This method is used in relatively thin overburden compared with the dragline dimensions. Thus, it is possible for the machine to spoil all the material to the side away from the coal deposit or what becomes the opposite side of the cut from the new highwall. All the material can be spoiled in this case using a maximum dragline swing angle of 60°. The dragline operating cost is greatly dependent on the angle of swing, as well as the percentage rehandle. The end-cut method has a very favourable dragline swing angle (maximum 90°), especially when compared with the side-cut method. However, the end-cut method is restricted to relatively shallow overburden thicknesses because of the limit on the dragline dumping radius. As the cut increases in depth, the waste pile increases in volume, until it can no longer be spoiled completely clear of the box pit.

The rehandle in these box-pit cuts is either 100% or nil, depending on whether the coal seam does or does not continue under the spoil pile. If the coal extends under the spoil pile because the box pit was started at a mining lease boundary, then it may be economical to recover the covered coal later. One method frequently used is the auguring technique. However, on occasion, this coal is lost.

2.4.2   SIDE CUT

The side-cut box pit is produced with the dragline positioned at the side of the pit, as shown in Figure 2.9. This method is used when the overburden is of such a depth that it is impossible for the dragline to use the end-cut position. The side position allows the spoil to be dumped at a greater distance from the pit, making it possible to widen the spoil pile at the maximum dumping height, as required.

Figure 2.9 shows that the swing angle can be increased to the theoretical maximum of 180° for certain parts of the cut. In practice, the highwall side of the pit would be allowed to lag behind the near side, reducing the maximum swing angle to approximately 130°. This is still considerably larger than for the end-cut method, but for thicker overburdens and this given machine size, it is the only way the box pit can be produced without the rehandle.

2.4.3   REHANDLE (END CUT)

An extension of the regular end-cut method incorporates spoiling from the end position to both sides of the cut. This method can be used to allow the end-cut position to be maintained as the overburden thickness increases. However, the attraction of reduced swing angles, as compared with the side-cut method, does not usually outweigh the additional material rehandle. It can also be used producing a regular end cut which was temporarily being produced in material of lower angle of repose. Thus, the end cut could be continued, with the excess spoil being placed on the highwall side. Figure 2.9 shows a plan and cross-sectional view of a box pit produced in this manner. Again, as with the regular end-cut method, the angle of the swing is seen to be a maximum of 90°.

Having produced this type of box cut, the material above the coal on the highwall side, which would be up to a maximum of 50%, must be rehandled in some manner. There are many methods available, including using a shovel/truck or loader/truck operation. However, the most common method is the use of dozers or dozer/scraper combinations. In this case, the material is then spread out on the highwall side to enable the dragline to rehandle it during later cuts. This offers the possibility of an advantage for this method. If the ground surface is pitted and uneven, the use of this extra spoil as a fill can enable the production of a good dragline floor for use in later cuts. This, then, could even reduce the later work of dragline pad production. Similarly, if the ground surface is unsuitable for the dragline bearing pressure in some areas, then selected material from the box cut can be spoiled on the highwall side.

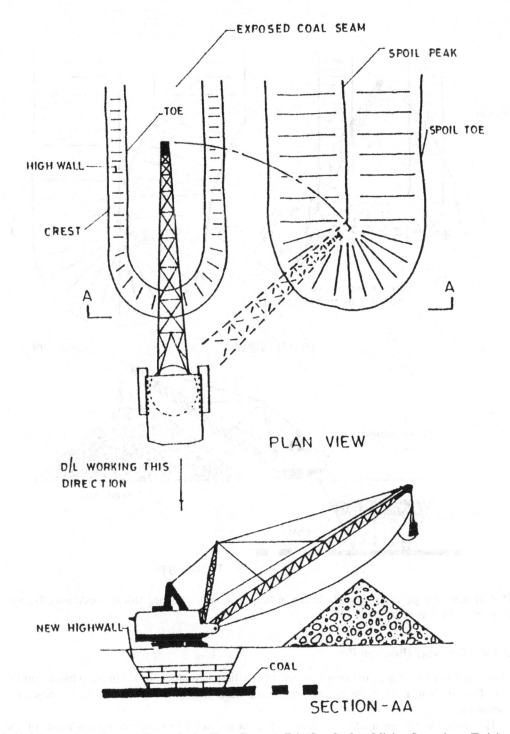

FIGURE 2.8 Box pit—end cut methods. (From Bucyrus-Erie Co., Surface Mining Supervisory Training Program, 1977. With permission.)

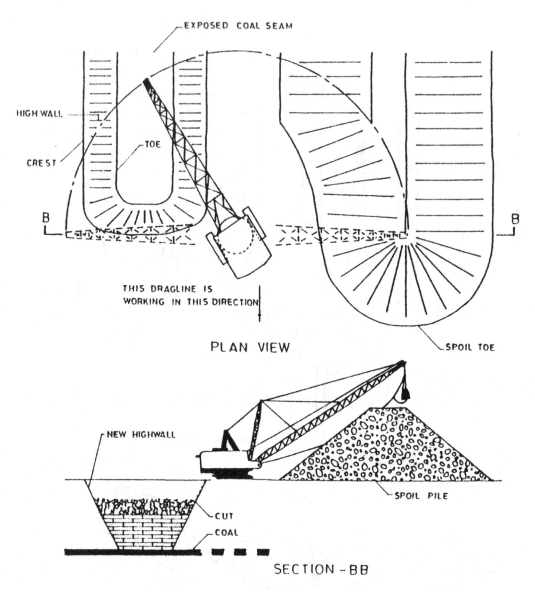

FIGURE 2.9 Box pit—side-cut methods. (From Bucyrus-Erie Co., Surface Mining Supervisory Training Program, 1977. With permission.)

2.4.4 REHANDLE (BORROW PIT)

This rehandle method does not entail the use of auxiliary equipment to aid in the rehandle operation. The aim is to spoil all the overburden on the waste side of the box cut by first producing a borrow pit.

The initial cut for this method produces a borrow pit parallel to, and on the waste side of, the proposed box pit, as illustrated in Figure 2.10.

The borrow pit encompasses the sum total of the rehandle. The volume of the rehandle depends solely on the amount of extra volume required to spoil the final box cut into a single pile. The final box cut itself can be an end or side cut, depending on the overburden cover. The side cut again provides the means of digging the maximum overburden depths.

Figure 2.10 shows the box cut being spoiled into the borrow pit. This figure suggests the dragline has deadheaded to the beginning of the borrow pit and is now returning in the same direction of

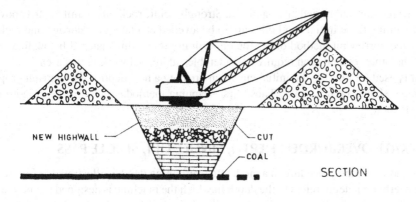

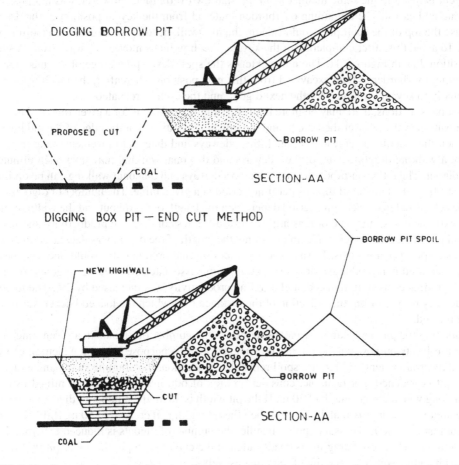

FIGURE 2.10 Box pit—end cut with rehandle, borrow pit, and digging box pit—end-cut method. (From Bucyrus-Erie Co., Surface Mining Supervisory Training Program, 1977. With permission.)

advance. This is certainly not essential. For most operations, the dragline would simply turn round and dig the box cut into the borrow pit with no deadheading. Suitable access would be designed into the cut from the appropriate end or, preferably, from both ends.[3]

In general, the larger material is kept near the base of the pile, whereas the clay would be dumped on top. It is frequently advantageous, and sometimes essential for reclaim purposes, to ensure that certain materials remain well buried after spoiling. Strongly acidic rocks are common just above the coal seams. Assuming these initial spoil piles are to be levelled at a later date during land reclamation, then only the borrow pit method offers the opportunity to get this material below final grade immediately. The other methods can require considerable, additional work in this area.

The box pit is used for making the initial cut when opening a new deposit. For a given property, only one such cut should be necessary. These first stripping methods, therefore, represent only a small part of the total.

2.5 MODERATE OVERBURDEN DEPTH, SINGLE LIFT, SINGLE PASS

A one-lift, one-pass dragline procedure in a moderate overburden depth is illustrated in Figure 2.11. The moderate overburden depth refers to the depth in which the machine is designed to operate with little or no spoil rehandle. The figure depicts a one-lift operation where the dragline operates from the natural ground surface.

Procedures are similar to those used in shallow overburdens except for two major differences. A keycut is made in the same manner as in the shallow overburden. However, it is not possible to excavate and cast all the remaining overburden material from the keycut position of the dragline, because the top of the spoil pile would ride up the highwall, needing subsequent rehandling of the spoil. To avoid this, after completion of the keycut, the dragline is moved sideways from Position 1 to Position 2 as in Figure 2.11. The move extends the spoil-developing range of the machine. The remaining overburden is then excavated from this new position. Thereafter, the machine is moved diagonally to keycut Position 3 for the next digout, and the cycle is repeated.

In some situations, there may be more than two dragline positions for a given digout. The beginning position is always over the keycut, and the finishing position is at the edge of the old highwall.

When the machine is crawler-mounted, the sideways and diagonal movements take little time. Where a walking dragline is used, the sideways and diagonal walking may take 5–15 minutes on each digout. The total non-productive time due to sideways and diagonal walking can be optimized by controlling the number of digouts per time period at a given mine. The number of digouts which can be completed in a given time period depends on the length of the digout and the width of the pit.

In extreme cases, this type of reasoning can result in loss of dragline productivity and increase in spoil grading costs. For example, in one mine, the length of the digout was decreased in an effort to reduce spoil grading costs. It was expected that dragline productivity would decrease because of the increased non-productive dragline-walking time associated with shorter digouts. As it happened, grading costs were reduced, and dragline productivity was increased by 2%. The increased productivity resulted from the reduction of the distance over which the loaded bucket was dragged being hoisted.

Widening the pit can be another means to reduce the number of digouts in a given time period. There are limits on how far the pit can be widened for this purpose because widening of the pit beyond a certain extent will require spoil rehandle. In addition, the dragline swing angle increases as the pit is widened. For example, consider an area of 600 m² which is to be mined using pits 600-m long with a digout length of 30 m. If the pit width is 30 m, the number of digouts to mine the entire area will be approximately 2,640. When the pit width is increased to 50 m, if this is feasible without resulting in unnecessary spoil rehandle, the number of digouts is reduced to 1,760. This is a reduction of 33%. If the dragline is deadheaded at the end of each pit, the wider pit will result in less deadheading time over the life of the mine as well.

With moderate depth overburden, the decision variables are the same as with shallow overburden, but the sideways movement of dragline on each digout complicates the decision-making process.

If the advantages and disadvantages of long and short digouts and wide and narrow pits are known, it may be difficult to arrive at an optimum decision. However, mine operators do take these decisions with varied results.

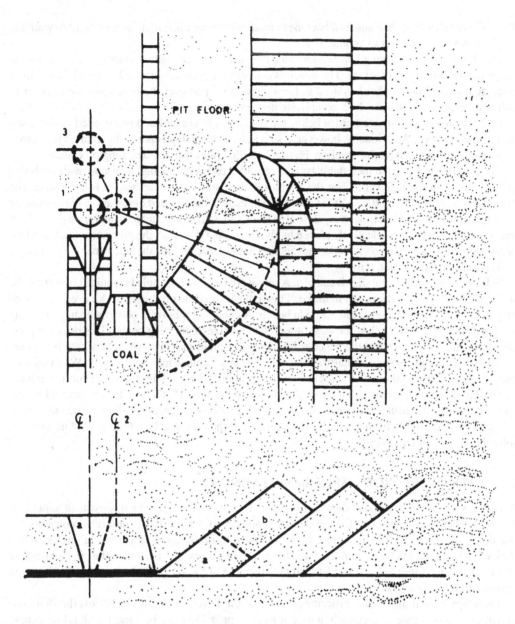

FIGURE 2.11 Dragline operating procedures in moderate depth overburden. (From Math Tech, Inc. Evaluation of Current Surface Coal Mining Overburden Handling Techniques, U.S. National Technical Information Service, PB-264 111 (1976).)

In a moderate overburden situation, the effect of pit geometry on productivity is more pronounced than in a shallow overburden situation. As the overburden becomes deeper, it becomes more difficult to keep the coal haulage roads open.

2.6 MODERATELY DEEP OVERBURDEN, TWO LIFTS, ONE PASS

Very often, the dragline works from a machine-supporting bench, rather than from the natural ground surface. The natural ground surface can consist of unconsolidated material which may be fairly deep and might not support the dragline when it is positioned close to the highwall edge.

The unconsolidated surface material becomes muddy in wet weather, making it difficult to walk the dragline from one position to another.

The dragline works from a bench—known as the established bench—which typically is 3–6 m below the natural ground surface. The bench is established over consolidated material. Sometimes competent shale from the bank is used to form a road or a pad over the bench surface in order to eliminate muddy conditions which complicate dragline walking.

On each pass of the pit, overburden below the bench height is excavated and spoiled as discussed in previous sections. Additionally, however, a side bench, that is the bench to be used on the subsequent pass, is excavated by the dragline. Figure 2.12 illustrates the side benching procedure.

The dragline works from the established bench and is turned to the highwall side, with the boom at right angles to the highwall. Then the unconsolidated side bench overburden is excavated. The boom is swung through a 180° angle, and side bench material is dumped onto the top of the existing spoil pile. This can be sometimes very desirable from a reclamation standpoint and is an example of one way in which spoil is selectively placed by the dragline. Of course, the procedure is adopted for production reasons, but the reclamation advantages are notable. In fact, the side benching procedure is also used for reclamation purposes.

Side benching has several disadvantages. A relatively large swing angle is required in cutting the side bench and spoiling the bench material. This reduces dragline productivity. An overhand chopping motion is required to excavate the side bench. A chopping motion is one in which the digging is done above rather than below the elevation of the machine-supporting bench. This does not present significant problems when the bench is high but may cause problems when benching is deeper.

When the side bench spoil is shallow, the reclamation advantage may be apparent. When the side bench spoil is fairly shallow on the surface of the spoil pile, the grading dozer will cut through the material and expose the underlying spoil. As shown in Figure 2.13, the side bench material on the spoil pile should be sufficient to prevent breakthrough by dozers during grading. If the side bench material is not desirable from a reclamation standpoint, then the side benching procedure may not solve the reclamation problem.

2.7   DEEP OVERBURDEN, SPLIT-BENCH MINING

A section view of this method is shown in Figure 2.14. On the first lift, which consists of about half the overburden depth, the operating procedure is similar to the previously described conventional practice. The top lift spoil is cast close to the highwall to minimize dragline swing angles. When the top lift stripping has been completed and the dragline has reached the end of the pit, the dragline is walked down a ramp to the lower lift and the stripping of the second lift proceeds in a direction opposite to that of the first lift.

There are several alternative procedures to make the keycut in the lower lift overburden. The keycut cannot be made immediately adjacent to the upper highwall because the dragline course would hit the upper highwall during rotations of the boom to cast the keycut spoil. The lower lift keycut has to be made at some distance from the upper highwall. To accomplish this, the top lift bench must be made wider than the pit.

An alternative procedure is to make the lower lift keycut using other equipment: typically, dozers and front-end loaders. Dozers make the keycut by pushing the overburden down into the pit, where it is loaded and placed using a front-end loader. When this procedure is followed, the lower lift keycut can be cut adjacent to the upper highwall. When the overburden is deep, this procedure is not suitable for safety reasons.

The highwall is cut on a slope of 65° to 75° from the horizontal. The dragline is closer to the second lift than to first lift. This can be seen in Figure 2.14 by comparing the dragline centrelines for Passes 1 and 2. This situation effectively increases the dumping radius of the dragline. For example, in a 25-m overburden depth with a 13-m bench depth and 65° highwall, the dragline would be 9 m closer to the spoil side on the second lift than on the first. This increases the effective dumping

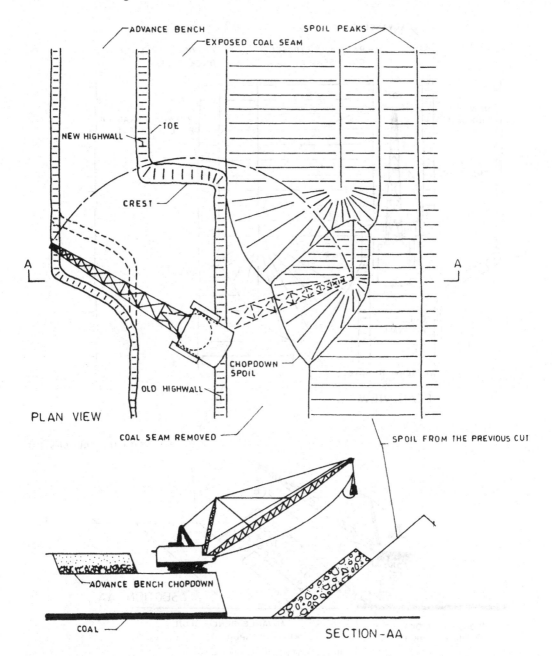

FIGURE 2.12 Dragline making chop down cut using an advance bench. (From Math Tech, Inc. Evaluation of Current Surface Coal Mining Overburden Handling Techniques, U.S. National Technical Information Service, PB-264 111 (1976).)

radius of the dragline, enabling no-rehandle stripping of overburden deeper than the design limit of the machine. The design limit of the dragline is defined as the maximum depth of overburden that can be excavated and spoiled without spoil rehandle. This limit is dependent upon the dumping radius of the machine, the distance of the machine centreline from the highwall during stripping, the angle of repose of the soil, the pit width, the spoil swell factor, and the highwall angle. For example, a maximum overburden depth of about 25 m can be handled without spoil rehandle by a dragline with a dumping radius of 75 m. A range extension of 9 m would increase this overburden depth to about 28 m. This is the main advantage of the split-bench method.

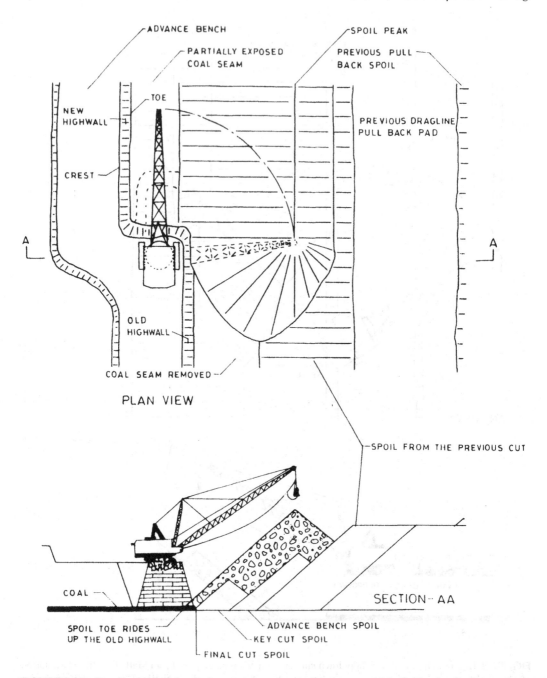

FIGURE 2.13 Dragline performing the last cut of an advance bench as part of the pullback stripping method. (From Math Tech, Inc. Evaluation of Current Surface Coal Mining Overburden Handling Techniques, U.S. National Technical Information Service, PB-264 111 (1976).)

The split-bench method suffers from several disadvantages. The deadheading time for the dragline is approximately double than that of the single-pass case because the dragline is deadheaded twice in each cut. Ramps must be constructed in each pit to enable movement of the dragline between the upper and lower benches. During the second lift stripping, the dragline is 13 m below the ground surface. So the bucket hoist heights during spoiling of second lift material are higher than usual. This may increase cycle time during second lift stripping operations.

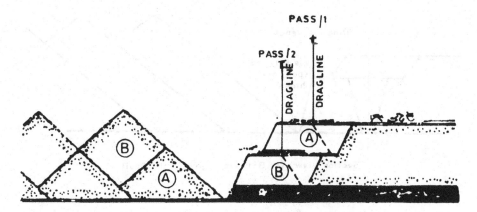

FIGURE 2.14 Section view showing the split-bench method with two dragline passes per pit. (From Math Tech, Inc. Evaluation of Current Surface Coal Mining Overburden Handling Techniques, U.S. National Technical Information Service, PB-264 111 (1976).)

The decision variables in split-bench mining are similar to those inside benching. But the determination of an optimum pit width, bench height and keycutting methods may be somewhat more difficult than inside bench mining.

In deep overburden situations, most mines prefer a one-pass extended bench method to the split-bench method.

After placing the keycut spoil against the existing highwall, the spoil is levelled by a dozer to form a level bench extension. After completing the keycut, the dragline moves to a position on this extended bench to complete the digout. The digout includes a portion of both the bench extension from the previous digout and the side bench. The width of the extended bench determines to what extent the side bench can be cut as the last activity in the digout. If the side bench cannot be cut, then the rehandle material is placed on the pits. Figure 2.14 shows section views of an extended dragline bench.

Usually there are two different approaches to the extended bench method—the high bench and deep bench. The high bench situation is illustrated in Figure 2.15. The alternative is to work the dragline from a bench whose depth is one- to two-thirds of the overburden depth. In 30-m overburden, the dragline bench would be approximately 10–20 m below the natural ground surface. The main advantage of this deep bench situation is that the spoil rehandle percentage decreases as the bench depth increases. For example, with an overburden depth of 30 m, a 15-m wide extended bench and a bench depth of 6 m, the rehandle is 30%. But with a bench depth of 15 m, the rehandle is 14%. These situations are illustrated in Figure 2.16.

During the rehandling operation, the spoil must be cut back to its natural angle of repose, which is usually between 35° and 40°. In some instances, the compaction of the extended bench spoil may allow oversteepening of the spoil during the rehandling operation, thereby reducing rehandle percentages. The degree of oversteepening depends primarily on the characteristics of the spoil, the method of spoil placement, and the height of the extended bench, as shown in Figure 2.17.

The deep bench method also suffers from several disadvantages in comparison with the high bench method. The average bucket hoist height is greater for the deep bench than for the shallow bench. Where the swing angles are sufficiently small, the overall swing and hoist time for the deep bench is determined by the hoist time. As a result, the total cycle time for the deep bench may be greater than that for the high bench.

A second disadvantage of the deep bench method is that as the bench depth is increased, the percentage of total overburden volume that must be excavated by an overhand chopping motion also increases. Many operators do not like this, as the dragline productivity in overhand chopping is less than in the conventional mode of operation.

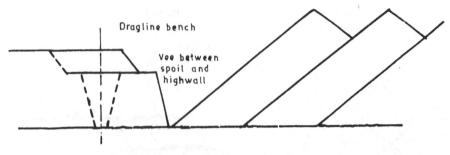

Section View: Vee Between Highwall and Spoil

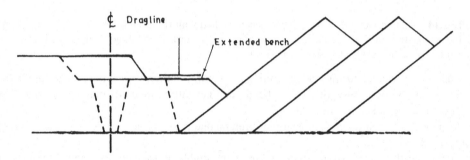

Section View: Vee Filled with Spoil to Extend the Dragline Bench

FIGURE 2.15   Section views showing an extended dragline bench. (From Math Tech, Inc. Evaluation of Current Surface Coal Mining Overburden Handling Techniques, U.S. National Technical Information Service, PB-264 111 (1976).)

2.8   DEEP OVERBURDEN, MULTIPLE LIFTS, ONE PASS (THE EXTENDED BENCH METHOD)

The primary objective of the extended bench method is to increase the effective range of the dragline by extending the dragline bench out into the open pit and filling the trough between the highwall and the adjacent angle of repose of the spoil pile. The dragline can then be moved out onto this extended bench beyond the edge of the existing highwall, thereby increasing the effective spoil-dumping radius of the machine. Ultimately, some or all of the extended bench material will have to be rehandled by the dragline to clear the spoil away from the highwall. The spoil rehandle requirement is a characteristic of the extended bench method. In most single-seam, single-dragline extended bench methods, the keycut material is used to extend the dragline bench. The main difference between the extended bench method and the side bench method is that instead of logging the keycut spoil, the boom is swung at approximately 120°, and keycut spoil is placed against the existing highwall in the direction of stripping. This procedure, known as leading the keycut spoil, is shown in Figure 2.18. Figure 2.19 illustrates a typical method of extending the dragline bench. Figure 2.19 shows a dragline on an extended bench.

The choice of pit width is also an important decision in the extended bench method. The rehandle percentage is largely independent of the pit width. Mine operations prefer wide pits because the wider the pits are the fewer pits in a given acreage. Widening of the pits also reduces the rehandle volumes per acre. However, wide pits increase dragline cycle time and spoil grading costs.

Extending the effective range of the dragline also yields other advantages compared to conventional operating methods. In conventional stripping of deep overburden using a long-boom machine, the spoil piles are very high, causing spoil slides into the pit. With the extended bench method, the

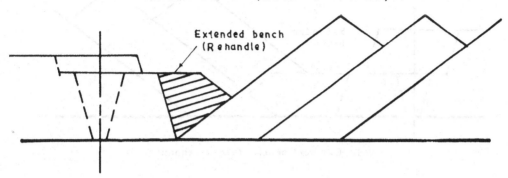

1. 100 foot overburden depth, 20 foot bench depth.
   Rehandle – 30%

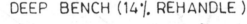

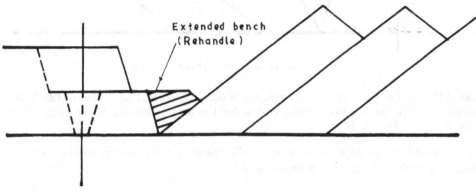

2. 100 foot overburden, 50 foot bench depth.
   Rehandle - 14%

FIGURE 2.16 Comparison of rehandle methods for high and deep benches. (From Math Tech, Inc. Evaluation of Current Surface Coal Mining Overburden Handling Techniques, U.S. National Technical Information Service, PB-264 111 (1976).)

spoil can be cast into the troughs between the spoil piles rather than on the tip of the pile closest to the pit. This results in better stability and better spoil grading.

The extended bench enables the use of a dragline with a steeper boom angle than is required if the bench is not extended. As the boom angle is increased in the dragline, the bucket size can be increased, and the machine radius can be decreased. Reduction in dump radius increases boom acceleration with radius boom swing time on each cycle. Additionally, because of the larger bucket, the volume of overburden material moved in each cycle is increased. This yardage advantage has to be compared with the volume of material to be rehandled in the extended bench method.

Most mines prefer to trade off bucket capacity for boom length. They prefer machines with long booms and moderately sized buckets. Some long-boom machines are capable of handling overburden depths up to 40 m without spoil rehandle. However, machine operators do not like to operate

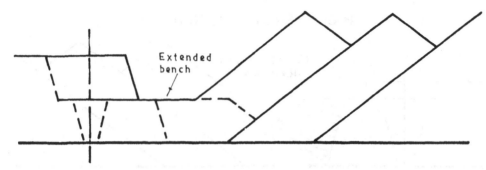

1. Extended bench: before rehandle

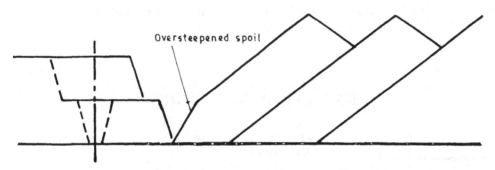

2. Oversteepened spoil: after rehandle

FIGURE 2.17 Section views showing extended bench and oversteepened spoil. (From Math Tech, Inc. Evaluation of Current Surface Coal Mining Overburden Handling Techniques, U.S. National Technical Information Service, PB-264 111 (1976).)

a large machine from the edge of a 40-m highwall, indicating that benching will be used in deep overburden regardless of the range of the dragline.

2.8.1 TWO-PASS EXTENDED BENCH METHOD

In extended bench mining, two passes can be made in each pit. This does not require the use of the overhand chopping procedure. The advantage of the two-pass method over the single-pass method is the reduction in rehandle percentages without loss of any productivity, as in overhand chopping. The disadvantages include increased dragline deadheading time, ramp construction, and ramping of the dragline between the two bench levels.

The first pass is made on an extremely high bench or on the natural ground surface. Top lift keyway material is placed against the lower highwall as the lead spoil to form the extended bench for the second pass. When the pit end is reached, the dragline is ramped down to the lower bench, and stripping continues in the direction opposite to the direction of the top lift stripping. The method is illustrated in Figure 2.20.

2.8.2 PULLBACK METHOD

The pullback method is an alternative approach to extended benching under similar pit conditions. It has additional use in multiple-seam applications. A typical pullback operation is illustrated in Figure 2.21. The pullback method can utilize a single dragline or two draglines. In the latter case, both machines are assigned a specific part of the operation.

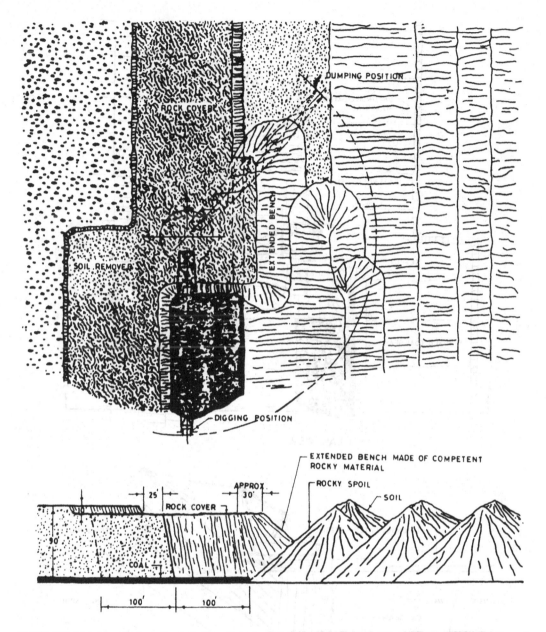

FIGURE 2.18 A typical method of extending the dragline bench. (From Math Tech, Inc. Evaluation of Current Surface Coal Mining Overburden Handling Techniques, U.S. National Technical Information Service, PB-264 111 (1976).)

The method involves first either side casting from the original ground surface or from an advance bench. Figure 2.21 shows the last cut being spoiled for the advance bench situation in the pullback method. In the pullback method, the dragline is unable to spoil all the overburden clear of the coal. As a result, this material rills over the coal seam at the toe of the spoil pile. This material has to be rehandled to completely uncover the coal.

The rehandle is performed with the dragline sitting on a prepared pad on the spoil pile itself. This pad is prepared using dozers or by the dragline removing the spoil peak using chop down. Final surface levelling is performed by a dozer. Figure 2.21 illustrates the pullback operation. The spoil pile is dug back away from the highwall and spoiled behind the dragline on the top of previous spoil.

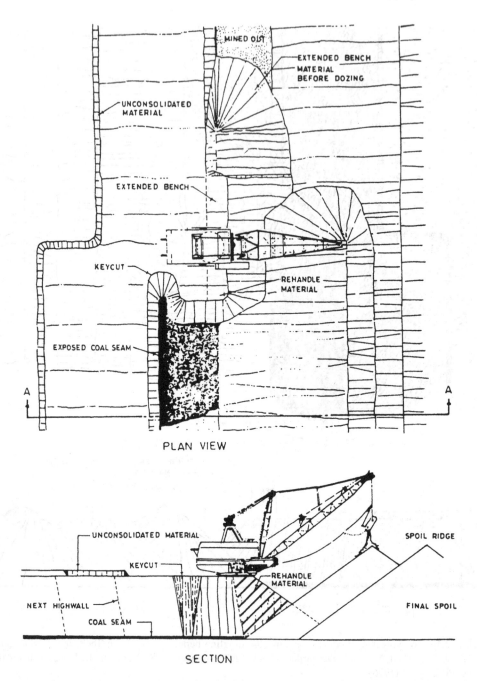

PLAN VIEW

SECTION

FIGURE 2.19 Dragline on an extended bench. (From Math Tech, Inc. Evaluation of Current Surface Coal Mining Overburden Handling Techniques, U.S. National Technical Information Service, PB-264 111 (1976).)

The spoil pile can roll over and cover the coal to a greater or lesser extent depending on the relative dragline and pit geometry. The rehandle can be theoretically calculated as the area of the spoil pile above a line intersecting the coal toe at the spoil pile, as shown in Figure 2.21. The mutual area, although it is removed in the pullback operation, is not rehandled, as it need not be removed during the initial spoiling operation.

The theoretical rehandle is not necessarily the most desirable from an economic standpoint. When the spoil pile slope has been pulled back to the top edge of coal, further rehandle exposes

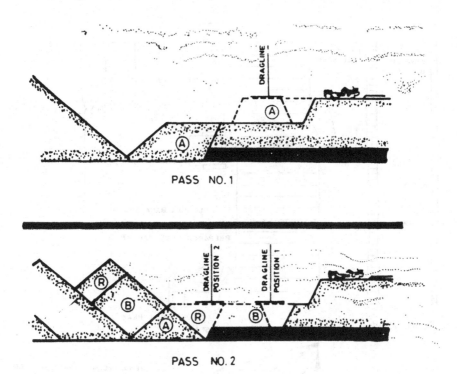

PASS NO. 1

PASS NO. 2

FIGURE 2.20 Section views showing a two-pass extended bench method. (From Math Tech, Inc. Evaluation of Current Surface Coal Mining Overburden Handling Techniques, U.S. National Technical Information Service, PB-264 111 (1976).)

proportionally less coal. There is a break-even point or an economic limit to the extent of uncovering of coal.

The main advantage of this method is to enable a dragline which has a limited operating radius to handle overburden covers of greater depth than would normally be feasible.

The pullback method can be performed with either one or two draglines. With a single dragline, the machine has to move periodically across to the spoil pile, either around the pit or on a section of extended bench.

The alternative is to use two machines: one permanently situated on the highwall, the other permanently on the spoil pile. The productivity of the two machines must be carefully matched so that they operate as a team.

In both these alternatives, the spoil pile stability is an important problem. Where spoil piles have inherently poor stability, it is inadvisable to place a dragline on the spoil pile. It would be better to use the extended bench method.

2.9 TERRACE MINING

This method enables deep overburden depths to be stripped. The operation usually incorporates three or more machines. Because of their height, both highwalls and spoil piles are stepped to maintain slope stability.

Figures 2.22A to C show cross-sections and plans of a typical mining sequence, one machine at a time. Figure 2.22A shows the first dragline starting the stripping sequence. This machine has a shorter boom and, consequently, a larger bucket than the two subsequent machines. Figures 2.22B and 2.22C show the other two machines and their relatively lighter workloads.

This method can be applied to strip greater and greater depths using any number of machines. However, as the overburden depth and the number of machines increase, so does the rehandle which can easily become more than 100%.

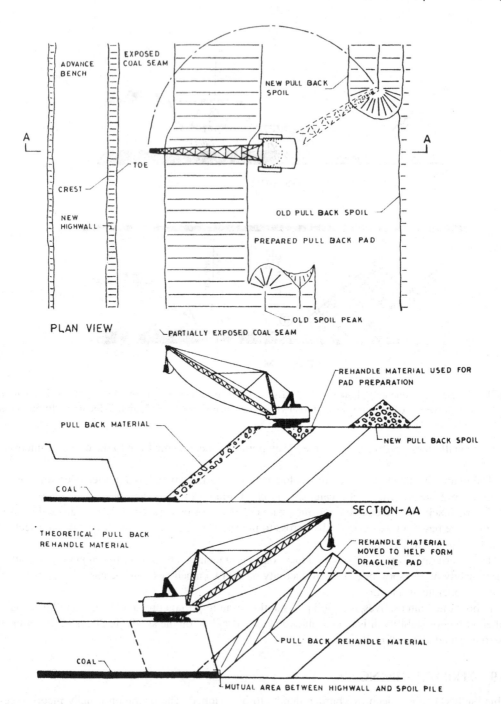

FIGURE 2.21 Dragline using the pullback method. (From Math Tech, Inc. Evaluation of Current Surface Coal Mining Overburden Handling Techniques, U.S. National Technical Information Service, PB-264 111 (1976).)

2.9.1 TANDEM MACHINE SYSTEMS

Tandem machine systems include the use of the dragline in conjunction with dozers, scrapers, front-end loaders, stripping shovels, bucket wheel excavators, and other draglines. In a single-seam mining situation, the tandem system most frequently used is a dozer/dragline combination. Other tandem systems are used in multiple-seam situations and are described in subsequent chapters.

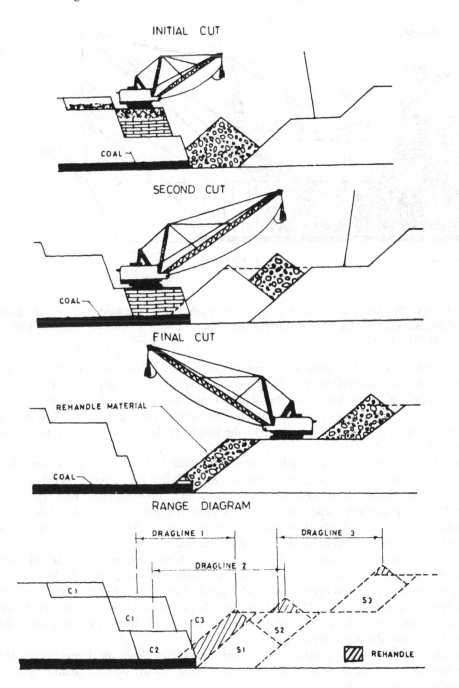

INITIAL CUT

COAL

SECOND CUT

COAL

FINAL CUT

REHANDLE MATERIAL

COAL

RANGE DIAGRAM

DRAGLINE 1   DRAGLINE 3

DRAGLINE 2

C1

C1   C3

C2   S1   S2   S3

REHANDLE

FIGURE 2.22 Terrace mining. (From Math Tech, Inc. Evaluation of Current Surface Coal Mining Overburden Handling Techniques, U.S. National Technical Information Service, PB-264 111 (1976).)

2.9.2 TANDEM DOZER/DRAGLINE STRIPPING

In many deep overburden situations, stripping is carried out using large dozers in tandem with relatively small, crawler-mounted draglines. These draglines typically have lengths of 35–45 m and bucket capacities of 7 or 8 m³. Under normal operating conditions, a machine with a 45-m boom can handle overburden up to about 15 m in depth, using a conventional no-rehandle procedure. But when operating in tandem with a dozer, these draglines can strip overburden 25–33 m deep, using the extended bench method.

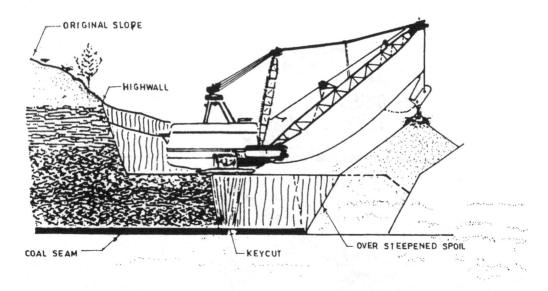

FIGURE 2.23 Dragline making keycut in second lift overburden. (From Math Tech, Inc. Evaluation of Current Surface Coal Mining Overburden Handling Techniques, U.S. National Technical Information Service, PB-264 111 (1976).)

In the extended bench method, maximum use of the force of gravity is made to move overburden. In the conventional situation using a single dragline, the top lift overburden is dragged down towards the pit, so that the overburden falls into the cut rather than interfering with the dragline bucket. This gravity principle can also be applied by using dozers to push the top lift overburden into the pit. So the dozers make the bench for the dragline, which then strips the second lift.

The stripping procedure for the top lift is shown in Figure 2.23. After the overburden is blasted, the material is pushed over the edge of the existing highwall into the open cut. As the depth of the overburden excavated by the dozer increases, the extended bench is automatically prepared by the dozer. The extended bench is essential as the range of the dragline is limited. The dozers compact the bench material fairly tightly. This allows for extreme oversteepening of the spoil during the dragline rehandle phase in order to minimize spoil rehandle percentage.

The dragline strips the second lift. The machine is positioned directly over the keycut outerline, as shown in Figure 2.23. The keycut is made at some distance from the highwall. To complete the digout, the dragline is moved out to the extended bench. During the rehandle phase, the spoil can be cut at an angle as high as 80°. The spoil can stand this angle, as it has been compacted. Because of this steep spoil angle, the rehandling is reduced from 15% to 5%. Figure 2.24 shows views of dragline on extended bench stripping second lift overburden.

The use of dozers in the top lift offers economic advantages. When the dozers operate level or downslope and push over fairly short distances, they are less costly than the small dragline. This may not be true for large draglines.

The important decision variables are the bench depth and the pit width. As the bench is deepened, the dozer push distance increases. At some point, the dozer has to push uphill over long distances, a procedure which can be costly. The decision involved is to determine at which point the dozer cost exceeds the dragline cost. This is explained in Figure 2.24. For shallow benches, the dozer costs are lower than those for the dragline as push distances are short and considerable amounts of overburden are moved downslope. At certain bench depths, dozer costs equal dragline costs, and this depth is the optimal depth for the top bench.

Similar reasoning can be used to determine the optimal pit width. The problem is complicated as pit width, stripping costs, spoil grading costs, and rehandling percentages are interrelated.

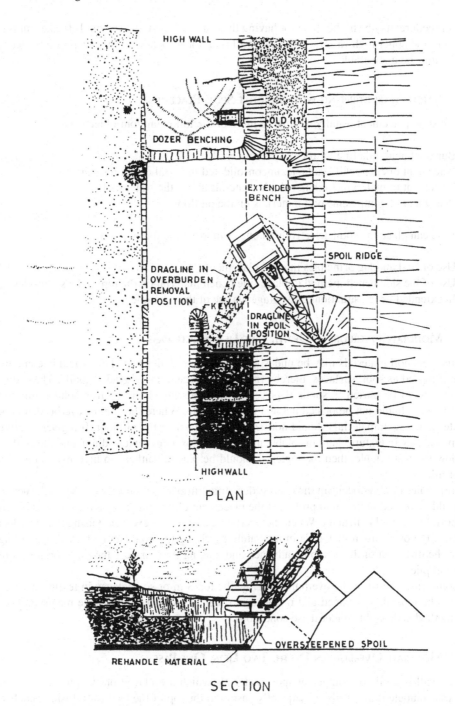

HIGH WALL

OLD H.T.

DOZER BENCHING

EXTENDED BENCH

SPOIL RIDGE

DRAGLINE IN OVERBURDEN REMOVAL POSITION

KEYCUT

DRAGLINE IN SPOIL POSITION

HIGHWALL

PLAN

OVERSTEEPENED SPOIL

REHANDLE MATERIAL

SECTION

FIGURE 2.24 Dragline on extended bench stripping second lift overburden. (From Math Tech, Inc. Evaluation of Current Surface Coal Mining Overburden Handling Techniques, U.S. National Technical Information Service, PB-264 111 (1976).)

There are alternative methods for making the keycut in the lower bench. Economic and safety factors determine the choice. When the dragline is used to make the keycut, the bench must be wider than the pit. Then the average dozer push distances are longer than when the keycut is made adjacent to the upper highwall. This results in a bench width equal to the pit width. But if the keycut is made adjacent to the upper highwall, and dozers and front-end loaders make the keycut, the costs of this procedure may be higher than for having the dragline make the keycut. If the keycut is made adjacent to the upper highwall, then the highwall will not be benched. In a deep overburden situation, this may be undesirable.

2.9.3   SINGLE-SEAM STRIPPING WITH SELECTIVE SPOIL PLACEMENT

Selective spoil placement usually has one or more of the following four objectives:

- Burial of acid, alkaline, or rocky spoil material
- Placement of chemically suitable, unconsolidated material on spoil surfaces
- Placement of materials high in soluble minerals above the water table
- Placement of competent spoil materials on the pit floor

A dragline can place the spoil selectively in one or more of the following ways:

- Use of lead and lag principles in spoil placement
- Use of the side bench method to place side bench material on the top of the spoil piles
- In extended bench situations, advantageous use of the rehandle material

2.9.4   MODERATE OVERBURDEN DEPTH, ONE LIFT, SINGLE PASS

Two kinds of spoil materials may be undesirable. Alkaline shales and clays would become impermeable if placed as spoil surfaces, thereby impeding revegetation of graded spoils. These types of materials should be buried in spoils if possible. Materials which are high in soluble minerals or trace elements should not be placed below the water table. Where the coal or overburden is below the water table, such materials should not be placed on or near the pit floor. If a given overburden stratum is clayey in texture, high in soluble minerals and exchangeable sodium, and the coal seams are below the water table, then the material should be placed neither too high nor too low in the spoil profile.

If soft shales or clays occur in the overburden strata immediately overlying the coal, such material should be buried in the spoil profile. If the placement of the material on the pit floor is permissible, burial may not be difficult. When the overburden strata are excavated using the layer loading technique, the overburden materials immediately overlying the coal can be buried by leading the spoil in the direction of the stripping advance and placing it on or near the pit floor, ahead of the main spoil pile.

In some cases, most of the overburden below the top 3 m or so is clayey in texture and high in sodium. The topsoil is salvaged and replaced in order to bury the undesirable material. The side bench method is desirable under these conditions.

2.9.5   MODERATE OVERBURDEN DEPTH, TWO LIFTS, ONE PASS

The backwall is a pile of competent spoil materials which are placed on the pit floor to buttress other, less competent materials subsequently cast onto the top of the backwall. Inside bench situations, construction of the backwall is simple. The dragline bench is usually cut deep enough so that all overburden strata below the elevation of the bench are competent. The bench surface is laid out at the interface of unconsolidated and consolidated materials. The keycut is made following the conventional practice, and this is the first activity on a given digout. Since all keycut materials are competent, they are used to build the backwall. This is done by leading the spoil at a boom swing angle of about 120°, so that the keycut spoils are placed on the pit floor, near the highwall, ahead of the remaining spoil materials which are cast behind and on top of the backwall. Sometimes, the backwall is called the lead spoil.

Acid material can also be buried using the lead spoil method. Acid materials can be rendered harmless by burying them, as both air and water are needed for acid production. Usually, the acid materials consist of slates and dark shales immediately overlying the coal. Where these do not comprise a large proportion of the overburden, they can be buried using the lead spoil principle. Placement of side bench spoil on tops of spoil piles gives additional insurance against acid production.

2.9.6 Deep Overburden, Two Lifts, One Pass (The Extended Bench Procedure)

In the extended bench method, successful spoil segregation and burial can be achieved if desirable spoil material with which to bury undesirable material is available. The lead spoil and side bench procedures can also be used to cover undesirable spoil materials with the rehandle material. The rehandle material should have desirable characteristics from a reclamation standpoint. Since the keycut material is used to prepare the extended bench, it may not always have desirable reclamation characteristics. This means that consolidated materials high in the bank are used to extend the bench. During the rehandle phase, these materials are placed on the surface of the spoil piles. A practical problem may arise in that during grading, a dozer may cut through the surface material. A second problem may arise because the use of top, unconsolidated material in the extended bench can cause stability problems.

2.9.7 Multiple-Seam Stripping with Draglines

Different approaches can be adopted for mining multiple seams with draglines. With a given set of equipment and physical conditions, the choice of the mining method is dependent on the following factors:

* Relative depths of the overburden and the interburden
* Dumping radius and height of the dragline
* Thickness of the coal seams being mined

One major difference between the single-seam mining system and the multiple-seam system is that the dumping height of the dragline is often important for the latter system and not for the former. An additional difference is that segregation and burial of undesirable materials are more frequently necessary in multiple-seam mining than in single-seam mining.

Multiple-seam mining requires multiple lifts. While stripping two coal seams, removal of the overburden requires at least one lift, and the removal of the interburden requires another lift.

Multiple-seam mining systems may be single or tandem machine systems. In subsequent sections, the use of tandem machines in multiple-seam situations will be viewed as advantageous than in single-seam situations.

2.10 SINGLE DRAGLINE, TWO-SEAMS, NONSELECTIVE SPOIL PLACEMENT

A single-dragline system stripping two coal seams is frequently used.

2.11 MODERATE OVERBURDEN AND INTERBURDEN DEPTHS

This method involves two lifts—one for overburden and the other for interburden and two passes. No rehandling of spoil is needed. The method is shown in Figure 2.25. On the first pass, the dragline is placed on the natural ground surface or a very shallow bench. The overburden is stripped and placed as in the conventional method. The top coal seam is mined closely, following the stripping operations. When the end of the pit is reached, the dragline is deadheaded back to the other end

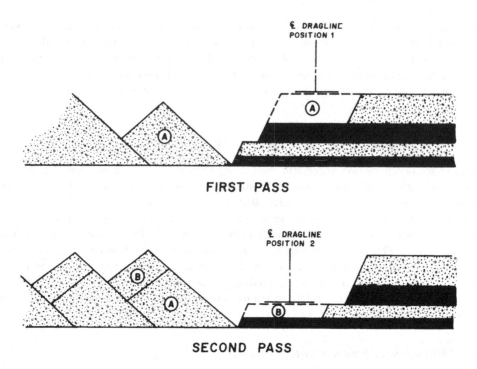

FIGURE 2.25 No-rehandle method of stripping two coal seams. (From Math Tech, Inc. Evaluation of Current Surface Coal Mining Overburden Handling Techniques, U.S. National Technical Information Service, PB-264 111 (1976).)

of the pit and ramped down to a position on top of the interburden. However, if the top coal seam can be mined closely in pace with the stripping of the overburden, then deadheading may not be needed. The machine can be ramped down after reaching the end of the pit and placed on top of the interburden. The bottom coal seam is then stripped in the direction opposite to that of the first pass.

The effective range of the dragline is increased while stripping the interburden because of the highwall slope angle. The interburden is stripped, and spoil is placed as in the conventional procedure. At the end of the pit, the dragline is ramped up to the top of the overburden, and a new top pass begins.

The same decision variables which are applicable to the single-seam situation are applicable in the multiple-seam situation. The only exception is that in multiple-seam mining it is difficult to control the depth of the bench on the second pass. The depth of the second pass is equal to the combined depth of the overburden and the top coal seam. This has an important implication, in that while stripping the interburden, the dragline should have sufficient dumping height so that all interburden material can be spoiled from a dragline position on the interburden. Additionally, the dragline should have sufficient range so that an extended bench is not needed in the second pass.

One of the disadvantages of this method is a high hoist height in the second pass. A second disadvantage is that in some cases final spoil piles may be very high, which may cause spoil slides. The interburden material is placed on top of the spoil piles. This may not be always desirable if the interburden material is not environmentally desirable. It may be costly or difficult to bury this material.

2.12   THE EXTENDED BENCH METHOD FOR TWO SEAMS

In a two-seam situation, when the dumping radius of the machine is short or where the overburden is thick, the use of the extended bench method can be efficacious.

A typical operating procedure is shown in Figure 2.26. The first pass on overburden is conventional. Some spoil material is cast close to the lower highwall to extend the second pass bench, and

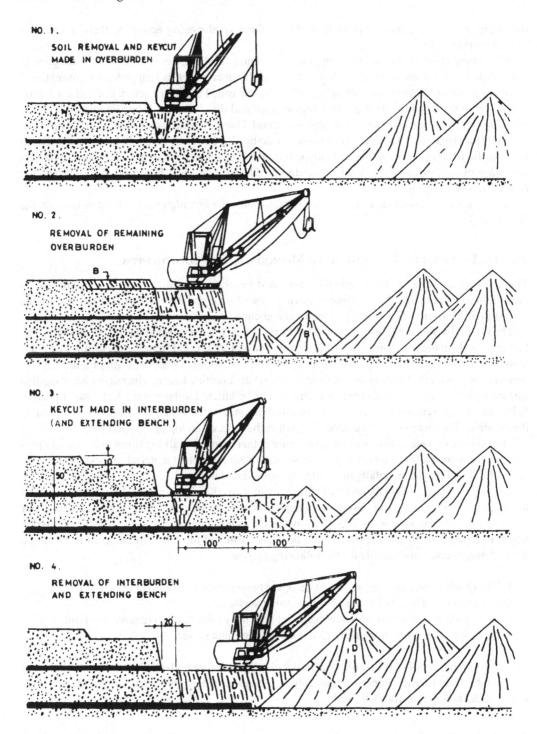

FIGURE 2.26 Stripping of two coal seams by a single dragline using an extended bench method. (From Math Tech, Inc. Evaluation of Current Surface Coal Mining Overburden Handling Techniques, U.S. National Technical Information Service, PB-264 111 (1976).)

the swing angles are usually greater than 90° in these spoil casting activities, thereby reducing dragline productivity.

After completion of overburden stripping and mining of the upper coal seam, the dragline is ramped down to a position on the interburden, and the interburden is stripped with conventional extended bench procedures. The rehandle material is excavated as the last activity of a given digout.

The extended bench method has both advantages and disadvantages when compared with the conventional two-seam no-rehandle stripping method. One advantage is the extension of the effective dragline range so that deeper overburden or interburden can be stripped. In the extended bench method, the interburden spoil material can be buried by the rehandle material. This is an important advantage of the environmental control standpoint. The extended bench method also improves the distribution of the spoil.

The disadvantages include requirements of spoil rehandle and high bucket hoist heights on the second pass.

2.12.1   THE ONE-PASS EXTENDED BENCH METHOD FOR TWO-SEAM STRIPPING

This method is similar to the single-pass extended bench method in single-seam mining. With multiple-seam situations, the dragline is always placed over the interburden and never over the overburden. All overburden is stripped by overhand chopping.

The method is illustrated in Figure 2.27. At first, a bench is established on the interburden. The keycut is made in the interburden by a dragline, but keycut spoil is led from the backwall. The interburden material has to be competent. Next, the surface overburden material (#2 in Figure 2.27) is removed by overhand chopping and is used to extend the dragline bench. The reason for using this surface material in the extended bench is that it will be utilized to bury the interburden material. Subsequently, the remaining overburden, the interburden, and the bench extension are removed by the dragline. The material of the extended bench is then placed on top of the spoil pile.

The main advantage of this one-pass procedure is that dragline-walking times, which are unproductive, are reduced. Since a backwall is constructed with competent material, the extended bench can be built with environmentally desirable, unconsolidated material.

The disadvantage of the method includes the use of overhand chopping. The one-pass method becomes more attractive as the overburden depth decreases relative to the interburden depth. For example, the method may work efficiently when the overburden depth is 14 m and interburden height is 25 m. However, if the overburden depth is 25 m and the interburden depth is 14 m, the method may become undesirable for the following reasons:

1 The dragline may not have sufficient dumping height to spoil the interburden when it works from a bench which is 25 m below the ground surface.
2 Even if the machine has sufficient height, it may not be desirable to remove overburden by chopping from a bench which is 25 m below the ground surface.

Under favourable conditions, the method works excellently, from both production and reclamation standpoints. The method is generally applicable to situations where the average overburden depth is less than the average interburden depth.

2.12.2   THE TWO-PASS SPOIL SIDE METHOD

The multiple-seam mining methods which have been discussed previously are feasible only when the dumping height of the dragline is large enough so that all interburden material can be placed on the top of the spoiled overburden material from a dragline operating on top of the interburden. In some situations, the dragline dumping height may not be sufficient to achieve the placement of interburden material over the overburden material.

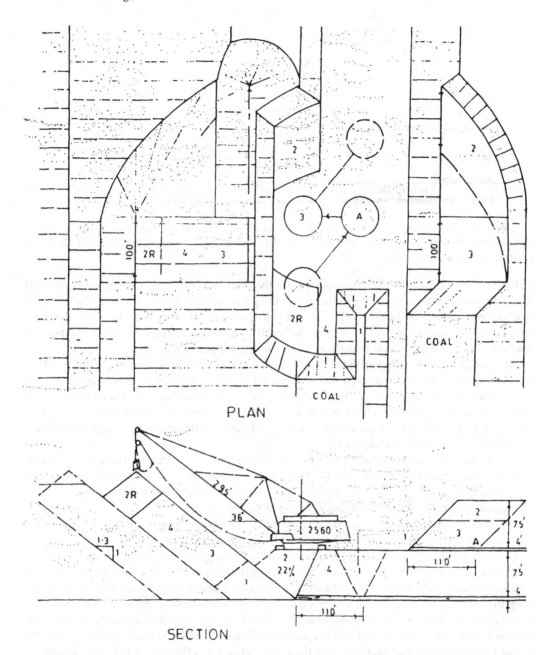

PLAN

SECTION

FIGURE 2.27 A one-pass extended bench method of stripping two coal seams. (From Math Tech, Inc. Evaluation of Current Surface Coal Mining Overburden Handling Techniques, U.S. National Technical Information Service, PB-264 111 (1976).)

The solution to this problem is to increase the effective dumping height of the existing dragline in the second pass by raising the height of the second pass bench. There are two possible ways to achieve this.

In the first method, a high bench is constructed over the spoil pile of the stripped overburden material. This is illustrated in Figure 2.28. The spoil pile can sometimes be oversteepened. Additionally, the bench can be extended towards the highwall side if the range of the dragline is not sufficiently long.

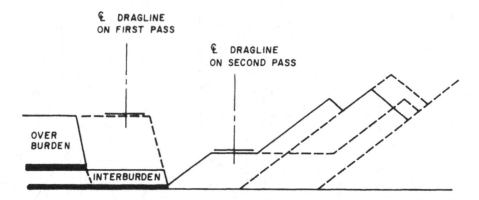

FIGURE 2.28 Bench is spoil pile from which the lower of two seams is stripped. (From Math Tech, Inc. Evaluation of Current Surface Coal Mining Overburden Handling Techniques, U.S. National Technical Information Service, PB-264 111 (1976).)

In the first pass, the dragline is positioned on the overburden and the overburden is stripped using conventional procedures. When the end of the pit is reached, a ramp is constructed and the dragline is moved across the spoil side of the pit and used to cut the initial bench on the spoil pile. Alternatively, dozers can be used to prepare the bench. After this upper coal seam is mined out, the machine is moved onto the bench on the spoil pile to strip the interburden.

When the dragline is stripping the interburden, its boom is typically at an angle of 60° to 90° to the highwall in contrast to the angle in the conventional practice, where the boom is roughly parallel to the highwall during the digging portion of the cycle. The interburden is excavated by chopping downward and dragging the bucket towards the dragline tub. The boom is then swung through an angle of 120° to 150°, to spoil the interburden material. Some of the interburden spoil can be used to build the bench for the next digout.

The primary advantage lies in the fact that the method increases the effective dumping height of the dragline so as to enable stripping of two coal seams rather than one. Additionally, in contrast to previously described methods, bucket hoist heights in the spoiling of interburden materials are reduced because the bench is raised above the top of the interburden.

The disadvantages include the need for chopping, the relatively large swing angle required for placement of interburden spoil, and the need to construct a bench in the spoil.

There are procedures to reduce or eliminate the need for chopping. One common procedure used to improve the productivity of the dragline in chopping is to fragment the interburden by blasting it very hard. A second method is to delay the shots from the spoil side of the interburden inward, thereby using the explosive to move the interburden material away from the lower highwall. This creates a pocket in the interburden. The dragline bucket is lowered into this pocket to digout the interburden. If rehandle is not otherwise needed, this blasting makes it necessary, as the interburden material will be cast by the blast into the trough between the lower highwall and the spoil pile. In most instances, rehandle is needed and the blasting technique does not create any additional rehandle.

A second procedure used to reduce the need for chopping involves keycutting the interburden prior to the main excavation activity. The dragline bucket enters the keycut to begin the digout. If the keycut is wide enough, it virtually eliminates the chopping method. However, as the keycut is narrower at the bottom than at the top, chopping may be necessary for the lower corner of the highwall. The keycut in the interburden can be made using dozers or loaders or the dragline.

Procedures for preparing the keycut using a dragline are as follows. In the second pass, the dragline is positioned on the interburden and the keycut is made for several digouts. A ramp is then constructed across the pit, and the dragline is moved over to the bench in the spoil pile and back along the pit to the starting point of the keycut. The interburden is then removed as described earlier. The productivity gains resulting from the reduction or elimination of chopping must be weighed against the dragline-walking time.

2.12.3 ELEVATED BENCH METHOD

The elevated bench method offers an alternative for eliminating underbench chopping and large boom swing angles needed to dispose of interburden materials from a position on a bench spoil. The method is illustrated in Figure 2.29. It shows a one-pass extended bench method where the dragline is always placed on the interburden. Some overburden is spoiled on top of the interburden to raise the dragline bench.

The major advantage of the method is the elimination of interburden chopping and reduction of boom swing angles during the spoiling of the interburden. The main disadvantages include a spoil rehandling requirement and the need for overhand chopping of the overburden. The actual rehandle

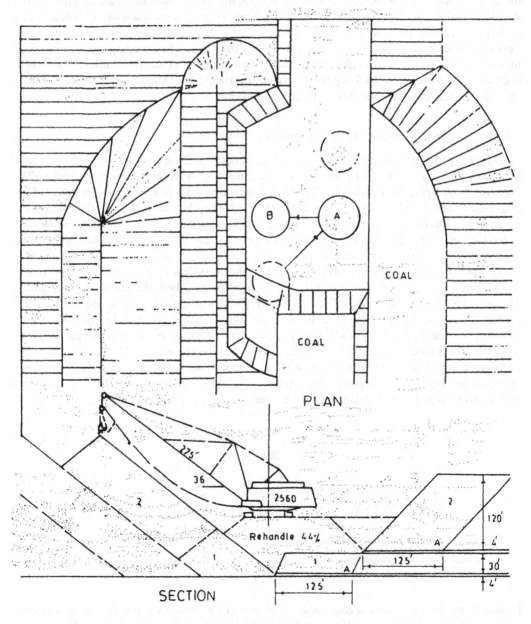

FIGURE 2.29 A method for increasing the effective dumping height of a dragline (From Math Tech, Inc. Evaluation of Current Surface Coal Mining Overburden Handling Techniques, U.S. National Technical Information Service, PB-264 111 (1976).)

percentage is smaller since some of the elevated bench material will have been moved by gravity. The spoil rehandle percentage decreases as the height of the elevated bench decreases.

In comparing the spoil bench and elevated bench methods, the tradeoffs can be obvious. In the elevated bench method, chopping off the interburden and large swing angles are involved. In the bench on the spoil method, rehandling of spoil and overhand chopping of the overburden are major disadvantages.

2.12.4   TANDEM MACHINE, TWO-SEAM CONDITION, NONSELECTIVE SPOIL PLACEMENT

In two-seam situations when tandem machines are used, one machine strips the overburden and the second machine strips the interburden. The tandem machine combinations are dragline/shovel, dragline/dragline, and shovel/shovel. A major advantage of the tandem machine subsystem is that each of the stripping machines can be tailored for a specific task. In most multiple-seam situations, the overburden depth is greater than the interburden thickness. The stripping capacity required for overburden exceeds that for the interburden. When stripping of the interburden is done by a machine needs more range and dump height than the overburden machine. Tandem machine system offers the highest efficiency for making these requirements.

2.12.5   THE TANDEM SHOVEL-DRAGLINE SYSTEM

The system is shown in Figure 2.30. The overburden depth is 20 m with interburden 6-m thick. Overburden is removed by a stripping shovel which has a 45-m boom and a 45-m³ bucket. The interburden is stripped by a dragline with a 55-m-long boom and a 10-m³ bucket. The range of the dragline exceeds that of the shovel so that the interburden can be spoiled beyond the ridge of the shovel spoils. The ratio of the shovel and dragline bucket capacities is 3.3:1, roughly the same ratio as the ratio of overburden depth to interburden depth. The rate of stripping advance for the two machines will therefore be approximately the same.

As explained earlier the main advantage of this tandem system over a single-machine system is it is economical. A single shovel or a dragline suitable to strip both the seams would require a 55-m-long boom and 45-m³ bucket. The capital cost of such a machine would exceed the combined capital costs of the smaller shovel and dragline. An additional advantage of the tandem machine subsystem is related to the rates at which the two coal seams are exposed. In this example, the stripping ratio for the upper seam is 28:1, while the ratio on the lower seam is 12.5:1. If a single machine is used to strip both seams, the lower seam would be exposed roughly twice as fast as the upper one, on a tonnage basis. This could complicate production scheduling in situations where the coal from both seams must be blended. A tandem machine can achieve a balanced production.

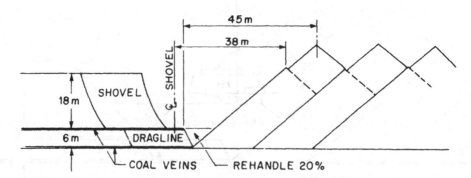

FIGURE 2.30 Two-seam tandem dragline/shovel method. (From Math Tech, Inc. Evaluation of Current Surface Coal Mining Overburden Handling Techniques, U.S. National Technical Information Service, PB-264 111 (1976).)

Ramping of the stripping machine between the overburden and interburden is not required when tandem machines are used since each machine works on one level only.

Disadvantages of the tandem system include higher inventory requirements for spare parts and lower availability compared to the single-machine case. Failure of either stripping machine may mean the failure of the whole system.

2.12.6   DOZER-DRAGLINE SYSTEM

When the interburden is relatively deep, all overburden is removed by D-dozers as shown in Figure 2.31. The second pass bench is extended and used by the dragline. After the upper seam has been loaded, the dozers are used to remove some of the interburden to make a level bench for the dragline.

Subsequently, the interburden is stripped by dragline. This can be done either from an extended bench or a bench in the spoil. Rehandle will be required.

The method has several advantages. Gravity is utilized to move overburden. The dragline always sits on the same bench level, thereby eliminating the need for ramping. The topsoil, which is the rehandle material, is placed on the top of the spoil pile, burying the interburden spoil.

When the overburden is thicker than the interburden, the dozer and dragline are used in tandem to strip the overburden, and then the dragline is ramped down to the interburden level. All interburden is then stripped by dragline.

2.12.7   TANDEM MACHINE-MULTIPLE DRAGLINE

Figure 2.32 illustrates that two draglines can be simultaneously used to strip two coal seams. One dragline strips the overburden, and the other dragline strips the interburden. The capacity of the two machines has to be matched with their stripping requirements so as to maintain their required rate of progress. The leading machine over the overburden has to maintain its lead ahead of the second machine.

The spoil of the leading machine does not abut the second seam. This allows the second machine to spoil the interburden on the pit floor without covering the seam.

2.12.8   SELECTIVE SPOIL PLACEMENT IN TWO-SEAM CONDITION

In most two-seam dragline situations, spoil rehandle material is the last material placed on the spoil pile during a digout. The rehandle material, if it is topsoil, is suitable for reclamation. Therefore, where rehandle is required for production requirements, segregation and burial of the spoiled interburden may have little effect on the operating procedure.

If spoil rehandle is not involved for production reasons, then a requirement for the burial of the interburden material may make it necessary. Figure 2.33 illustrates a situation where an extended bench was not required for the range of the dragline. Nonetheless, in order to satisfy the requirement that the interburden spoil be buried, the rehandle material was used.

2.12.9   STRIPPING OF UNSTABLE OVERBURDEN MATERIAL

Some surface coal mines are located in swampy lowland areas. Draglines are used for overburden removal. Unstable overburden consists of a thick layer of mud. Conventional dragline procedures cannot be used in such situations in order to avoid spoil slides into the pit.

The first step in the special mud handling procedure is shown in Figure 2.34. The dragline works on an established bench (not the extended bench) and the boom is turned to the highwall side. The mud in the overburden is scooped out by the dragline. The boom is then swung through a 180°

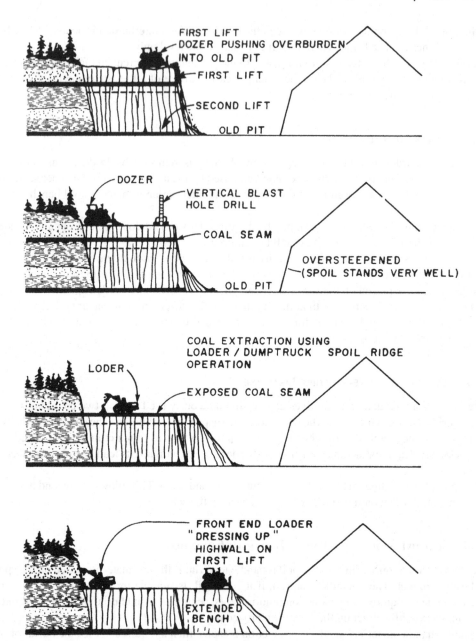

FIGURE 2.31 Removal of overburden by dozer in tandem dozer/dragline stripping of two seams. (From Math Tech, Inc. Evaluation of Current Surface Coal Mining Overburden Handling Techniques, U.S. National Technical Information Service, PB-264 111 (1976).)

angle, and the mud is placed in the vee between the first and second spoil piles that are adjacent to the pit.

After the excavation of the mud hole has been completed, the hole must be filled with competent spoil material to force the machine-supporting bench for the next pass. Competent material is excavated from the overburden bank, and the boom is turned 90° to the highwall side to deposit the material in the mudhole.

Thereafter, the dragline is moved out to the edge of the extended bench and the bucket is lowered into the vee between the first and second spoil piles, and the mud which has been placed recently

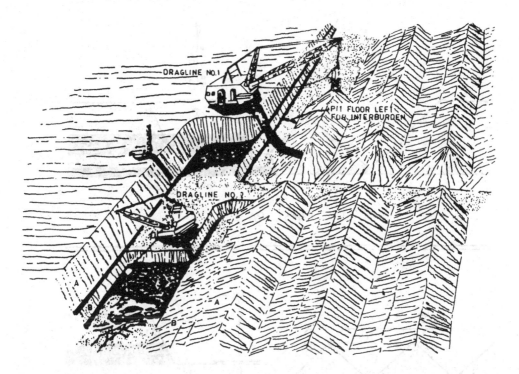

FIGURE 2.32 Multiple seam—tandem machines. (From Math Tech, Inc. Evaluation of Current Surface Coal Mining Overburden Handling Techniques, U.S. National Technical Information Service, PB-264 111 (1976).)

is excavated and cast further away from the pit to rest in the vee between the second and third spoil piles. The movement prevents spoil from sliding into the pit.

2.13 MULTIPLE-SEAM SYSTEMS

Up to this point the discussion has cantered primarily on two-seam systems. It remains now to describe a multiple-seam system mining, say five seams at the one operation.

Figure 2.35A shows a dragline taking the top cut off a five-seam operation using the simple side casting method. As in all these diagrams the dimensions are representative of actual practice, and specific values are not usually significant. In this case, however, the operation is more unusual. These particular drawings illustrate a 40-m pit width being taken, the top cover being 35 m, the coal seams being in the range 2- to 4-m thick, and the parting thicknesses in the range of 6–9 m. The overall operation has a height in excess of 60 m.

The various partings, other than the uppermost, are taken using a second machine situated on the spoil pile produced with the upper overburden. Figure 2.35B is a cross-section illustrating the first parting being taken. Figure 2.35C illustrates the second parting being taken by another machine. In preparation for this operation, the spoil pile top had to be levelled. Figure 2.35D shows a pit cross-section illustrating the third parting being dug, while Figure 2.35E shows the final parting and some rehandle being taken.

Spoiling dragline swing angles are approximately 90° for the top cut and 180° for the partings, other than the last parting. Here, the final material, including some rehandle, is spoiled at 90° directly on the working pad beside the dragline.

Rehandle varies with pit geometry and the reach capabilities of the first dragline. The rehandle will be further increased by material sloughing and falling from the bucket of the draglines while they take the upper partings. The majority of the rehandle is necessary in order to take the lowest

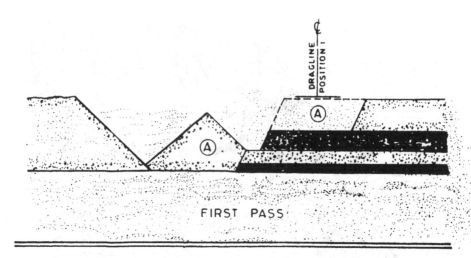

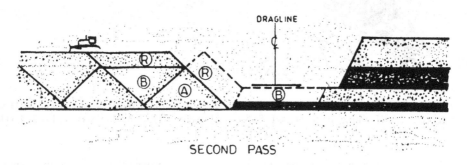

FIGURE 2.33 Use of rehandle material to bury spoiled interburden material. (From Math Tech, Inc. Evaluation of Current Surface Coal Mining Overburden Handling Techniques, U.S. National Technical Information Service, PB-264 111 (1976).)

seam. The economics then balance on whether the fifth seam is worth mining. However, in certain locations, government agencies may require that this seam be taken, even if it is not economical.

For properties having these types of coal resources, this method offers a means of strip mining that can handle large bank heights at relatively low overall strip ratios.

The coal seams are successively mined after each stripping operation. Inclines are provided at suitable intervals. For the upper seams these access roads take the form of temporary bridges off the spoil pile across to the coal. These bridges are formed where required by the dragline dumping extra spoil, followed by suitable dozer work. The frequency of access roads should be such that two accesses are always between each dragline stripping operation.

A series of multiple seams with thin interburdens can be stripped by dragline-dozer combination, as shown in Figure 2.36.

The top overburden is extracted by dragline placed on the highwall. The subsequent interburdens are extracted by dragline placed on the spoil or with a dozer.

2.13.1 HORSESHOE MINING SEQUENCE

The "horseshoe" mining sequence is a method of dragline operation that can be used to mine one end of a two-seam cut with a single machine. The stripping method used would be double pass with either rehandle or no rehandle.

This mining sequence is illustrated in Figures 2.37A through 2.37D. In this case the end of the pit is assumed to have no access, but the method can be equally well used when end access is provided.

STEP 1: DIGGING MUD HOLE & CASTING TO SPOIL

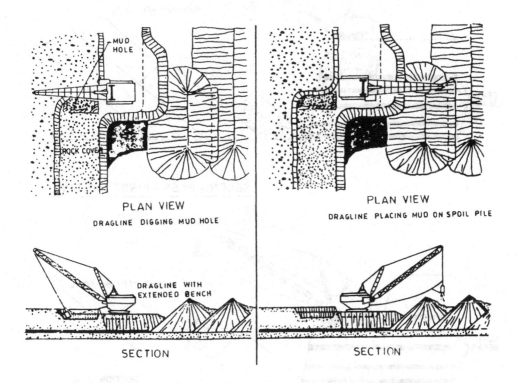

PLAN VIEW
DRAGLINE DIGGING MUD HOLE

PLAN VIEW
DRAGLINE PLACING MUD ON SPOIL PILE

SECTION

SECTION

STEP 2: REHANDLING MUD (OPT.)

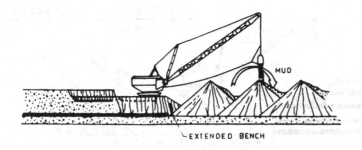

FIGURE 2.34 Dragline mud-handling procedure. (From Math Tech, Inc. Evaluation of Current Surface Coal Mining Overburden Handling Techniques, U.S. National Technical Information Service, PB-264 111 (1976).)

The dragline move to the spoil pile may require ramps, but parting stripping need not be held up until all the upper coal seam is mined because of the alternate haulage route via the end access.

Figure 2.37A shows the first step—stripping of the upper seam. As this operation proceeds the upper coal seam can be mined out. Upon completing this initial cut the dragline is deadheaded around the end of the cut and along the top of the prepared spoil pile to commence the parting stripping (Figure 2.37B).

When the upper seam has been mined out, the parting can then be stripped. Any waiting period here can usually be used for some scheduled maintenance. Figure 2.37C shows the parting spoil being taken. Mining of the lower coal seam follows after the dragline.

Figure 2.37D shows the parting spoiling complete, and the dragline deadheaded back to the opposite area of the pit. This section may well be taken by an alternative machine, especially if this

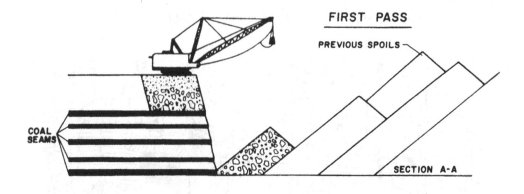

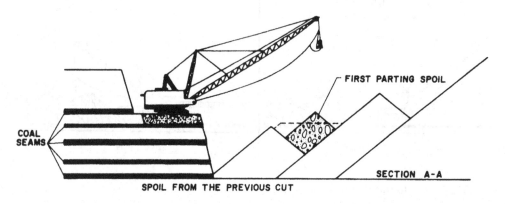

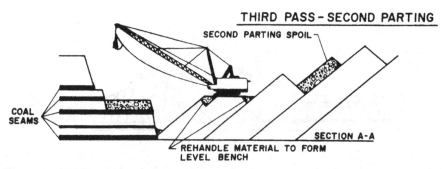

FIGURE 2.35A–C   Multiple seam mining—five seam operations. (From Bucyrus-Erie, Co., Surface Mining Supervisory Training Program, 1977. With permission.)

is just one end of a larger pit with only one incline. In this case the dragline would not be in a position to repeat the sequence on the other arm.

2.13.2   Steep Slope Mining Systems

Three mining methods are commonly used in steep slope areas. These are the conventional contour mining method, the haulback method, and the mountaintop removal method. The environmental damage caused by past use of the conventional contour mining method, primarily erosion, sedimentation, landslides, and aesthetic degradation, was severe and extensive. As a result, use of the method is now prohibited by law in all states.

FOURTH PASS — THIRD PARTING

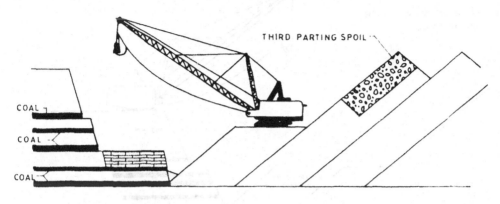

FIFTH PASS — FOURTH PARTING

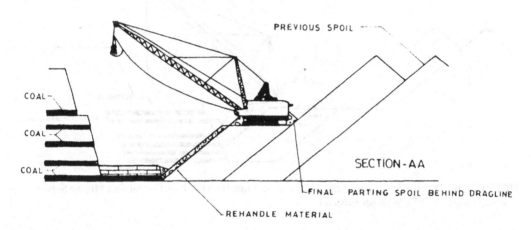

FIGURE 2.35D, E Multiple seam mining—five seam operations. (From Bucyrus-Erie, Co., Surface Mining Supervisory Training Program, 1977. With permission.)

There have been many dramatic changes in steep slope mining practices over the past several years, and there will be more changes in the future. Spoil haulage trucks, until several years ago used only at the largest mines, are now a virtual necessity for mining in compliance with the reclamation laws in all states. Although construction equipment is used for overburden removal and spoil placement at an estimated 97% of the mines, it is likely that draglines will be used to mine large mountaintop areas.

Although only three basic mining methods are in use in steep slope areas, there are many variations of each method. Additionally, state reclamation laws cause still further variations of the basic mining methods.

A description of the sequence of procedures used in mine development, mine operation, and reclamation is presented next. Following that are discussions of conventional contour, haulback, and mountaintop removal mining methods.

2.13.3 GENERAL SEQUENCE OF MINING AND RECLAMATION OPERATIONS

Many mine planning and development activities are common to all mining methods in three of the four central Appalachian states. Prior to mining, sediment basins are constructed in selected

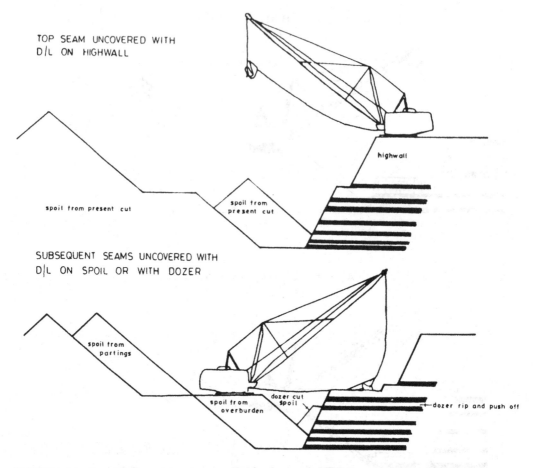

TOP SEAM UNCOVERED WITH
D/L ON HIGHWALL

highwall

spoil from present cut

spoil from
present cut

SUBSEQUENT SEAMS UNCOVERED WITH
D/L ON SPOIL OR WITH DOZER

spoil from
partings

spoil from
overburden

dozer cut
spoil

dozer rip and push off

FIGURE 2.36 Dragline/dozer operation, seven seams. (From Bucyrus-Erie, Co., Surface Mining Supervisory Training Program, 1977. With permission.)

drainage ways. The capacities and methods of construction for these basins generally are regulated by the states. If hollow fills are to be used, the hollows to be filled are selected and the storage capacities of the hollows are computed. If spoil is to be placed downslope of the coal seam elevation in other than hollow fill areas (this is prohibited only on very steep slopes), the legal limit for the position of the spoil toe is marked on trees downslope of the coal seam elevation in Kentucky and Tennessee. Areas upon which spoil is to be placed are generally scalped of trees, brush, and topsoil prior to placement of spoil on the area. The vegetation and soil are windrowed at the projected position of the toe of spoil.

Where hollows are to be filled, more extensive preparation measures are used. In addition to scalping the area, a drainage way is cut in the hollow and is then filled with rocks. After the fill has been placed, this rock-lined trench will serve as the drainage channel for the fill area.

Coal haulage roads are built to the coal crop line. In the past, these roads were typically narrow, winding, poorly surfaced, and were purportedly a major source of sedimentation. Today, in most states, the roads are better constructed. Drainage from the roads is controlled by ditches and culverts, as specified by reclamation regulations. Nonetheless, even today, roads at some of the smaller mines are poorly constructed and maintained. In some cases, bad road conditions caused by winter freeze-thaw cycles result in the closing of mines during winter months. Roads conditions are especially bad at some of the small mines. Because the acreage at many of these mines is small, there is not enough room to construct a road that has a gradual grade. As a result, roads at some small mines are very steep and winding. A little rain makes such roads impassable.

DRAGLINE STRIPS OVERBURDEN

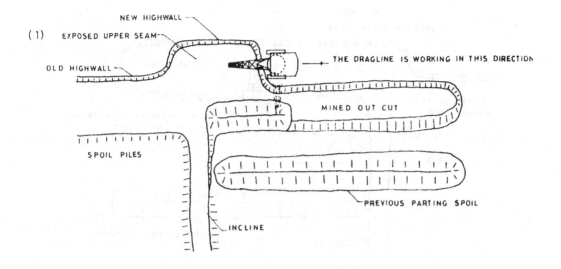

DRAGLINE DEADHEADS AROUND END OF CUT

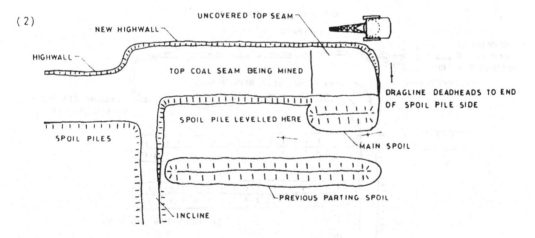

FIGURE 2.37 Horseshoe mining sequence. (From Bucyrus-Erie, Co., Surface Mining Supervisory Training Program, 1977. With permission.)

After site preparation and haul road construction have been completed, mining activities begin. Drill benches are constructed using dozers, overburden is vertically drilled and blasted, overburden is removed, coal is loaded out and hauled to the tipple by contract truckers, and the area is reclaimed. Overburden removal and placement methods and reclamation practices vary widely and are described in subsequent sections of this chapter.

2.13.4 ONE-CUT, SINGLE-SEAM CONVENTIONAL CONTOUR MINING

Conventional contour mining, called the shoot-and-shove method by some mine operators, is based on the use of dozers and end loaders for overburden removal and placement. At one time, this was

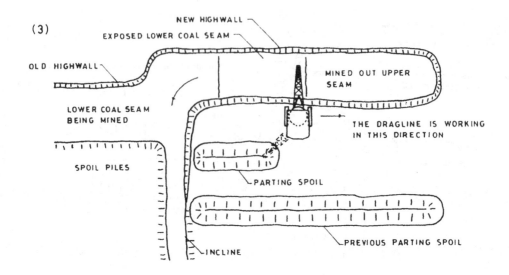

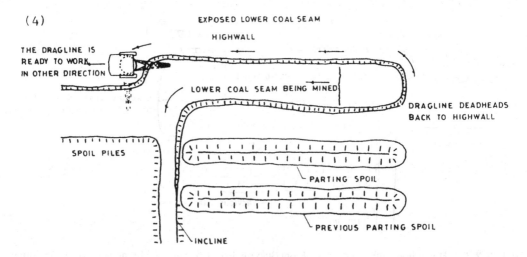

FIGURE 2.37 This figure is the continuation of the Figure 2.37. The step (3), dragline strip parting- shows parting spoil being taken, dragline working direction, previous parting spoil is under the parting spoil, and mining of the lower coal seam. Step (4) shows the completed parting spoiling. The direction of dragline is ready to work in other direction near the highwall (left side) as the lower coal seam is being mined. It also shows parting spoil is on top of previous parting spoil.

the only surface coal mining method used in steep slope areas, but now it accounts for only a relatively small percentage of total production. Mining usually commences as soon as the coal haulage road has been extended to the coal seam cropline. Mining advances in such a way that pits become longer as mining progresses.

Typically, only one cut, averaging 60–80 ft in width, is made into the hillside. Mine operators use long-established rules of thumb to decide upon pit widths, or, more correctly, to decide upon the height of the final highwall. A commonly used rule is that the final highwall height should be 1 ft for every inch of coal thickness. Thus, if the coal seam is 50 in. (127 cm) thick, the final highwall should be 50 ft (15.2 m) high. If the ground slope angle is 2:1, the resultant pit width would be 100 ft (30 m). Generally, ground slopes are steep in hollows (inside curves) and more gradual on points (outside curves). For this reason, when using the aforementioned rule of thumb, the pits are generally narrowed in hollows and widened on the points.

For a given ground slope angle, the average stripping rates can be reduced by narrowing the pit. In the authors' opinion, this procedure, known in the industry as creaming, is used by some conventional contour miners as a means of maximizing profit per ton of coal mined. The obvious disadvantage of the use of this procedure is that coal which could profitably be mined is left in place.

Overburden is excavated in blocks that are usually 250–1,500 ft (76–457 m) long. The length of the block is determined in part by the number of dozers that are used for overburden removal. The first step in the overburden removal process is the construction of the drill bench. Dozers are used to cut a solid bench and make a fill bench; i.e., an extension of the solid bench made from excavated material. The drill bench can be cut only as deep as the unconsolidated or semiconsolidated material near the ground surface. If this material is thin, a wide drill bench cannot be made, and the overburden must be drilled in two or three lifts. Sometimes, however, the overburden is drilled in one lift with the result that the lower outside corner of the overburden is not well fractured. The result is the huge sandstone boulders that are a commonplace at conventional contour mines.

At some mines, federal safety inspectors require the construction of a safety berm above the highwall. The purpose of the berm is to catch falling rocks. The berm is cut on a 1:1 slope using dozers. Where the unconsolidated surface material is thin, construction of the berm causes problems for mine operators.

Overburden blast holes are drilled on 10- to 15 ft (3- to 4.5 m) centres and are loaded with bulk or bagged ANFO. Blasting is delayed by row from the outcrop in towards the highwall; i.e., the outermost row of holes is shot first. This is done so that the explosive charge will move some of the overburden downslope, to rest on the ground surface below the elevation of the coal seam. Large boulders, termed flyrock, are sometimes cast far downslope when this blasting technique is used.

Subsequently, dozers are used to push the shot overburden down the hill. If the area has not been scalped prior to mining, trees are knocked down and buried in the spoil. Eventually, these trees will rot, leaving voids in the soil. These voids may eventually contribute to the instability of the fill.

Removal of overburden in this manner is relatively inexpensive since the dozers push downhill over short distances. According to mine operators, methods involving haulage of spoil are far more expensive than the dozer push method. They are understandably reluctant to use haulback methods.

It is difficult for the dozers to work right next to the highwall, so front-end loaders are generally used to square up the highwall. Spoil excavated by the loader is carried to the edge and dumped over, or is stacked for eventual use as backfill material.

If the overburden is drilled in two lifts, a second drill bench is constructed. The lower lift is drilled and blasted, and more spoil is pushed downslope by the dozers. At some mines, the overburden in the outer half of the pit is cleared first and half of the coal is loaded out. The front-end loader can then strip the remaining overburden from the front, turning and stacking the spoil on the fill bench for eventual use as backfill material.

When all overburden has been removed from a block, a fill bench and long slope outslope are left permanently in place. The uncompacted spoil rests on the steep slopes at its natural angle of repose. Landslides sometimes are the result of these factors in combination with the heavy rainfall typical of the area. Tension cracks in the fill bench are early indicators of spoil movement.

After all the coal has been removed from a block, the block is backfilled, as shown in Figure 2.38. The backfill is sloped back towards the highwall at an angle of 2–4° as a means of reducing runoff and erosion on the spoil out slope. In theory, ditches parallel to the highwall are constructed so that water which collects near the highwall will be carried to controlled discharge areas. In practice, water collects near the highwall, seeps into the spoil, and possibly contributes to slope instability.

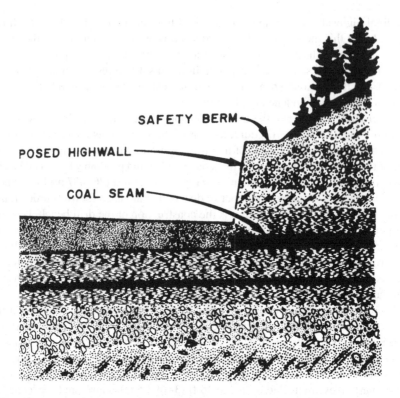

FIGURE 2.38 Conventional contour mining final grading. (From Math Tech, Inc. Evaluation of Current Surface Coal Mining Overburden Handling Techniques, U.S. National Technical Information Service, PB-264 111 (1976).)

After backfilling, the bench area is graded. The bench and spoil out slope are then fertilized and seeded using hydroseeders. With the exception of the lower third of the out slope, which is difficult to reach with the hydroseeder, revegetation efforts are usually successful. Erosion on the spoil out-slope is nonetheless a problem (Figure 2.38).

Although the reclaimed bench may have greater use potential after mining than before, the use potential of adjacent land may be reduced due to the danger of slides, increased erosion and sedimentation, and degraded aesthetics.

The major environmental impacts of conventional contour mining are landslides, erosion, sedimentation, and aesthetic degradation, which results from exposed highwalls and eroded outslopes. Reduction of the environmental effects of conventional contour mining is needed. Modified methods, such as the slope reduction and parallel slope fill methods, were proposed years ago as a means of reducing erosion and slope instability, but they were used only on occasional experimental bases and reportedly did not work well. These kinds of modifications, although widely reported in the literature, were never used on a production basin in the past, nor are they used today.

In devising research programmes whose goal is the reduction of the environmental impact of conventional contour mining, economics are of paramount importance. In conventional mining, dozers push downhill or on level ground for one-way distances averaging less than 100 ft (30 m). The capital and operating costs of this method are relatively low. No spoil haulage trucks are required. Mine planning requirements are minimal. If a modified method could be devised wherein the amount of spoil placed downslope was significantly reduced, but the costs of mining and reclamation were similar to those of conventional contour mining, then research will have served a useful purpose.

The solutions devised are the full and partial haulback methods, but according to mine operators, haulback is as much as 60% more costly than conventional contour mining. Both capital and operating costs are affected. Since haulback mining is believed to be a means of reducing the environmental impacts of steep slope mining to acceptable levels, research should be conducted to devise a haulback method that has the same cost characteristics as the conventional contour method.

2.14   OTHER CONVENTIONAL CONTOUR MINING SITUATIONS

In some situations, multiple seams are mined by the conventional contour method, or two cuts are taken on a single seam. There are three kinds of multiple-seam mining. The first of these, which apparently occurs infrequently, is the case where two seams appear in the first cut highwall.

Where slopes are gradual or coal is of very high quality, two cuts are sometimes made in conventional contour mining. After the coal has been loaded out the first cut, second-cut overburden is blasted, pushed, and carried into the open first cut. It should be possible through modification of operations procedures to reduce the amount of spoil pushed downslope in this situation.

The haulback method, as the name implies, involves haulage of spoil laterally back along the bench, where it is placed on the pit floor. The method is now widely used to comply with regulations that prohibit or limit the downslope placement of spoil and require that the final highwall be completely or partially buried.

Unlike the conventional contour method, which is essentially the same wherever practised, haulback methods are state- and site-specific. Four parameters are important in defining haulback situations; these are the percentage of spoil that can be placed downslope, the permissible height of the reduced final highwall, the number of cuts made into the hillside, and the number of coal seams mined per pit. The values of the first two of these parameters are controlled by state laws, the third by mining economics, and the fourth by overburden and coal stratigraphy.

In steep slope areas, no spoil can be placed downslope, except in hollow fill areas. Maximum permissible height of the reduced highwall is 30 ft (9.1 m). Some spoil can be pushed downslope, the amount depending on the ground slope. In very steep areas, 28° and above, no spoil can be placed downslope. Reduction of the final highwall may be required. A determining factor is the competence of the highwall. In areas less than 28°, a limited amount of spoil can be pushed downslope—the straight-line distance from the outer edge of the solid bench to the toe of spoil cannot exceed 50 ft (15.2 m)—and complete highwall burial is required. An exception to the latter rule is made where a previously stripped area is mined. Then the final reduced highwall can be as high as that left after the original mining operation.

2.14.1   SINGLE-CUT HAULBACK MINING OF SINGLE SEAMS

Figure 2.39 shows an artist's conception of a single-seam, single-cut haulback method based on the use of dozers and loaders for overburden removal, and rock haul trucks for spoil haulage and placement. Many features shown in the figure are common to all haulback situations.

A drill bench is first cut by dozers and overburden is vertically drilled, using procedures identical to those used in conventional contour mining. Haulback blasting procedures are different from conventional procedures, however, because spoil must not be cast downslope by the blast. This is true even where limited amounts of spoil can eventually be placed downslope. Operators of most haulback mines visited during the field survey delayed the shots so that the overburden would be lifted but not moved outward during blasting. A special haulback blasting procedure had been devised by one West Virginia company. This procedure consists of delaying the shots in curvilinear "rows" so that overburden is thrown laterally back into the open pit by the blast. According to a company official, several years were required to develop and perfect the method, which now works well.

The initial block cut is generally located adjacent to a hollow which is to be filled with spoil during mining. This method of opening the initial block depends on prevailing reclamation regulations. Dozers push spoil from the first block directly into the hollow in conventional fashion.

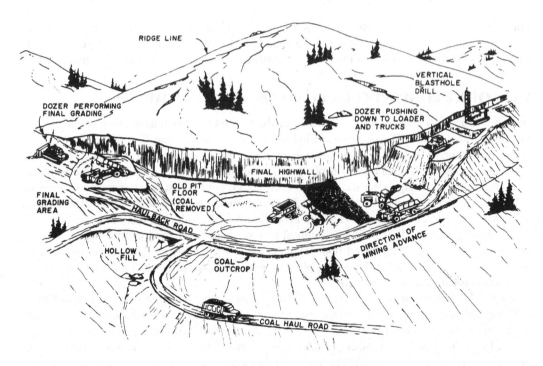

FIGURE 2.39 Haulback mining in a single seam. (From Math Tech, Inc. Evaluation of Current Surface Coal Mining Overburden Handling Techniques U.S. National Technical Information Service, PB-264 111 (1976).)

When a working room has been established, loaders and trucks may also be used to dump spoil into the hollow. As this is done, dozers may be used to work the outslope of the fill to about a 2:1 slope. The hollow fill must be constructed in lifts, from the bottom up. This necessitates truck haulage of first-block spoil down the hollow to the projected position of the toe of the fill.

After a sufficient length of the pit has been opened, the haulback operation begins. In a full-haulback situation, in which spoil cannot be placed downslope, a dozer works on the shot overburden, pushing it down to an end loader. Typical loader capacity is 12 cubic yards (9.17 m^3). According to one engineer, with the exception of only the largest loader built, dozers must push to the loader because the loader does not have breakout force sufficient to excavate the sandstone overburden without the dozer assist. This is but one of the factors that makes haulback mining more costly than conventional contour mining.

It is easy to segregate spoil materials when this method is used. First, the dozer pushes all soil down to the loader, which loads it into haul trucks. Typical truck capacity is 35–50 tons. Usually, two trucks work with one loader. The soil is hauled back a relatively long distance and is placed on top of graded soil. Next, the bulk of the remaining overburden is pushed down to the loader, loaded into the trucks, hauled back, and placed against the highwall. Finally, the overburden directly overlying the coal seam is loaded, hauled back, and placed on the pit floor near the highwall, where it subsequently will be covered with other spoil. Not all mine operators use this method, but it may not be expensive to do so.

Stripping procedures are somewhat different in a partial haulback situation, where some spoil can be pushed downslope by dozers. In this case, after blasting, dozers push some spoil downslope in conventional contour fashion. The amount of spoil placed in this manner is fairly small, however, and in the authors' opinion, poses no major landslide or erosion problems. It does, however, eliminate the possibility of using topsoil as dressing on reclaimed areas, since the soil is pushed downslope. After the permissible amount of spoil has been placed downslope, the previously described haulback procedures are used.

Occasionally, in situations where some spoil can be placed downslope, after the initial block has been opened, no further spoil will be placed in hollow fill areas. More typically, however, because of the 30% spoil swell factor, as mining progresses, it will be necessary to store additional spoil in a hollow. This is always the case where spoil cannot be placed downslope. Operating procedures used to deal with this problem vary. One mine operator opens a long initial block, storing all the first-block spoil in a large hollow fill area. Then, for a while as mining advances, all spoil is hauled back and placed on the solid bench; none is placed in hollow fill areas. Because of the spoil swell, the spoil eventually "catches up" (this is the mine operator's phrase) with the stripping and loading operation. This means that the length of the open cut gets progressively shorter as mining advances. When the open cut reaches a minimum acceptable length, another hollow fill is started. With some planning, this point will be reached directly adjacent to the hollow, which is to be filled.

Another design includes a hollow fill area at least every 3,000 ft (914 m) along the contour. Although mining proceeds in only one direction, spoil is hauled to hollow fill areas in two directions, ahead of the block being stripped and behind it. Suppose, for example, that mining is proceeding in an easterly direction. One hollow fill area would be located west of the active pit. Spoil from the active pit would be hauled to this area in a westerly direction back along the contour on the road used by coal haulage trucks. The second hollow fill area would be located east of the active pit, in an area which had not been stripped. In order to enable haulage of spoil to this fill area, an advance road would be constructed on the contour ahead of the active pit area. In practice, spoil is hauled the shortest possible distance. Sometimes this is to a position on the solid bench, sometimes to the advance hollow, and sometimes to the "retreat" hollow. The hollow fill areas are deliberately spaced no more than 3,000 ft (914 m) apart so that the haul distance to a hollow will never exceed 1,500 ft (457 m).

Spoil haul road profiles are dependent primarily on the following factors:

- Number of overburden lifts
- Haulage distance
- Height to which spoil is stacked on the solid bench

If overburden is drilled and stripped in one lift, the end loader will work from a position a few feet above the top of the coal seam. Haul trucks approach the loader on relatively level ground. If, on the other hand, two or three overburden lifts are used, when the top lift is stripped, the loader will work on the upper bench and trucks will have to be ramped up to this loader.

At southern West Virginia full-haulback operations, the coal haulage road is constructed on the solid bench as shown in Figure 2.40. The road is not built on the fireclay pit floor, however, but is built up using competent spoil materials. The road, which is 15–20 ft (4.5–6 m) wide, is inclined downward towards the highwall so that runoff from the road will flow into the drainage ditch shown in the figure. Sumps are located in the ditch at intervals along the contour. Water that collects in the sumps is taken off the bench through culverts that have been installed under the haul road. The culvert discharges to a rock-lined drainage way. This is a good drainage-control system.

Some mine operators like to have two active pits at each mine. In haulback mining, this is easily accomplished by mining outward in two directions from the initial block. If sufficient stripping equipment is available, both pits will be worked simultaneously, with the spoil being hauled back in the direction of the initial block. Except for the sharing of coal loading and haulage equipment, these are merely two separate haulback operations. In some cases, if there is not enough stripping equipment to work two pits simultaneously, the haulback can still be worked in two directions by shuttling equipment between the two pits. Coal is loaded from one pit while the other is being stripped, and vice versa. Some have called this latter procedure the block cut method; others have called it the modified block cut method, although no one seems to know what has been modified. Practically speaking, it is just single-cut haulback mining.

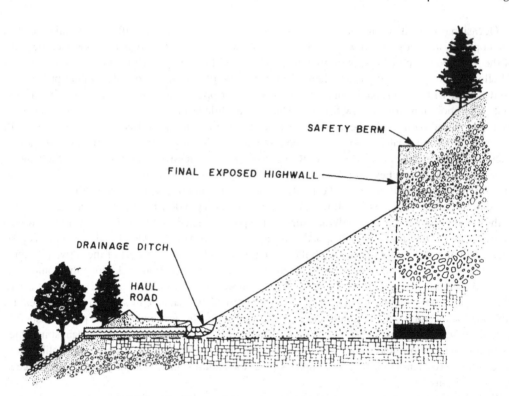

FIGURE 2.40A Reclamation with no spoil downslope (section). (From Math Tech, Inc. Evaluation of Current Surface Coal Mining Overburden Handling Techniques, U.S. National Technical Information Service, PB-264 111 (1976).)

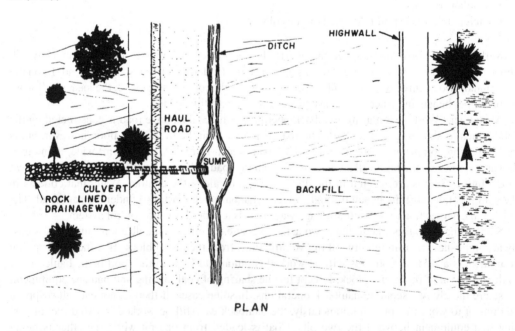

FIGURE 2.40B Reclamation with no spoil downslope (plan). (From Math Tech, Inc. Evaluation of Current Surface Coal Mining Overburden Handling Techniques, U.S. National Technical Information Service, PB-264 111 (1976).)

Mining proceeded in only one direction at each of the 11 field survey haulback operations, but several operators were considering the use of the two-directional method.

Backfilling and grading activities are an integral part of the haulback mining process. After spoil has been dumped on the solid bench by the spoil haulage trucks, dozers are used to grade the spoil. One method entails grading of spoil on the solid bench to about a 2:1 slope, the maximum slope that can be navigated by a large dozer. After grading, topsoil may be spread over the spoil surface. This is done at 3 of the 11 survey haulback operations. Fertilizer, seed, and sometimes mulch are applied by hydroseeder (Figures 2.40A and 2.40B).

BIBLIOGRAPHY

1. *Surface Mining Supervisory Training Program* (Bucyrus-Erie Co., 1977).
2. Fluor Engineers, Inc., Optimal Dragline Operating Techniques, U.S. National Technical Information Service, DE83-006980 (1982).
3. Math Tech, Inc., Evaluation of Current Surface Coal Mining Overburden Handling Techniques (Washington, DC: U.S. National Technical Information Service, 1976), PB-264.

3 Reclamation and Revegetation of Mined Land

3.1 INTRODUCTION

Federal and state reclamation laws and regulations sometimes require, or tend to imply, that mined land be restored to premining use. This has been understood to require restoration of the mined land to approximate premining topography, re-spreading of all or most of the original soil material, and re-establishment of the same or similar vegetative cover. The goal of reclamation should be to establish a permanently stable landscape that is aesthetically and environmentally compatible with surrounding undisturbed lands. Postmining land use should be one that contributes most effectively to the productive capacity and stability of the greater ecosystem, of which the mined land is one component. The size of the ecosystem depends upon practical considerations at each site. In areas where land ownership is predominantly private, each privately owned block could comprise an ecosystem. Where the ownership is public, the ecosystem may be much greater in extent. We cannot consider each mined area as a separate entity but, rather, as one component of the ecosystem in which it occurs. Reshaping the mined land to a topography that contributes best to the stability and productive capacity of the entire area can be more important than reshaping to premining contours. Revegetation should use species that will contribute the most to the stability and utilization of the entire system. The land should be reclaimed so that it is suited for as many alternative uses as may be practically feasible.

The postmining productive capacity depends upon the inherent properties of the soil and/or soil materials replaced in the root zone. The actual productive potential of a reclaimed soil depends upon its topographic location and the degree of restoration of specific root zone properties that were disrupted in the mining and reclamation processes. A guiding concept was that soils existing prior to mining could be replaced on the mined area. With a minimum change in properties that influence land productivity, the soils of the mined land should approach the productivity of the original soils at the site. Even if the exact thickness of topsoil and subsoil were replaced, two soil properties that likely would not be the same as those premining are

- Size, shape, and volume of pores
- Number and kind of soil microorganisms

The latter can be avoided by direct, immediate replacement of soil without stockpiling. The number of microorganisms in the stockpile decline due to the lack of oxygen.

Some natural soil horizons are not particularly suitable for use in reclamation, so in some instances, substitution of soils for both the sub- and the topsoil may be necessary. Some cases exist in which the bank-run mine spoil or soil mixes that could be created with specific mining operations are as adequate as the subsoil and could be provided at much less cost to the mine operator.

The potential productive capacity of a reclaimed soil is dependent upon the inherent properties of the soil and/or spoil materials in the root zone. Among the many inherent chemical properties that affect plant growth are

- pH
- Toxic elements
- Cation exchange capacity
- Nutrient availability

The most important inherent physical factor is the texture which determines water-holding capacity and the entrance and internal movement of air and water.

These soil properties have formed in soils during the long period of their development. The properties are partially or totally destroyed during mining and reclamation. They must be re-established before the potential productivity of the soil can be reached.

Of the in-site properties, bulk density is used most frequently. It is often erroneously considered the only necessary measure of root zone physical properties. Bulk density is simply the weight of soil per unit volume, and as such can be used to estimate the total pore space in the soil. Pore size distribution is important because measurements of the space occupied by the smaller pores give an estimate of the available water-holding capacity of the soil, while measurements of the larger pores (macropores) give an estimate of the volume of pore space available for conducting water and air through the soil.

The root zone material must be regraded with proper compaction. The restoration of all surface mined land would require regrading, if only to level the spoil banks. The grading, however, should be performed in such a way that compaction is minimized.

Several mine operators have made progress in developing methods of handling the soil in order to minimize compaction. The use of a bucket wheel excavator and hauling by trucks with base-level dumping that avoids traffic on top of the soil being replaced have given encouraging results. The main difficulty in the compaction problem is the use of scrapers, especially in cases in which the lifts are thin and the equipment must make repeated passes over the area. Making the lifts 3 m or more thick is helpful.

Handling the soil when it is dry is very critical in reducing compaction. Soil that is excavated when wet is likely to be puddled during excavation and is not likely to dry while in a stockpile. Puddling is then likely to recur when the soil is regraded. Wet soil is vulnerable to over compaction, even with minimal compactive forces. Handling of top and subsoil materials should be planned for seasons when the soil has the lowest moisture content.

The bottom of the soil zone used by plants such as corn is in the range of 40–72 in. (1–2.5 m). Corn plants can extract significant water or nutrients below a depth of 48 in. (1.3 m) on a few soils with thick subsoils and well-developed soil structure. Root-restricting layers occur in some prime farmland soils and have presented a special problem—they may affect soil productivity. Such horizons should be considered a part of the root zone provided for in soil reconstruction plans.

Restoration of the mined land to its original contour can place some restrictions on the final soil slope. Some surface land has poor natural soil drainage. Most regraded mine soils have slower permeability after mining than before regrading, and it is likely that if the regrading does not provide for better surface drainage than the original surface did, the reconstructed soil will be wetter than the original. Productivity might be much lower than the original soil because of higher wetness of the soil. Slopes of 2–3% are desirable.

3.2   RECLAMATION OF SURFACE MINED LAND IN AUSTRALIA

Reclamation planning is integrated with mine planning from the earliest conceptual stages so that reclamation constraints can be accommodated in the mining plans and vice versa. One important requirement is the development of a flexible reclamation plan which will accommodate changes in mining operations.[1] Specific attention is paid to the physical and chemical characteristics of the soil and to the catchment size, drainage density, and channel capacity in landform design. Greater attention is given to quality control in the selection, stripping, and use of topsoil.

Drainage stability erosion control and pre- and postmining land use are primary considerations of the reclamation plan. The need to construct a suitable landform within the confines of an operable mining plan is important. To achieve this, an interaction between environmental personnel and the mine planning group during the planning stages is required with respect to equipment scheduling, overburden volumes, swell factors, and locations for haulage ramps and roads.

The topography consists of a narrow ribbon of highly textile floodplain, no more than 2–3 mi wide. The floodplain is intensely irrigated and used for dairying and the production of oilseed crops. Rising away from either side of the floodplain are gently rolling foot slopes with relatively infertile soils that are used for low-intensity beef cattle production on native pastures. Isolated areas of more fertile volcanic soil types have been used for grape growing and area crop production. Surface coal mines are located on this type of land.

The objective of reclamation is to return mined land to its previous capacity and productivity for beef cattle grazing. This requires constructing stable landforms and establishing maintenance for pastures. More vigorous and productive exotic grass and legume species are used in place of the native species because the seed of the native species is difficult to procure commercially, germination percentages are poor, and the plants are slow to establish. Native grass and legume species do not provide sufficient protection from soil erosion during the early stages. Native trees are planted in areas ranging from small woodlots serving as windbreaks and stock shelter to sizable areas of forest on steeper slopes.

The following is a discussion of typical reclamation planning procedures.

3.2.1 MATERIAL CHARACTERIZATION

Overburden material is analyzed from exploration borehole scrapes taken during the planning stage. This helps determine fertilizer requirements on reclaimed pastures and identifies any potentially toxic or otherwise intractable layers which may have to be selectively handled. Table 3.1 presents a typical physical and chemical analysis of the two principal overburden types: a poorly laminated argillaceous grey mudstone and a massive brown-grey lithic sandstone. The main physical difference between the two is that the mudstone weathers rapidly upon exposure to form a reasonably acceptable substitute soil or subsoil, while the sandstone is strongly resistant to weathering. Both materials are extremely deficient in nitrogen and phosphorous, indicating the need for heavy fertilizer application. Phosphorous is present in the rocks in the form of apatite or hydroxy apatite which becomes increasingly plant available as weathering progresses. Analysis of spoils more than 20 years old has shown levels of plant-available phosphorous sufficient to sustain plant growth, but not in toxic concentrations.

The principal deficiencies in the materials are high pH and high salinities. Salt levels are rarely sufficient to inhibit plant growth but do cause dispersion of clays leading to surface crusting and sealing, which impedes moisture penetration. Sulfur levels are quite low, except in the northern end of the valley.

3.2.2 LANDFORM DESIGN

In the development of a final land surface, establishing a lateral drainage system is essential. The drainage system should be compatible with the off-lease drainage pattern, minimize flow into mine workings, and provide containment and treatment options to satisfy pollution control requirements. Interference between the drainage system and access to mine workings should be minimized.

The ultimate objective is to use major coal haulage roads and ramps as the principal drainage channel. The drainage density of the surrounding land, particularly those with similar slopes, vegetation, and land use, guide the determination of appropriate catchment sizes. One of the main objectives of the drainage design for the mine as a complete system is separating runoff coming from undisturbed areas from the drainage which may be contaminated by mine workings and spoil piles. All water flowing from reclaimed surfaces is contained and treated in sedimentation dams before release.

Attaining satisfactory slopes on reformed spoil piles is an important requirement. Influencing factors include the erosion potential of the reformed surface in relation to slope length, upslope catchment and maximum anticipated runoff, soil materials used for topdressing, and vegetation density.

TABLE 3.1
Overburden Characteristics

Parameter	Sandstone	Mudstone
Particle size (%)		
Sand (2.0–0.2 mm)	14.2	12.8
Fine sand (0.2–0.02 mm)	48.8	33.6
Silt (0.02–0.002 mm)	9.4	19.6
Clay (0.002 mm)	27.6	34.0
Dispersal index	4.2	3.3
Moisture characteristics		
Field capacity (pF 2.0)	21.5	21.5
Wilting point (pF 4.2)	9.0	9.0
pH (1:5)	8.5	8.0
EC (mS)	0.16	0.12
Water-soluble cations (meq/100 g)		
Na	0.50	0.30
K	0.60	0.30
Ca	0.10	0.20
Mg	1.70	1.70
Exchangeable cations (meq/100 g)		
Na	1.0	0.5
K	1.8	0.6
Ca	4.8	5.1
Mg	12.0	13.7
Trace elements (ppm/0.5 ml HCl extract)		
Zn	9.0	2.1
Ca	1.1	0.9
Pb	n.d.	0.5
Mn	38.0	10.5
Mo	1.0	1.0
Cu	0.2	0.4

From Gordon, R. M. and J. G. Hannen. Reclamation of surface coal mines in the Hunter Valley, New South Wales, Australia, in *Innovative Approaches to Mined Land Reclamation* (Carbondale, IL: Southern Illinois University Press, 1986), pp. 525–559. With permission.

n.d. = not detected.

Natural slope angles in the vicinity of several old mines were surveyed and the median angle was found to be 10°. This angle has been accepted as a standard. In practice, reformed slope angles range between 6° and 10°. A minimum slope of 0.5° is used to ensure free drainage.

In the development of the reclaimed landform, computer models are used to simulate mining and beef-filling operations over the operating life of the mine. In generating landforms, the computer operates within the defined constraints of slope angle, overburden swell factors, and elevation. The landform is designed to visually blend with the surrounding topography, to be stable in the long term, and to suit the proposed postmining land use of beef cattle grazing.

3.2.3 USE OF TOPSOIL

Before any mining operation is conducted, a detailed soil survey is undertaken. These surveys are conducted in increments, covering the area to be mined during each 5- to 7-year open-cut approval.

Soil profiles are examined at exposed areas such as cuttings and gullies, and by means of coring. Soil types are classified according to horizon, colour, texture, structure changes, and pH. In addition, representative samples are collected for laboratory analysis.

A topsoil stripping plan is prepared which indicates the required stripping depth or the thickness of soil to be removed within each portion of the mine site, including areas to be used for out-of-pit spoil emplacements and washery reject disposal. Stripping is indicated in increments of 15 cm, about the limit of accuracy at which the dozers and scrapers are able to operate.

Soils covering most of the rolling foot slopes where the surface mines are located have developed more or less in-site from sedimentary rocks of the coal measures. They are relatively infertile duplex types in which a shallow sandy or loamy A horizon overlies a medium to heavy clay B horizon. The organic-rich A horizon is frequently <10 cm thick. On the lower parts of slopes, this is often underlain by a bleached, hard-setting A_2 horizon, which is quite infertile, highly dispersible, and of no value for reclamation purposes. The B horizon is also usually dispersible, often saline, and of little value for reclamation. Table 3.2 presents physical and chemical analyses for a typical sodic soil profile.

The optimum thickness for spreading topsoil over reformed spoil piles is 10 cm. Over quite significant portions of the mine lease, none of the soil horizons are suitable for reclamation purposes because of high salinity, a high proportion of stone in upper horizons, or the absence of A_1 horizon due to erosion. To compensate for this problem, stripping depths are increased on better soil types in the lease.

Suitable topsoil material may not always be available, however, to top dress all reformed spoil piles and other disturbed areas. The available material must be used in the most beneficial way so as to rapidly establish dense vegetation on areas susceptible to erosion, such as steeper slopes and

TABLE 3.2
Analysis of Typical Sodic Soil Profile

Parameter	A1 Horizon (0–10 cm)	A2 Horizon (10–40 cm)	B Horizon (+40 cm)
Particle size (%)			
Sand	22	21	6
Fine Sand	57	62	19
Silt	8	9	2
Clay	11	8	72
pH (1:5)	6.2	6.3	8.3
EC (mS)	0.06	0.04	0.05
Water-soluble cations (meq/100 g)			
Na	0.1	0.1	0.2
K	0.1	0.1	0.1
Ca	0.1	n.d.	n.d.
Mg	1.8	0.3	0.1
Exchangeable cations (meq/100 g)			
Na	0.2	0.4	1.9
K	0.6	0.6	1.9
Ca	3.0	3.6	3.4
Mg	1.9	3.4	4.4

Source: From Gordon, R. M. and J. G. Hannen. Reclamation of surface coal mines in the Hunter Valley, New South Wales, Australia, in *Innovative Approaches to Mined Land Reclamation* (Carbondale, IL: Southern Illinois University Press, 1986), pp. 525–559. With permission.

n.d = not detected

the channels of newer watercourses. Preference is given to establishing trees directly into the spoil material. Some native trees establish and grow better in some spoil materials than on sites that have been topdressed with soil. The main use of topdressing with soil is to provide a better physical environment for the initial germination of pastures and as a source of native seed which helps increase the diversity of species in the final pasture. The relatively thin layer of applied topsoil means that vegetation depends upon the underlying spoil materials for long-term growth and survival.

3.2.4 RECLAMATION PROCEDURES

Standing timber is cleared in strips ahead of mining operations before the topsoil is cleared. To avoid dust generation problems, the length of the strips is limited to a short distance ahead of the highwall. Self-loading scrapers are the most commonly used for removing topsoil. In some instances, the topsoil is removed by bulldozers and transported by trucks.

Soil is stockpiled at strategic locations around the mine for subsequent recovery and for spreading over reformed spoil piles. However, after reclamation has commenced, the topsoil is directly transported to the areas being reclaimed.

In mines that utilize trucks and shovelers for waste handling, the desired landform can be roughly constructed as the material is dumped. Afterwards the bulldozers can build the landform by trimming the spoil piles. Dragline spoil piles require considerably more shaping.

The first step in the reshaping program is to erect slope profiles or survey pegs that indicate the required slopes in accordance with the landform plan. Slots are then bulldozed to achieve the working slope at selected points over the area to be reshaped.

At several mines, graded banks have been progressively introduced and have proven successful as an erosion control measure on longer slopes. The banks are designed with channel gradients of 0.5%, though it is believed that gradients could be increased to 1 %. Differential settlement of the spoil leads to overtopping and bank failure. A limitation on the use of graded banks is the availability of suitable discharge points.

In most instances, scrapers used for topsoil spreading utilize material transferred from a point directly ahead of the cut or material reclaimed from stockpiles. Scrapers are usually worked on the contour to avoid creating tracks that may become sites for gully erosion after a heavy rain. Scraper movements over previously top-dressed surfaces are minimized to avoid over compaction of the soil. Topsoil is spread at a nominal thickness of 10 cm.

3.2.5 REVEGETATION METHODS

Valleys experience a warm temperature climate with summer-dominant but winter-effective rainfall. The average annual rainfall is 0.6 m. Summer rainfall commonly occurs as short duration, high-intensity storms, often causing localized flooding and soil erosion damage.

The climate of the region presents particularly difficult conditions for pasture establishment as it favours warm-season species. However, the unreliability of spring rains and the hot, dry, early summer make their establishment in early spring a difficult process.

After some research, a technique was developed involving sowing in the early autumn to take advantage of the relatively reliable late summer and autumn rains, using a mixture of cool- and warm-season species sown simultaneously. The cool-season grasses and legumes germinate and establish quickly to provide early protection of the surface from soil erosion. The warm-season seeds remain dormant in the soil until the following spring or early summer when they germinate and establish through the mulch provided by the hayed-off, cool-season grasses. These warm-season species then develop into permanent, productive pasture.

All cultivation in preparation for sowing is carried out on the contour to retard surface runoff and promote moisture infiltration. The usual practice is to deep-rip the ground to a depth of 0.6–1 m using a dozer-mounted ripper with tines spaced about 1 m apart. This is followed by contour cultivation with a conventional chisel plough to a depth of about 22 cm. The purpose is to produce a very coarse, rough seedbed that retards the formation of a surface crust or seal on the soil with deep furrows for moisture retention.

Inoculation of legume seed by Rhizobium occurs via tractor or truck-mounted broadcasters immediately prior to sowing. Sowing is usually timed for the early autumn season (March to April) when soil moisture levels are high. The cool-season species (Barrel medic and ryegrass) germinate to provide a quick-growing cover crop, protecting the surface against erosion. The warm-season species germinate from the early spring onward to provide a Rhodes grass- and alfalfa-based long-term pasture.

3.2.6 TREE PLANTING

Sites for tree planting are chosen based on their physical suitability for trees, aesthetic appearance of the reclaimed surface, and their usefulness in providing stock shelter. Favoured locations include ridgelines adjacent to major watercourses and dams, and in scattered woodlots.

The areas selected for tree planting are pegged out, and topsoil is not placed on them, although normal cultivation work is performed. The sites are then deep ripped on the contour a single time and are spaced at 4–5 m. The rip lines are allowed to settle for 4–6 weeks prior to planting.

Planting methods use tubed seedlings 6–9 months old, which are raised in a commercial nursery. Some mines collected the seed of native tree species growing on the lease and gave them to a nursery to raise the tubed stock.

Seedlings are planted out May through June. Planting is carried out by hand, with seedlings being placed about 5 m apart; several slow-release fertilizer pellets are placed into the hole with some water.

3.2.7 PASTURE MANAGEMENT

A management program is implemented to ensure the persistence of a productive pasture and tree lots on the reclaimed land. To maintain a balance of legumes and grasses in the pasture, livestock grazing is required as soon as possible after the pasture has become established. Annual fertilizer topdressings are applied by aircraft or by ground broadcaster twice during the year.

Erosion damage is promptly repaired. Weeds can usually be controlled by appropriate pasture and grazing management, but spot spraying is necessary.

3.3 REVEGETATION OF A SURFACE MINED LAND IN MONTANA

Reclamation of coal-mined land in Montana is oriented towards rebuilding the land to its pre-mining structure, function, and use. The state regulations require the establishment of suitable, permanent, diverse vegetative cover capable of funding livestock and wildlife, limiting soil erosion, and persisting under prevailing environmental conditions. Restoring productivity of the land to provide a renewable resource requires coherent functioning of soil and plant systems. These systems are interdependent, and their development during and following reclamation is of particular concern.[2]

Plant and soil successions on mined land contain elements of both primary and secondary succession. The substitute lacks normal microorganism populations, cycling, and biogeochemical balances, but it has water-holding capacity and can perform soil functions. Plant succession on abandoned spoil is not the same as on abandoned land. Some early successional stages associated with soil formation can be achieved in reclamation proves by top soiling, fertilization mulching, and the use of seeding mixtures of climax and subclimax native and introduced plant species.

Concerns about reclamation success have included lack of diversity, declining productivity, and marked compositional characteristics. Difficulty in establishing adaptive native grass, forb, shrub, and tree species has posed a serious problem. Other problems affecting revegetation include alkalinity, salinity, excessive litter accumulation, insect and pathogen damage, and soil nutrient deficiencies.

Plant succession on seeded mine spoil is strongly influenced by the following variables:

- Species seeded
- Initial seeding success
- Cultural practices
- Weather

At the end of the first growing season on seeded mine spoil, annual grass and forbs provided about 90% of the vegetation cover. Seeded perennial grasses, forbs, and shrubs provided the other 10%. By the end of the second growing season volunteer and seeded annual and biennial species were dominant and provided most of the cover. Perennials, especially introduced grasses, increased in cover to about 40% of the total, depending on the success of stand establishment. By the third growing season, perennials had assumed complete dominance. Perennial cover stabilized at about 40% in the fifth growing season. Litter cover and moss increased rapidly to more than 95% and 5,500 kg/ha, respectively, after three growing seasons.

On sites revegetated, introduced perennial grasses have commonly achieved dominance despite simultaneous seeding of native species. The most successful species have been crested wheatgrass, smooth bromegrass, and alfalfa. When seeded at moderate rates (<100 PLS/m^2), these species produced most of the cover and yield following establishment. Over time they have tended to exclude other species. Because of the competitive ability of introduced species, the advantages of seeding mixtures, as well as topsoil containing native species, were quickly lost.

Several native species have proven valuable in reclamation. Rosana western wheatgrass, Critane thick spike wheatgrass, and Wytene four-wing saltbush have occasionally performed well. These species have persisted on seeded raw spoils for 8 years when seeded without cover crops, crested wheatgrass, and other introduced grasses.

Repeated fertilization was probably an important factor in the failure of some native species to establish and persist. On several seeded stands in which introduced species showed vigorous stands, repeated fertilization, and favourable growing conditions resulted in high accumulation. The thick mulch layer delayed pH development and decreased vigour and yield. On one site seeded, in 5 years the yield decreased from 2,200 to 750 kg/ha. Wide C/N ratios on such sites indicated that the nutritional status of the spoil soil had been lowered.

Final preliminary observations indicate that successful reclamation in semi-arid lands in Montana will not be easily or quickly achieved. Reclamation law requires establishing permanent diverse vegetative cover which must be composed of predominantly native species. Native species, particularly shrubs and forbs, have not been successful in reclamation seedings. Species diversity has remained or become low in recut mined lands in comparison to native rangeland and old, naturally revegetated spoil. Variations in yield, despite normal climatic conditions, suggest that adequate soil functioning was not attained. Management practices following stand establishments, such as grazing, prescribed burning, and proper fertilization may improve successional development. Greater seeded stand diversity should occur through proper seeding formulation as well as species enrichment from topsoil and native plant communities.

3.4 FACTORS AFFECTING NATURAL REVEGETATION OF COAL MINE SPOIL BANKS IN OHIO

Strip mining spoil banks cover a large area of land surface in Ohio. These spoil banks erode severely and, in many instances, increase silting and the acidity of nearby streams. In addition to erosion, downslope displacement and slumping of surface soil can occur on unvegetated areas.

Spoil banks are also characterized by extreme acidity.[3] A pH of 4.0 or less is common and is classified as toxic to most species in this area. Along with the high acidity are high levels of iron, aluminum, and manganese and a low supply of magnesium, calcium, potassium, and phosphorous. Toxic levels of aluminum, iron, manganese, copper, zinc, and nickel have been reported in the acidic soils of spoil banks. Calcium, potassium, nitrogen, and phosphorus ions may be deficient in soils of low pH. The soil and surface temperatures play important roles in the vegetation developments on strip mine banks where surface temperatures exceeding 60°C were reported. High surface temperatures cause high evaporation rates of soil water, reducing water availability and limiting plant establishment.

Past researchers have indicated that once the acid-tolerant species became established, the spoil banks became richer in organic and microorganisms and could support less tolerant species. On slopes exposed to the southwest, soil temperatures were too high for seedling establishment. Surface conditions, rather than pH or soil content, were more important in plant establishment on raw spoil. Factors that affected moisture in the environment, such as evaporation due to wind exposure and high surface temperatures, were important. In extremely acidic areas with a pH of 3.5 or less, no plants became established. Soil particle size (a fraction of the sample smaller than 2 mm diameter) <20%, soil pH <3.5, and a specific of 3 mm or higher were limiting factors for the invasion and growth of natural vegetation.

A special research project was conducted to determine whether soil conditions in both vegetated and unvegetated areas of coal mine spoil banks were correlated with the establishment of natural vegetation. The research site was a former coal strip mine, abandoned some 30 years ago. It was chosen because there was a distinct boundary that separated vegetated "islands." This zonation of natural vegetation can be attributed to soil acidity. In areas where pH is below 4.0, no vegetation existed.

A soil of 4.0 or less is usually considered a toxic level for plants in this region. Tolerance to soil acidity varies from species to species, so it is reasonable to assume that low pH values on this spoil bank excluded many intolerant species. Therefore, pH can be considered a major factor limiting the species that will become established.

Exchangeable aluminum ion concentrations of 1.0 meq/100 g or more are considered toxic to plants. In open sites significantly higher concentrations of aluminum were observed, whereas in the vegetated and open sites, toxic aluminum concentrations were found. Exchangeable manganese, magnesium, and calcium concentrations and water-soluble potassium concentrations for the soil extracts were near or below critical values given for deficiency levels. Other researchers reported that iron, copper, and zinc may also reach toxic levels on spoil banks. Nitrogen and phosphorus may be unavailable to plants in the spoil bank; however, nutrient levels in the soil may not be important to an establishment on bare spoil for species that tolerate such conditions. Successful transplants were made on the bare spoil, indicating that mature plants can tolerate the higher concentrations of aluminum and lower concentrations of potassium.

Mature plants can survive the low pH, high aluminum concentrations, and low nutrient levels, and seeds can germinate under these conditions. No vegetation was found on much of the spoil bank, because no viable seeds exist in the soil, conditions were not suitable for seed germination, or seedlings could not survive.

Field germination data revealed that on the soil surface, no seeds germinated in the open sites, whereas seeds did germinate in the vegetated sites—buried seeds germinated in the open sites, whereas the surface seeds did not.

Soil moisture and soil temperature data indicated that water availability may be a problem at the soil surface. The effect of low soil moisture on plants varies with species tolerance to moisture stress and the textural properties of soil.

Where vegetation is absent, much of the water from rainfall may be lost in runoff. Evaporation of soil water may be greater in unvegetated areas. On warm days surface temperatures of the open spoil material can be much higher than soil temperatures measured below the surface. Shading the soil with plant cover and leaf litter decreases soil temperature, which in turn reduces evaporation and increases soil moisture.

Variation in soil temperature may explain field germination results. Buried seeds are not exposed to the extreme surface temperatures and low soil water levels and may be able to germinate. Once the buried seedlings in open sites break through to the surface, they may not be able to survive the high temperatures and the low moisture content. Such conditions did not prevail in vegetated sites, and seed germination and seedling survival do not appear to be inhibited.[4]

Robinia pseudo-acacia trees were planted on the bare spoil bank in an attempt to reclaim the site. Cores of these trees indicated that they were 12–15 years old. Leaf litter accumulated at the base of these trees and altered the edaphic conditions enough to allow seeds to germinate and for acid-tolerant species to become established. These plants in turn accumulated more leaf litter, and the spoil bank became richer in organic matter and could support less tolerant species. In this way, the vegetated islands have grown to their present size.[5,6]

3.5 VEGETATION DEVELOPMENT ON OLD AND ABANDONED LEAD AND ZINC MINE WASTES

Establishment of a stable vegetative cover has been shown as the only feasible method of physically stabilizing metal-mine wastes and reducing pollution effects. Artificial vegetation has been attempted, but little is known about the processes that accompany natural revegetation of contaminated sites. Patterns of natural successional development can serve as models for artificial revegetation where the goal is to grow stable, well-established vegetation without extensive use of fertilizers and irrigation.

Ecosystem response to the deposition of metal-mine waste occurs at different levels. Individual organisms may exhibit specific sensitivity to metals which can be translated to a population-level response in the evolution of process-tolerant vegetation. Compositional changes have been observed.

A research project was conducted on old and abandoned mill tailings of lead and zinc mines in Wisconsin. Some physical and chemical properties of the tailings are given in Table 3.3.

The mine flora was composed of primarily native species, many of which are characteristic of droughty waste places. No significant trends can be discerned in either species richness or variety. A slight decrease in dominance is observed with the increasing age of the site. Tree abundance was related to the mine age. Aggregation of vegetation was extremely pronounced at each of the

TABLE 3.3
Physical Properties of the Tailings Material[5]

Date of Abandonment	(1900) Badger Mine	(1928) Middle Mine	(1945) Thompson Mine
% Gravel by weight	64	59	60
Textural class	gls[a]	gls	gsl[b]
pH	7.3	7.8	7.9
Total N (μg/g)	1,400	1,200	750
Available P (μg/g)	33	1.5	6.7
Available K (μg/g)	118	34	21
Total Zn (μg/g) $\times 10^{-4}$	1.6	1.7	2.1
Total Pb (μg/g)	450	540	760
% Soil moisture	—	6.6	7.1
% Organic matter	—	0.6	0.9

[a] Gravelly loamy sand.
[b] Gravelly sandy loam.

sites. Numerous bare areas existed interspersed with well-vegetated zones. The open areas supported sparse populations of annuals. The closed-turf areas were dominated by perennial grasses and sedges and perennial forbs.

Cover changes with time were observed. Little vegetation establishment occurs in the first decade after abandonment. The long lag in initial colonization can be attributed to the lack of species capable of tolerating or evolving a tolerance to extremely harsh sites. Evolution of tolerance is known to be a function of species sensitivity, metal concentrations, and time.

The initially slow rate of cover development was followed by rapid growth due to an increase in the frequency of turf-forming sedges and grasses. These species are among the earliest but represent only a small contribution to total cover.

Subsequent vegetation development proceeds slowly. The virtual absence of cover change in 30 years suggested that vegetation has expanded to the limits of available resources. Inputs of new propagates and clonal expansion may be balanced by high mortality due to the vigours of the harsh and nutrient-poor site. This will result in a stable cover pattern.

As woody vegetation continues to increase in importance, the tailings environment is altered. Increased shading, litter accumulation, and the improved moisture begin to ameliorate the severe environment. Shifting dominance and increasing cover are accompanied by changes in spatial pattern.

Aggregation is highest at the outset of colonization. Individual plants are clustered together and separated by large areas of bare tailings. However, the intensity of aggregation continues to diminish with successional development. Despite the trend of declining aggregation, the pattern of vegetation cover at even the oldest mine is extremely patchy.

3.6 REVEGETATION AT THE USIBELLI COAL MINE, ALASKA

The Usibelli coal mine has been operating in Alaska for three decades. The company has been involved in all phases of surface coal mining from exploration to postmining reclamation. After an approximate permit area boundary (based on the identified coal resource) is outlined, environmental studies are initiated. Studies include vegetation, wildlife, surface and subsurface hydrology, soils, and cultural resources. The objectives of the vegetation studies are to determine and describe the existing vegetation and the amount of area this covers. The data are used to determine what vegetation types will be disturbed and what plant sources are available for reclamation.[6]

Vegetation inventory data include the following:

- Percentage of cover by vegetation
- Mosses, lichens, and litter
- Density of shrubs >20 cm (8 in.) tall
- Age, height, and dbh (diameter at breast height) of trees and tall shrubs
- Height and basal diameter of dominant woody species <3 m (10 ft) tall

A list of vascular plant species and moss and lichen genera is prepared. The sampling methods include reporting every plant species at 50 cm (1.5 ft) spacing along a 20-m (66-ft) transect. This is converted to percentage of cover. Stems of woody plants are counted for density estimates in belt transects (rectangles) beside the transects. Enough of these transects must be sampled to ensure that the total living vascular cover is examined with an acceptable degree of precision. The successional and ecological conditions associated with the various vegetation types are described.

The mine is located in the boreal forest zone of interior Alaska. Its coal leases contain almost 20 different vegetation types ranging from low shrub communities to dwarf tree forest types. Grasslands are dominated by bluejoints. The most important tree species in the boreal forest vegetation types include white spruce, black spruce, and paper birch.

The topography consists of flat ridges with adjacent steep slopes where streams are actively downcutting. Along the slopes, vegetation is influenced by landslides. Plant species growing under conditions of slope and aspect are identified for potential use in reclamation. Plant species that occur in steep south-facing, windblown slopes before mining are those that would succeed on the south-facing slopes after mining. Plant species occurring in early successional sites are the most likely candidates in reclamation.[7]

One of the main objectives of reclamation is to establish a diverse, self-reproducing plant community. Successful reclamation is usually among various options. Grass is usually the fastest, most reliable plant type for establishing protective cover on the mined soil. However, heavy grass cover can slow down natural succession on the site and compete with woody plant establishment. Moderate grass cover may help conserve subsurface soil moisture.

The reclamation plan is developed during the mine planning process. Site conditions such as steep slopes and permafrost are considered in both the engineering and the biological aspects of reclamation. The mined area is backfilled and graded to restore approximately the original contour. Reclamation guidelines require that topsoil (an A horizon) be replaced over the graded overburden. Some soil profiles in Alaska do not have A horizons or other material suitable for use as a plant growth medium.

Grass species are the primary plant species used for initial soil stabilization. Woody species are then planted from local transplants or cuttings or seedlings so as to provide diversity and sometimes deeper rooting, which also contributes to soil stabilization. Leaf fall from woody plants may help increase the organic matter and nutrients in the soil. Leaves of alder (*Alnus*) contain high levels of nitrogen. Woody species are usually selected based on their local availability, rapidness of growth, tolerance of various conditions, and usefulness for wildlife or other postmining land use. The reclamation plan includes plant species, planting techniques, amendments such as fertilizer or inoculation and their rates of application, and general location for different plant communities.

The reclamation plan must include a postmining bond-release standard. Before disturbing the land, the mine operator posts a bond sufficient to cover reclamation costs at any given point. If a mine operator fails to meet reclamation requirements after mining, then the state maintains funds to reclaim the existing disturbance. At Usibelli, the reclamation must meet certain standards after 10 years.

The biological objective of these bond-release standards is to ensure that a diverse, self-reproducing plant community capable of supporting postmining land use is established. Bond-release standards can be developed in several ways and possess three parts. The standards include total living vegetation cover, woody species density, and plant species diversity. A simple method of creating a standard is to develop a technical standard in which the mine operator agrees to meet a certain level of the given parameter, such as 70% (level) living plant cover (parameter). This technical standard can be justified by data from existing revegetated areas, test plots, existing native vegetation, or a combination of these. The reclamation plan must be designed to meet these standards at the end of the bond-release period.

After the mining permit is approved, the area is mined according to the mine plan. After mining is completed in an area, the pit is backfilled and graded to approximately the original contour. The slopes are sometimes furrowed for reducing runoff and for accumulating moisture and nutrition.

The site is usually aerially seeded. The seed mix consists of grasses adapted to the area, including Arctared red fescue, Manchur smooth brome, common foxtail, and reed canary grass. Two annual species are also used to establish ground early cover. Some mines frequently are found to have seed mixes of 20 species with different mixes for different conditions.

The mine personnel regularly transplant young woody plants from nearby vegetation into the reclaimed areas to assist natural succession toward boreal forest communities. This ensures that the plants are adapted to climatic conditions and the rooting zone contains the microbial communities, especially mycorrhizae, needed for proper plant growth.

BIBLIOGRAPHY

1. Gordon, R. M. and J. G. Hannen. Reclamation of surface coal mines in the Hunter Valley, New South Wales, Australia, in *Innovative Approaches to Mined Land Reclamation* (Carbondale, IL: Southern Illinois University Press, 1986), pp. 525–559.
2. Sindelar, B. M. Successful development of vegetation on surface mined land in Montana, in *Ecology and Coal Resource Development*, Vol. 2 (New York: Pergamon Press, 1979), pp. 550–556.
3. Bell, T. J. and I. A. Ungar. Factors affecting the establishment of natural vegetation on a coal strip mine spoilbank in southeastern Ohio, *Am. Midl. Nat. 105*:19–31 (1981).
4. Skousen, C. A. *A Chrono Sequence of Vegetation and Mine Soil Development on a Texas Lignite Surface Mine* (Washington, DC: U.S. Bureau of Mines, 1988), IC 9184, pp. 79–88.
5. Kimmerer, R. W. Vegetation development on a dated series of abandoned lead and zinc mines in Southwestern Wisconsin, *Am. Midl. Nat. 111*:332–341 (1984).
6. Elliott, C. and J. D. McKendrick. Strip mine reclamation and Alaska's big game wildlife, *Agroborealis 23*(1):41–43 (1991).
7. Helm, D. J. From boreal forest to reclaimed site: Revegetation at the Usibelli coal mine, *Agroborealis 23*(1):45–50 (1991).

4 The Acid Mine Drainage Problem from Coal Mines

4.1 INTRODUCTION

Acid drainage from underground coal mines and coal refuse piles is one of the most persistent industrial pollution problems in the United States. Pyrite in the coal and overlying strata, when exposed to air and water, oxidizes, producing ferrous ions and sulphuric acid. The ferrous ions are oxidized and produce a hydrated iron oxide (yellow boy) and more acidity. The acid lowers the pH of the water, making it corrosive and unable to support many forms of aquatic life. The iron oxide forms an unsightly coating on the bottom of streams, and further limits the ability of aquatic life to survive in streams affected by acid mine drainage (AMD).

Before coal is mined, very little of the pyrite is exposed to the conditions necessary to produce acid drainage. The mining and coal cleaning process exposes the pyrite to surface or ground waters and allows pyrite oxidation to occur. A ton of coal containing 1% pyritic sulphur has the potential of producing 33 lb of yellow boy and over 60 lb of sulphuric acid. However, the rate of acid production varies, and abandoned mines and refuse piles can produce acid drainage for >50 years. The drainage, if discharged into surface streams or ponds, constitutes an extensive, expensive, and persistent environmental problem. Federal law now requires that water discharge from active coal mines have a pH between 6 and 9, and the law places limits on the total iron, manganese, and suspended solids. Controlling AMD from active mines usually requires expensive water treatment and the necessity of handling very large volumes of water.

4.2 CHEMISTRY OF FORMATION

The oxidation of iron pyrite (FeS_2) and the release of acidity into waters draining through coal mines can be represented by the reaction sequence given in Table 4.1. Fe^{2+} is released in the initiator reaction either by simple dissociation of the iron pyrite or by oxidation of the pyrite by oxygen. After the sequence has been initiated, a cycle is established in which Fe^{2+} is oxidized by oxygen to Fe^{3+}, which is subsequently reduced by pyrite, thereby generating additional Fe^{2+} and acidity.[1]

Two possible oxidants for iron pyrite are available, namely oxygen and ferric iron. The reduction of Fe^{3+} by pyrite both in the presence and in the absence of oxygen at 0.20 atm showed no difference in the rate of the reaction (the parallelism exhibited by the two straight lines is indicative of equivalent reaction rates) or in the rate of change of soluble ferrous iron. The rate of the reaction is relatively rapid; for example, at pH 1, 50 min was required for the reduction of 50% of the initial ferric iron concentration by approximately 3 m^2 of pyrite per litre of solution. In the absence of ferric iron, no oxidation of pyrite was observed even after 1 week. Further evidence for the slowness of reaction 1 is the fact that pyrite can be used as a reasonably inert electrochemical electrode. Hence, the major oxidant of iron pyrite is ferric iron, as indicated in the propagation cycle.

The rate of oxidation of Fe^{2+} by oxygen in abiotic systems is a function of pH.[2] These results were obtained in pH-buffered systems (using $HClO_4$ or CO_2 and HCO_3^-) under a constant partial pressure of oxygen and, at pH values >4.5, were found to be compatible with the kinetic relationship

$$\frac{-d\left[Fe^{2+}\right]}{dt} = k[Fe^{2+}][O_2][OH^-]^2 \tag{4.1}$$

TABLE 4.1
Reactions Responsible for Pyrite Oxidation

1. $FeS_2 + \frac{7}{2}O_2 + H_2O \rightarrow Fe^{2+} + 2SO_4^{2-} + 2H^+$

2. $Fe^{2+} + \frac{5}{2}H_2O + \frac{1}{4}O_2 \rightarrow Fe(OH)_3(s) + 2H^+$

3. $Fe^{2+} + \frac{1}{2}O_2 + H^+ \rightarrow Fe^{3+} + \frac{1}{2} + H_2O$

4. $FeS_2 + 14Fe^{3+} + 8H_2O \rightarrow 15Fe^{2+} + 2SO_4^{2-} + 16H^+$

Stage 1		
Mechanism	{Reaction 1:	proceeds both abiotically and by direct bacterial oxidation
	{Reaction 2:	proceeds abiotically, slows down as pH falls
Chemistry		pH above approximately 4.5; high sulphate; low iron; little or no acidity
Stage 2		
Mechanism	{Reaction 1:	proceeds both abiotically and by direct bacterial oxidation
	{Reaction 2:	proceeds at the rate determined primarily by activity of *T. ferrooxidans*
Chemistry		approximate pH range of 2.5–4.5; high sulphate; acidity, and total iron increasing; low Fe^{3+}:Fe^{2+} ratio
Stage 3		
Mechanism	{Reaction 3:	proceeds at rate totally determined by activity of *T. ferrooxidans*
	{Reaction 4:	proceeds at rate primarily determined by rate of reaction 3
Chemistry		pH below approximately 2.5; high sulphate, acidity, total iron and Fe^{3+}:Fe^{2+} ratio

Source: From Kleinman, R. L. P. Biogeochemistry of acid mine drainage and a method to control acid formation, *Min. Eng.* 33(3):300–305 (1981). With permission.

where $k = 8.0 \times 10^{13}$ L^2 mol^{-2} atm^{-1} min^{-1} at 25°C. At pH values below 3.5, the reaction proceeds at a rate independent of pH, i.e.,

$$-d\left[Fe^{2+}\right] = k'\left[Fe^{2+}\right]\left[O_2\right] \tag{4.2}$$

where $k = 1.0 \times 10^{-7}$ atm^{-1} min^{-1} at 25°C. The reaction has previously been reported to be first or second order with respect to Fe^{2+}, depending upon the ionic medium; in the presence of ligands which form relatively strong complexes with Fe^{3+} (pyrophosphate, fluoride, and dihydrogen phosphate), the rate is first order in $[Fe^{2+}]^2$. The actual rates of the reaction, however, are of the same order of magnitude as those reported here in the lower pH range. The half-time of the reaction in this acidic pH region is approximately 1,000 d, reflecting the slowness of the oxygenation reaction when compared to the rapid oxidation of pyrite by ferric iron. Since reaction 2 is significantly slower than reaction 1, the oxidation of Fe^{2+} by oxygen appears to be the rate-limiting step in the propagation cycle. It is irrelevant in this model, whether pyrite or marcasite, the orthorhombic polymorph of FeS_2, is considered to be the sulphide source; reaction 2 continues to be the rate-controlling reaction.

Field investigations of the oxidation of Fe^{2+} in natural mine drainage waters were conducted in the bituminous coal regions of West Virginia. Observations of the rate of this reaction in these acidic streams (at pH values closer to 3) showed that it proceeded considerably more rapidly than the laboratory studies at pH 3 predicted.

Many agents indigenous to these natural mine streams have been cited in the literature, in various circumstances, as displaying catalytic properties in the oxidation of Fe^{2+} by oxygen. The catalytic effects of sulphate, iron(III), copper(II), manganese(II), aluminium (III), charcoal, iron pyrite, clay particles and their idealized counterparts, alumina and silica, and microorganisms were investigated and compared in synthetic mine waters in our laboratory. Of these, microorganisms appeared to exhibit the greatest effect in accelerating the oxygenation of Fe^{2+}. Comparisons between the rates of oxidation of Fe^{2+} under sterile conditions after inoculation with untreated and with sterilized natural mine water showed that microbial mediation accelerates the reaction by a factor $>10^6$.

If the reaction scheme describing the oxidation of pyrite and the formation of acidity (reactions 1–3) is considered, it appears that the oxygenation of Fe^{2+} is the rate-determining step and that in natural acidic systems the reaction is greatly accelerated by microbial mediation. A similar mechanism is probably applicable to acidic leaching processes in other mines, such as copper mines, where iron is also invariably present.[3] If so, copper sulphides are oxidized by ferric iron and the resultant ferrous iron is reoxygenated, again with the aid of microorganisms, to form additional iron(III).

In a coal mine, pyrite degradation may occur in the self-accelerating cycle identified in Table 4.1 with pyrite being oxidized by ferric iron. Of the two ferric iron-producing equations shown in Table 4.1 (Equations 1 and 2), only ferrous iron oxidation is a significant source of ferric iron. Assuming that influent mine waters have an average total iron concentration of 0.5 mg/L, that ferrous iron oxidation is the only ferric iron source, and that no ferric iron is hydrolyzed, 93 turns of the pyrite degradation cycle (Equations 2 and 3, in Table 4.1) must occur to result in the release of 300 mg/L total iron commonly observed in mine drainage. If the average mine water residence time is 10 months, each pyrite degradation cycle can thus take no longer than 3.2 days. Ferric iron hydrolysis and precipitation (Equation 4) will naturally increase the number of cycles required and thus decrease the theoretical time available per cycle.

The iron oxidation-pyrite degradation cycle identified in Table 4.1 will proceed slowly at pH <4.5 in the absence of catalysis. This is because at pH <4.5, the cycle is rate-limited by the ferrous iron oxidation rate. Catalysis of ferrous iron oxidation can be biologically or abiotically mediated. The effect of chemical and physical catalysts on abiotic ferrous iron oxidation is a catalysis factor of <30. The iron bacterium *Thiobacillus ferrooxidans* significantly catalyzes iron oxidation with catalysis factors >300 under laboratory conditions. However, the optimal activity of this organism is at pH <3.5. Thus, no mechanism exists for rapid catalysis of the iron oxidation-pyrite degradation cycle in the pH range 4.5–3.5. This chapter describes the effect of a filamentous iron bacterium on catalyzing the formation of the environment necessary for optimal *T. ferrooxidans* activity and thus for pyrite degradation.

4.2.1 THE ROLE OF BACTERIA

The major reactions responsible for pyrite oxidation are summarized in Table 4.1. These four reactions reflect the current understanding of acidification mechanisms. The actual reaction process occurs in a multistage sequence dependent upon the activity of *T. ferrooxidans* and solution Eh and pH. During the first stage of this process, fine-grained pyrite is oxidized either by *T. ferrooxidans* or by air, with equal amounts of acidity produced by the oxidation of sulphide to sulphate (reaction 1) and by the hydrolysis of Fe^{3+} (included in reaction 2). The most reactive pyrite is the framboidal form due to the presence of pyrite granules <0.5 μm in diameter. The ability of *T. ferrooxidans* to accelerate the rate of pH decline is important, for each rainfall potentially interferes with the initial build-up of acid. During this stage, it is possible to forestall acidification by adding alkalinity to the reaction system; if alkalinity exceeds acidity, the only major downstream effect is an increase in sulphate concentration. Thus, adding crushed limestone or other sources of alkalinity to freshly exposed pyritic material can stop or postpone acidification. Once acidity changes to significant alkalinity, it becomes much more difficult to return an acid-producing system to stage 1, although the fall of pH is moderated as it approaches 4.5 by a decrease in the rate of reaction 2. For example, at $[Fe^{2+}] = 5$ ppm, and pH $= 5.5$, the half-time for reaction 2 is 3 d; at pH 4.5, 300 days are needed for the oxidation of half the initial Fe^{2+}. This is due to the second-order dependence of the reaction on OH^- activity, as given by rate law:

$$-d[Fe^{2+}] / dt = k \left[Fe^{2+} \right] \left[OH^- \right]^2 \cdot PO_2 \qquad (4.3)$$

where $k = 8.0 \ (\pm 2.5) \cdot 10^{13} L^2 \, mol^{-2} \, atm^{-1} \, min^{-1}$ at 25°C.

As abiotic oxidation of Fe^{2+} slows, *T. ferrooxidans* takes on its primary role of oxidizing Fe^{2+}, thereby allowing reaction 2 to continue producing acidity and ferric hydroxide.[4] This initiates stage 2 of the reaction process. Once again, it is possible for the pH to stabilize in this region, though this will usually occur only when permeability or the amount of exposed pyrite surface area is low. The pH decline otherwise continues to the third stage, where acid production is most rapid.

At pH <3, the increased solubility of iron and the decreased rate of $Fe(OH)_3$ precipitation results in increased Fe^{3+} activity. Stage 3 begins as Fe^{3+} activity becomes significant at a pH of approximately 2.5; a vicious cycle of pyrite oxidation and bacterial oxidation of Fe^{2+} results from the combined effects of reactions 3 and 4 (Table 4.1). The rate of reaction 3 exerts primary control on the cycle by limiting the availability of Fe^{3+}, the major oxidant of pyrite. The steady-state activity of Fe^{3+} is determined by the combined effects of bacterial oxidation of Fe^{2+}, reduction of Fe^{3+} by pyrite, and the formation of associated ferric sulphate and hydroxy complexes. Stage 3 includes the oxidation of both framboidal and coarse-grained pyrite.

The transition to stage 3 can be seen in the graph of total dissolved iron and acidity vs. pH which has been compiled from a series of experiments conducted in simulations of a coal refuse pile. Acidity was determined in the laboratory by titrating with NaOH to pH 8.3 after boiling with samples of hydrogen and is expressed as milligrams per litre of $CaCO_3$ as described in the American Public Health Association Standard Methods. The advantage of small laboratory simulations is their rapid acidification as compared to the months necessary for acidification of coal refuse piles in the field. The three stages of acidification can be observed in the laboratory data and have also been observed in the field.

An Eh pH analysis indicates that pyrite and $Fe(OH)_3$ cannot theoretically coexist at acid pH. The fact that they are commonly found together at the source of acid drainage demonstrates that pyrite oxidation is controlled by kinetic processes rather than by equilibrium chemistry.

The rise in Eh that accompanies stage 2 (due to the oxidation of Fe^{2+} by *T. ferrooxidans*) controls the manner in which Eh levels off as stage 3 is reached and steady-state cycling begins between reactions 3 and 4. The Eh limit is determined by the pyritic material and the activity of *T. ferrooxidans*; in laboratory experiments, stage 3 occurred at roughly 670 mV, which represents an $Fe^{3+}:Fe^{2+}$ activity ratio of approximately 0.02. This suggests that stage 3 may be initiated at relatively low Fe^{3+} activities.[5] The pH boundary of approximately 4.5 between stages 1 and 2 does not appear because neither the mechanism nor the rate of pyrite oxidation is affected by the transition from abiotic to bacterial oxidation of Fe^{2+}.

It should be noted that relatively high Eh is also possible at near-neutral pH due to the rapid abiotic oxidation of Fe^{2+} and the low concentrations of iron in solution (on the order of 10 ppb). At higher iron concentrations and lower pH, Eh rises due to the activity of *T. ferrooxidans*. In general, the ratio of ferric/ferrous concentrations for all species also stabilizes at about 1:2 as stage 3 is reached, although drainage values can be much higher as reaction 3 proceeds without reaction 4. The difference in the iron activity and concentration ratios is primarily due to the fact that most aqueous iron in this system is present as sulphate and hydroxy complexes, which are included in the iron concentrations, but not in the Fe^{3+} and Fe^{2+} activities.

Once stage 3 is reached, acid production can only be reduced by slowing reaction 3. It is traditionally done by limiting the available oxygen, but the more direct route of inhibiting *T. ferrooxidans* is also possible.

Acid mine drainage is a dilute solution of sulphuric acid and iron sulphate with iron in ferrous and/or ferric form. Treatment of the AMD involves neutralization of the acid with a suitable alkali, oxidation of ferrous iron to the insoluble ferric form, and removal of the precipitates by a sedimentation process.[6] Even though one basic treatment process is applied, many options are available in each unit process or subprocesses.

4.3 CONVENTIONAL NEUTRALIZATION PROCESS USING LIME

In the conventional process, the five basic treatment steps are:

- Equalization
- Neutralization (mixing)
- Aeration
- Sedimentation
- Sludge disposal

A flowsheet for a typical system is shown in Figure 4.1.

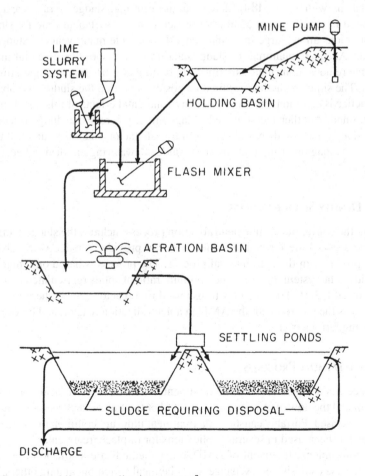

FIGURE 4.1 Conventional lime neutralization process.[7]

A constant flow with only small variations in quality is desired so as to minimize needed controls and operator attendance.[7] The mine drainage is collected in large holdings or equalization basins or in large sumps within the mine. The holding basins should have a capacity for 2–3 days' flow. Usually, 12–24 hr of flow is maintained in the holding basins to equalize flow and quality to the treatment facility. From the holding basin, the mine drainage flows by gravity to the treatment plant. The holding and settling basins are usually of earthen construction.

Lime is used as the alkali for neutralization of the acid in almost all large treatment facilities; however, the selection between quicklime and the hydrate is determined by cost and availability.

Aeration is used for oxidizing ferrous iron to the ferric form, which is less soluble in water. Ferrous iron is much more soluble than the ferric iron. The solubility of ferrous iron occurs in the pH range of 9.3–12.0. Ferric iron is much less soluble and begins to precipitate as hydroxide at a pH of 4.0, with minimum solubility occurring at about pH 8.0. Economic advantage is obvious in removing iron in the ferric form at the lower pH as less lime is needed for neutralization to the pH level needed to maintain minimum iron solubility (8.0 vs. 12.0).

The oxidation of ferrous iron is commonly included in the AMD treatment processes. The oxidation is pH dependent, with the reaction proceeding rapidly at a pH above 8.0. At this pH level, iron oxidation becomes dependent upon the availability of oxygen. Theoretically, one unit weight of oxygen is required for seven unit weights of ferrous iron to be oxidized.

After neutralization and oxidation of ferrous iron, the subsequent step is sedimentation. Settling of iron hydroxide and other suspended solids is accomplished in settling basins. The basins must have a capacity of at least 12 hr of clear water detention above the sludge storage zone. When small settling basins with 12- to 48-hr detention and minimal sludge storage capacity are used, two units operated in parallel are a common practice to allow sufficient time for sludge removal. If plant site conditions allow, large impoundments that provide many years of sludge storage can be advantageous. Adequate planning for sludge handling and disposal is essential in all treatment processes. Sludge removal contributes significantly to both construction and operating costs of the AMD facilities. The simplest method for final disposal is to pump the sludge into abandoned deep mines. This practice is common, but the overall environmental effects of this disposal method have yet to be determined. Another method used is lagooning, wherein the sludge thickens naturally. Eventually the sludge must be disposed of in a more satisfactory manner, such as burial in a reclamation project. Among other methods used for sludge dewatering are drying beds, vacuum, and pressure filtration.

4.3.1 HIGH-DENSITY SLUDGE PROCESS

One variation in the conventional lime neutralization process includes the sludge recirculation process. This process uses lime for neutralization and can provide a sludge with substantially less volume than is produced in the conventional process. The process is based on a high sludge recirculation rate within the system, in which the optimum ratio of solids recirculated to solids removed is in the range of 20:1–30:1. The sludge is transported to a reactor vessel where the lime slurry is added. The slurry is then mixed with the AMD in a neutralization reactor, and here aeration is provided for oxidizing ferrous iron.

4.3.2 OTHER TREATMENT PROCESSES

Many other processes are available for the treatment of AMD, wherein acid concentration is the primary problem and the flow of mine water is low. Other treatments such as soda ash, caustic soda, or limestone can be used. Portable caustic soda treatment units are common in surface mine operations. Limestone has been used in several applications for in-place treatment.

Besides the conventional treatment of AMD, other methods are available to produce water of higher quality: reverse osmosis, ion exchange, and chemical softening, among others.[7]

4.4 CHEMICAL TREATMENT

Where the formation of AMD cannot be prevented or its discharge cannot be controlled, chemical treatment is necessary before the mine water can be discharged. Lime treatment is the most commonly used system. The chemistry of the lime treatment process is as follows:

$$Ca\,(OH)_2 + H_2SO_4 \rightarrow CaSO_4 + 2H_2O \tag{4.4}$$

$$Ca\,(OH)_2 + FeSO_4 \rightarrow Fe(OH)_2 + CaSO_4 \tag{4.5}$$

$$Ca(OH)_2 + Fe_2(SO_4)_3 \rightarrow 2Fe(OH)_3 + 3CaSo_4 \tag{4.6}$$

The choice of an alkali in the neutralization process is influenced by factors such as cost, suitability, relativity, availability, ease of use, and sludge volume. Each of these factors should be carefully evaluated because each alkali requires significantly different equipment, and the alkali selected may affect the design of other processes in the overall system.

4.4.1 LIME

Lime is a general term that, by definition, encompasses only burned forms of limestone. The two forms of lime of particular interest in AMD treatment are quicklime and hydrated lime.

Quicklime (CaO) is produced by the calcination of limestone. Limestone generally consists of 50–90% calcium carbonate. Based on chemical analysis, quicklime may be divided into three categories:

- High-calcium quicklime—containing <5% magnesium oxide
- Magnesium quicklime—containing 5–35% magnesium oxide
- Dolomitic quicklime—containing 35–40% magnesium oxide

Quicklime is available in different standard sizes, but the major types used in AMD treatment include ground lime and pulverized lime. The ground lime is produced by grinding the larger-sized material and screening of the fine size. A typical size is −#8 mesh and 40–60% −#100 mesh. Pulverized lime is produced by intense grinding to produce a size between −#20 mesh and 85–95% −#100 mesh. Quicklime is usually delivered in bulk carloads or trucks and then transferred to a storage silo.

For efficient use in mine drainage neutralization the quicklime must be slaked. The slaking process must be carefully controlled and requires daily attention. Well-slaked quicklime offers a low unit cost per gram of acidity neutralized. Major disadvantages include high capital investment for a slaker, grit removal, close operational control, and the danger from possible severe burn injuries to personnel.

Hydrated lime (Ca(OH)$_2$) is the most commonly used alkali for neutralizing AMD in existing treatment plants when lime consumption rates are low or when the cost of a slaking system is prohibitive.

An air classification system is used to produce the fineness necessary to meet the process requirements. A common size is 75–95% −#200 mesh. The commercial hydrated lime is purer than quicklime.

Storage and handling of lime in AMD treatment plants require careful attention. Most often bulk lime, either quick or hydrated, is more efficient and economical to use. The bulk lime is delivered by truck and conveyed by mechanical or pneumatic systems into weather-tight bins or silos. Bagged lime is delivered loose or palletized in a truck or boxcar and generally handled by hand truck or forklift to storage. Bagged lime should be stored in dry areas. Hydrated lime is normally packaged in multiwall paper bags. Moisture can permeate the liner and cause caking. Dry storage is important. The use of bulk lime can yield considerable savings over the use of bagged lime, not only in initial cost but also in reduced labour costs for less handling and for other advantages, including faster loading, elimination of losses from broken bags and spillage, better housekeeping by modern handling systems, and less dust hazard to employees.

Bulk lime is commonly delivered to the treatment plant by truck. The blower truck is the fastest, most common, and most economical. The lime is blown from the truck directly to silo storage via a pipeline. Conventional steel or concrete bins and silos can be used for lime storage as quicklime, and hydrated lime is not corrosive. A steel silo with a cone bottom is the most popular.

The flowability of lime is important inertia in storage bin design. The flowability varies from good, for pebble and granulated quicklime, to erratic, for hydrated and pulverized quicklime. Lime can absorb moisture quickly, forming a sticky, soft cake that can reduce flowability in the bins. Hydrated lime tends to form ratholes. In order to avoid the problems inherent to lime flowability, special design in bins, external vibrators, internal antipacking and antiarching devices, and live bin bottoms are used.

Currently AMD treatment plants utilize dry lime in a liquid suspension or slurry before feeding it into the raw water, but the new practice is to feed dry hydrated lime directly to the acid water, thereby eliminating a slurry feed system. Dry lime feed has been used when the drainage streams are small and mildly acidic, the lime requirements being <0.1 kg/1,000 L. For drainage systems requiring larger amounts of lime, for example, 0.36–0.48 kg/100 L, a special feeding arrangement such as a larger aeration or a flash mixer is used for achieving complete mixing.

The efficiency of an AMD treatment plant depends on the speed and accuracy with which the lime is fed to the process. Various types of feeders available for use include:

- Volumetric feeder
- Screen feeder
- Oscillating hopper
- Belt feeder
- Rotary paddle
- Vibrating feeder

As explained earlier, slaked lime is often used in AMD treatment plants. The term "slaking" refers to the combination of varying proportions of water and quicklime which yields milk of lime, a viscous paste of lime. Continuous slakers have largely replaced manually operated batch slaking. The variables exerting influence on slaked hydrated lime include:

- Reactivity of the quicklime
- Particle size and gradation of quicklime
- Optimum amount of water
- Temperature of water
- Distribution of water
- Agitation

Concentrations of lime solids in the milk of lime slurry vary between 5% and 20%, with 10% being very common. Lime concentrations can be checked for specific gravity by a hydrometer. Automatic pH control systems for the feeding of lime solutions into the flash mix tank are commonly used.

The diluted paste or slurry is transferred to the neutralization tank. Scaling can become a serious problem in the transfer process. Lime is only slightly soluble. At the saturation level the solubility of lime is 1.7 g/L. With an increase in temperature, the solubility of lime decreases to 0.55 g/L at

90–100°C. It is economically desirable to carry lime in a much more concentrated form such as suspension. Because lime slurry has a pH higher than 12.0 it softens the water and calcium carbonate is precipitated. Calcium carbonate forms a dense, hard scale. If left unattended, the scale will build up and clog pipes. No foolproof solution exists; however, some corrective measures can be taken. Scaling can be minimized by using sodium hexametaphosphate, which softens the slaking or dilution water so the calcium carbonate does not scale up in the pipes.

The feeder location should be such that the slurry flows by gravity directly into the mixing tank or the solution. Scaling at the end of the slurry pipe can be avoided by discharging slurry through an air gap into the open solution tank. Heavy-duty flexible rubber hoses are preferred over metal pipes. The components of the slurry feed system such as piping, valves, pumps, and configuration should be designed so that they provide quick-disconnect, easy-to-assemble fittings and valves; avoid confined spaces; and enhance flow patterns. Slurry feed systems have used heavy-duty plastic pipe (Schedule 80), flexible rubber hoses, and galvanized or stainless steel pipes. The pipe material choice is determined in consideration of economics, anticipated problems, type of slaking water, and ease of assembly.

A flushing system is not essential but highly recommended for systems with frequent shutdowns; for example, AMD plants not operating daily or operating only in wet weather. A minimum flow velocity of 1.0–1.2 m/sec should be maintained within the slurry loop.

The type of valves used in the slurry feed system determines the degree of maintenance. Pinch valves have proven the most successful. In fully closed or fully open situations, ball or plug valves can be used.

A tight control on the metering system is essential for the bleed-feed slurry system. The material used in the tank construction should be resistive to the high pH of the slurry and suitable from a structural standpoint. The configuration is determined by the slope available. Tanks exceeding 9,500 L (2,500 gal) should be fitted with baffles set 90° apart to prevent vortex formation. Adequate agitation in slurry storage tanks must be provided.

AMD treatment plants do not need a sophisticated slurry feed control system. Usually the drainage flow is constant in quantity and quality because equalization basins are used. The system can be efficiently controlled by the pH of the flow, leaving the neutralization tank. This pH signal can control the dry lime feeder, a lime solution feeder, or a control valve. Another type of lime slurry feeding involves proportioning the slurry flow in response to the pH signal.

Pumps used for lime slurries generally fall into two categories: controlled volume and centrifugal. Centrifugal pumps are more popular for a wide range of slurry flows as they are inexpensive and their standard design incorporates flow patterns that lend themselves to easy slurry transfer. Two configurations have been frequently used, as shown in Figures 4.2 and 4.3.

4.4.2 LIMESTONE

Considerable interest has arisen in the possible use of limestone for the treatment of AMD. Limestone equals about 30% of the cost of quicklime or hydrated lime. Two types of limestone are found: the dolomitic and the high calcium. The high-calcium type has potential application in AMD treatment.

A typical chemical composition of limestone is

	Calcium Limestone % Composition
Calcium oxide (CaO)	53.00–56.00
Magnesium oxide (MgO)	0.12–3.11
Calcium carbonate ($CaCO_3$)	92.66–98.60
Silica dioxide (SiO_2)	0.10–2.89

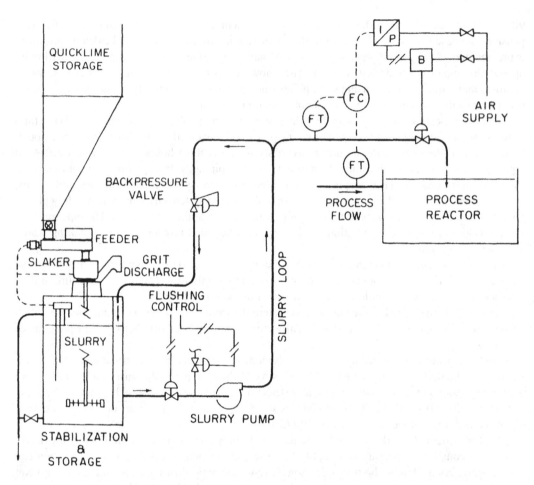

FIGURE 4.2 Slurry feed with pH control loop.[7]

The reaction of limestone with the components of AMD can be represented by the following equations:

$$CaCo_3 + H_2SO_4 \rightarrow CaSo_4 + H_2CO_3 \tag{4.7}$$

$$3CaCO_3 + Fe_2\left(SO_4\right) + 6H_2O \rightarrow 2CaSO_4 + 2Fe(OH)_3 + 3H_2CO_3 \tag{4.8}$$

$$3CaCO_3 + Al_2\left(SO_4\right)_3 + 6H_2O \rightarrow 3CaSO_4 + 2Al(OH)_3 + 3H_2CO_3 \tag{4.9}$$

To utilize limestone effectively as a neutralizing agent, certain quality criteria must be met:

- Minimum particle size, preferably a −#325 mesh
- High calcium content
- Low magnesium content
- High specific (surface) area

For the treatment of ferrous iron, limestone use is also possible, but limitations include slow reactivity of limestone and its inability to increase the pH above 7.0. At a pH level of 7.0, the slow rate of

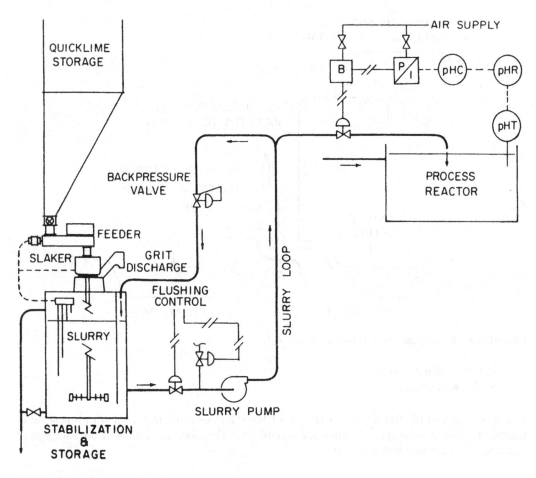

FIGURE 4.3   Slurry feed by flow proportioning.[7]

oxidation of ferrous iron is a deterrent. Excessive aeration time is needed for complete oxidation of the ferrous iron.

The design approaches for limestone treatment of AMD include the following:

Ferrous Iron Conc. (mg/L)	Response to Limestone Treatment
0–50	Effective treatment may be achieved without pre- or postneutralization iron oxidation
50–100	May be effectively treated but requires postneutralization aeration and significant retention time
>100	Potential treatment is uncertain

Important factors to be considered in designing a limestone treatment process include:

- Specifications for limestone grade, size, and hardness
- Mode of operation (mixer or rotary mill)
- Supplementary reagent requirements (live, polymers)
- Operating pH and aeration requirements for ferrous iron oxidation

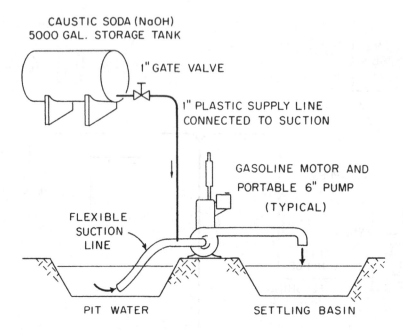

FIGURE 4.4   Portable caustic soda feed arrangement.[7]

- Recycle volume ratio
- Sludge settleability

If iron is mostly in the ferric form (at least a 4:1 ferric:ferrous ratio) treatment with limestone is feasible, but limestone utilization efficiency is only 32%. Therefore, the economic advantage over hydrated lime treatment may not exist.

4.4.3   CAUSTIC SODA

Caustic soda can be utilized to treat AMD for low-flow, mildly acidic drainage problems. Generally, it is used for surface mine drainage problems in remote areas. The treatment process consists of a horizontally mounted 38 m^3 (10,000 gal) storage tank and a flume-type chemical feeder. In surface mining operations caustic soda is frequently used to treat pit water. A schematic layout is shown in Figure 4.4. The disadvantages of caustic soda use in AMD include high cost, dangers in handling, and special design requirements. Sodium hydroxide produces an excellent effluent quality that is low in suspended solids, turbidity, and iron content.

Soda ash (Na_2CO_3) is rarely used to treat large-flow AMD facilities. Due to its high cost and limited availability, soda ash has been used for the treatment of low-flow drainage that contains little ferrous iron such as would occur in some surface mines. It is selected primarily for convenience rather than cost efficiency.

4.5   IRON OXIDATION

Acid mine drainage usually contains significant iron concentrations that result from the oxidation of pyritic minerals present in coal seams. The effect of the process of oxidation on pyrites produces ferrous sulphate and sulphuric acid. These salts readily dissolve in water forming the AMD.

Iron is initially present in the ferrous (Fe^{2+}) form. Ferrous iron can be oxidized to ferric iron which is much more soluble and hence can be precipitated as a hydroxide to effluent quality levels below the allowable pH of 6.0. The minimum solubility of ferric iron occurs at a pH of 8.0 (Figure 4.5), while ferrous iron does not reach minimum solubility until the pH approaches 11.0.

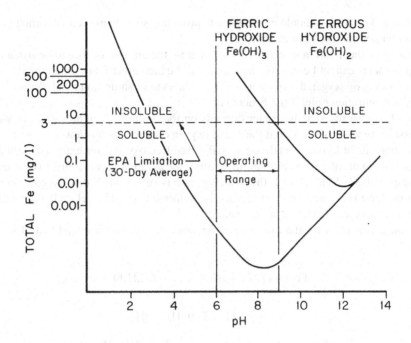

FIGURE 4.5 Solubility of ferric and ferrous iron at various pH.[7]

At the maximum allowable discharge pH of 9.0, ferrous iron is soluble to about 4 mg/L, which exceeds the discharge limitations for new sources. Therefore, in most AMD treatment systems, it is imperative to oxidize any ferrous iron to the ferric form and then to remove it at lower system pH. Methods used for this oxidation include:

- Mechanical aeration
- Chemical oxidation
- Biological systems

4.5.1 AERATION SYSTEMS

Ferrous iron, when exposed to oxygen, oxidates to ferric iron. The rate of oxidation depends upon the ferrous iron concentration, the dissolved oxygen concentration, and the pH of the solution. At pH values higher than 6.0, the reaction occurs according to the following rate equation:

$$\frac{-d}{df}\left(Fe^{2+}\right) = k\left(Fe^{2+}\right)\left(O_2\right)\left(OH^{-2}\right) \tag{4.10}$$

The reaction is a first-order reaction with respect to the ferrous iron and the dissolved oxygen concentration. The oxidation rate decreases as the concentration of ferrous iron or oxygen decreases. The reaction rate is second order with respect to the hydroxyl ion (OH^-) concentration for pH values >6.0. The reaction rate increases 100 times for each one-unit rise in pH above 6.0. The rate of ferrous iron oxidation is extremely slow at a pH of <3.0, slow in the pH range of 3.0–6.0, moderate to fast in the pH range of 6.0–8.0, and rapid above this point. At a pH level of 5.0, ferrous iron will precipitate as ferric hydroxide sufficiently to meet the effluent limits at a pH of 5.0. At this pH value, however, the oxidation rate for ferrous iron is slow. Until the pH is 8.0 or greater, the oxidation rate does not increase until the pH value reaches 8.0 or higher. When iron occurs mostly in the ferrous form, aeration processes are most efficiently operated within a pH range of 8.0–9.0, when oxidation takes place in a matter of minutes. At this stage, the controlling parameter for the design of the aeration unit becomes a function of the oxygen transfer efficiency and not the chemical reaction of oxygen and iron. The aerator should be designed to provide dissolved oxygen saturation in the aeration basin with maximum oxygen transfer.

The capacity of the aeration system is determined by the amount of iron to be oxidized. If the oxygen requirements cannot be met, the oxidation will be incomplete. The oxidation rate increases as the concentration of oxygen dissolved in water increases to its saturation point. Aeration capacity in excess of the saturation point is not beneficial.

The rate of oxygen transfer into water depends on the initial oxygen deficiency of water. It is easier to dissolve oxygen by aeration if the initial oxygen concentration is lower.

The oxidation rate of ferrous iron depends on the dissolved oxygen concentration, with the maximum rate occurring at saturation. For optimization of the aeration process, these two mechanisms must be compromised. At a pH >8.0 the oxidation rate is sufficient so that oxygen concentration near saturation level is not necessary. If aeration is performed at a pH level <8.0, then a fairly high level of dissolved oxygen should be maintained.

The chemical equations for the oxidation of ferrous iron to ferric iron and hydrolysis of ferric iron are

$$Fe^{2+} + 1/4O_2 + H^+ \rightarrow Fe^{3+} + 1/2H_2O \tag{4.11}$$

$$Fe^{3+} + 3H_2O \rightarrow Fe(OH)_3 + 3H^+ \tag{4.12}$$

These chemical equations indicate that 1 kg of oxygen will oxidate 7 kg of ferrous iron under ideal conditions. In these reactions, 1 mol of acidity (as H_2SO_4) is formed for each mole of ferrous iron oxidated. Sufficient alkalinity must be added to neutralize the extra acid formed and to maintain optimum pH conditions.

The aeration system chosen must meet the oxygen demand for ferrous iron oxidation. The theoretical oxygen demand for any mine water can be calculated from the following equation:

$$O_2 = QW \times Fe \times 5.16 \times 10^{-4} \tag{4.13}$$

where
O_2 = theoretical O_2 demand (kg O_2/hr)
QW = AMD flow rate (L/sec)
Fe = Initial concentration of Fe^{2+} (mg/L)

Atmospheric air contains about 21% of oxygen by volume. Only a fraction of the oxygen in the air that comes in contact with water, called oxygen transfer efficiency, is actually absorbed by water. This fraction differs for each aeration system and operating conditions. The total air needed to supply the theoretical oxygen quantity demanded can be calculated by the following equation:

$$Qa = \frac{6.324 \times O_2}{E} \tag{4.14}$$

where
Qa = total air demanded (m³/min)
O_2 = theoretical oxygen demanded (kg/hr of oxygen)
E = oxygen transfer efficiency (as %)

Oxygen transfer efficiencies (E) range from 3% to 25% depending upon the type and size of the aerator and the depth of submergence.

The aeration system must also be capable of keeping the ferric hydroxide solids and unreacted reagent in suspension. If the mixing is insufficient these solids will settle at the bottom of the basin.

Settled solids will reduce the aeration volume and aeration time, causing incomplete ferrous iron oxidation. Thus, the aerator must be designed to meet both oxygen and mixing requirements.

Aeration processes that dissolve atmospheric oxygen in mine drainage water can be classified into four types:

- Mechanical surface aeration
- Submerged turbine aeration
- Cascade aeration
- Diffused air aeration

The aeration basin should be efficiently designed for complete oxidation of ferrous iron without requiring excessive aeration time. Important parameters are basin plan, depth, and inlet and outlet structures. Aeration basins are excavated in the ground and levelled with riprap or asphalt. They also can be constructed as concrete or steel structures.

Mathematical models have been developed to predict aeration times for the oxidation of ferrous iron at varying pH ranges. The nature of mine drainage is variable, and the effects of the other dissolved ions on the reaction are not well known. Therefore, laboratory tests are the most reliable way to optimize the aeration system design. The tests should be conducted on a sample containing the maximum expended ferrous iron concentration and at the lowest anticipated operating pH level in the AMD.

The detention time needed for ferrous iron oxidation must be multiplied by a safety factor for the design of the operating aeration basin. The capacity of the aeration basin is determined from the following formula:

$$V = Q \cdot D_t \cdot f \qquad (4.15)$$

where
  V = volume (m³)
  Q = flow (m³/sec)
  D_t = detention time (sec)
  f = safety factor

The aeration basin must be designed to efficiently accommodate the aerator, and the entire volume of the basin should be well mixed and aerated.

Besides aeration, chemicals have been used for iron oxidation. Ozone and hydrogen peroxide have been used.

Theoretically, 1.0 kg of ozone will oxidize 2.3 kg of ferrous iron. The same amount of acid is released during ozone oxidation as during aeration. At 86% ozone utilization, 1.0 kg of ozone will oxidize 2.0 kg of ferrous iron. The advantages of ozone treatment over lime aeration include:

- The oxidation reaction is efficient and quick.
- Close process control needed in the lime treatment process is not required.
- Neutralization to pH 6.0 is all that is needed.
- The sludge produced by the limestone-ozone reaction is denser than lime sludge, reducing sludge handling requirements.

Hydrogen peroxide can be used when specific conditions exist:

- Alkaline mine drainage (pH higher than 6.0) with 10 W oxygen requirements for iron oxidation values
- Need for a supplemental oxidizing source when the system is overloaded with iron and expansion is impossible

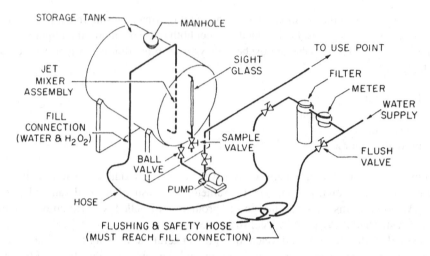

FIGURE 4.6   Hydrogen peroxide feeding system.[7]

The chemical reaction of ferrous iron with hydrogen peroxide is given by the following equation:

$$H_2O_2 + 2Fe^{2+} + 2H^+ \rightarrow 2Fe^{3+} + 2H_2O \tag{4.16}$$

According to the above equation, 0.45 kg of H_2O_2 will oxidize 1.5 kg of ferrous iron. A hydrogen peroxide feeding system is shown in Figure 4.6.

4.5.2   BIOLOGICAL OXIDATION

Bacteria capable of oxidizing ferrous iron exist naturally in most acid mine drainages. As explained earlier, these bacteria, *T. ferrooxidans*, act as a catalyst in the formation of AMD. These bacteria obtain their carbon and nitrogen requirements from inorganic sources and their energy from the oxidation of ferrous iron. Researchers have proposed several methods to utilize their ability to oxidize ferrous iron in mine drainage. Dispersal growth systems have not been successful. Fixed growth systems, such as trickling filters and rotating biological contractors (RBCs), are effective in supporting the bacterial growth necessary for oxidation in a trickling filter.

The RBC is an aerobic treatment device consisting of a series of four plastic discs mounted on a horizontal shaft. This assembly is placed in a trough through which the wastewater flows, submerging almost half of the surface area of the discs. The discs rotate slowly on the shaft, causing the biological growth on the discs to alternately contact the air and water (Figure 4.7).

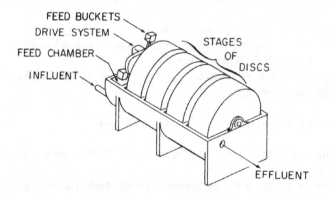

FIGURE 4.7   Rotating biological contactor.[7]

Experimental data have been obtained at two peripheral disc velocities, 0.32 and 0.17 m/sec, and at five hydraulic loadings ranging between 110 and 440 e/d/m². A linear relationship has been observed between ferrous iron removal and stage retention time. A faster disc velocity produced better iron oxidation at any constant hydraulic loading. At a given disc velocity, increased hydraulic loading resulted in an increased effluent ferrous iron concentration, even though the ferrous iron oxidation rate also increased. The RBC can produce an effluent containing <10 mg/l ferrous iron at loading rates up to 88 kg Fe^{2+} applied per 1,000 m² of disc surface.

Tests have been conducted on AMD flowing overland for more than 1 mi before reaching the treatment facility, thereby exposing the water to stream ecology and ambient air temperatures. Observations during the winter indicated that the effect of low temperatures on the biological oxidation process is not significant. At 0.4°C, the lowest temperature recorded, the removal efficiency decreased by only 10% below the efficiency observed at 10°C, the initial mine water temperature. This effect can be compensated for by a lower hydraulic loading.

It may not be possible to remove ferrous iron to 1.0 mg/e with RBC. However, this degree of removal by a biological system is not necessary. An effluent of 10 mg/L of ferrous iron is adequate for RBC. The effluent will require neutralization which will provide enough aeration to oxidize any remaining ferrous iron.

Design of a four-stage single-shaft RBC system can be made using the following equations. The equations for disc area surface area can be determined from:

$$A = \frac{FoQ86.4}{L} \tag{4.17}$$

where

A = disc surface area (m²)
Fo = initial ferrous iron concentration (mg/e)
Q = flow (p/sec)
L = ferrous iron loading (kg/d/1,000 m²) (calculated from Figure 4.8)

The designer chooses the desired effluent ferrous iron concentration from the RBC system listed on the vertical axis in Figure 4.8. A line is drawn horizontally to intersect the curve, and then from the point of intersection, a vertical line is dropped to the horizontal axis. This point on the horizontal axis gives the maximum ferrous iron loading (L) that will yield the desired effluent. This value of L and the initial iron concentration (Fo) and the flow (Q) are applied to the equation, yielding the necessary disc area.

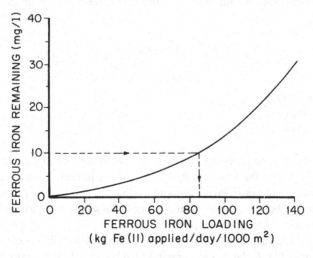

FIGURE 4.8 Design procedure for a four-stage rotating biological contactor.[7]

4.5.3   OXIDATION RATE

The oxidation rate of ferrous iron is dependent upon the ferrous iron concentration at any point in time. The reaction can be expressed as:

$$\frac{-d}{dt}(Fe) = K(Fe) \tag{4.18}$$

The integrated form of the equation is

$$Fe = Fe_0 e^{-kt} \tag{4.19}$$

where Fe_0 is the initial ferrous iron concentration and Fe is the ferrous iron concentration at time t.

The slope of the graph, log Fe vs. time, which is a straight line, determines the constant K, the reaction rate.

$$K = Slope \times 2.303 \tag{4.20}$$

The oxidation rate test should be performed on fresh samples of mine drainage to minimize any natural oxidation of the ferrous.

A small sample is taken and carefully preserved for analysis to determine the initial ferrous iron concentration. Lime or caustic soda is added to raise the pH to the desired level. Air is supplied through an aeration stone.

Thickeners and separator units receive clarifiers underflow with a high percentage of solids. They are used to produce an acceptable overflow and to store sludge to produce denser underflows.

Tilted-plate gravity settlers are sometimes used instead of clarifiers in instances in which very limited space is available. They are inclined-plate, shallow-depth settling devices that perform the same function as a clarifier but occupy only one-tenth the space.

4.6   SLUDGE DEWATERING AND DISPOSAL

Sludge is costly to handle, dewater, and dispose. In AMD treatment plants, sludge disposal presents the most recurrent and demanding problem. Environmental regulations categorize sludge as a potential pollutant. In AMD treatment plants, two effluents are produced: treated water and sludge.

The volume of sludge to be disposed depends on the chemical composition of mine drainage and the neutralization method used. Commonly, the sludge volume is 5–10% of the daily flow through the treatment plant.

$$V = Qt_d \tag{4.21}$$

where
   t_d = detention time (min)
   V = volume of sludge (m^3)
   Q = design flow (m^3/min)

Additional capacity must be provided for sludge accumulation, which is estimated to be 5–10% of the average daily volume treated. Sludge storage volume requirements in a settling basin depend on the frequency of sludge removal. Ponds without sludge removal facilities should allow sufficient volume for sludge storage between withdrawal operations. In some cases, basins have high volumes to hold sludge for the life of the treatment plant.

Many AMD treatment plants use circular-shaped clarifiers for liquid-solids separation when limited land area is available. A clarifier is a gravitational liquid-solids separator having the primary objective of producing clean overflow regardless of underflow solids content. Sizing of the clarifier is made from the following formula:

$$V_{RR} = Q / A \qquad (4.22)$$

where

V_{RR} = rise rate (m/min)
Q = design flow rate (m³/min)
A = cross-sectional area of clarifier (m²)

Two types of earthen ponds are commonly used: settling ponds and impoundments. Settling ponds are small and designed permanently for settling with periodic sludge removal. The sludge is disposed of or recycled to utilize the untreated portions of the neutralizing agent. In settling ponds, surface baffles prevent short circuiting. Submerged baffles are used for sludge confinement. The two types of baffles are illustrated in Figure 4.9a and b.

Many factors are considered in calculating the size of sedimentation basins. Important factors include:

- Detention time
- Sludge removal
- Disposal method
- Mode of operation

The basin volume can be calculated, using the detention time and design flow, with the following equation:

$$V = Qt_d \qquad (4.23)$$

where

V = volume of settling pond, without sludge storage (m³)
Q = design flow (m³/min)
t_d = detention time (min)

The efficient design of inlet and outlet is important in providing quiescent conditions for good settling pond performance. Inflow should be uniformly distributed over as much of the pond width as possible. Multiple inlets or a continuous width, multiple V-notch box weir offers an efficient design. This reduces the inlet flow velocity, reducing the possibility of washout or resuspension of

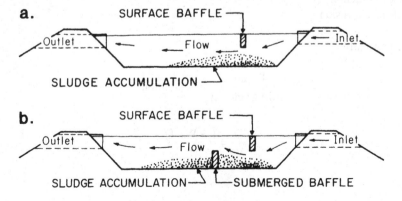

FIGURE 4.9 (a) Surface baffled pond.[7] (b) Combination surface and submerged baffled pond.[7]

solids. Bad inlet design can cause short circuiting, reducing the detention time and removal efficiency of the settling pond. Dead areas of noncirculating water can be created in the settling pond, creating channels of flow within the pond. Uniform distribution of influent across the width of the settling pond inhibits isolated mounding of settled particles within the basin, maximizing sludge storage volumes. Similarly, outlet openings should also be multiple or continuous to maintain low exit velocities. High exit velocities create turbulence that can resuspend settled solids, causing poor effluent quality.

Baffles, placed selectively in a settling pond, prevent short circuiting. Two types of baffles, surface and submerged, are used. Clays and silty clays are excellent compounds for providing an impermeable layer. Plastic membrane liner can be placed in areas where fill material is unsuitable. Soil conditions must provide stable support for pond embankment foundation, along with necessary imperviousness to the passage of water. Mixtures of coarse- and fine-textured soils like gravel-sand, sand-clay, and sand-silt provide good stability and resistance to the passage of water. Layers of clay and silty clay are often used to ensure tight contact between embankment and foundation. The basin bottom should not be founded on bedrock or on stony, rocky soils. A thick layer of relatively impervious consolidated material can form the most suitable foundation.

The embankments forming the sidewalls of the basin should be constructed of impervious materials similar to those used in the basin's liner. The design of the embankment elements, such as the slope of the sidewalls, height, and crown width, should provide stability. The inside slope of the embankments should be protected against erosion caused by wave action. Piling of a 0.61-m wide collar of riprap at the expected upper level on the embankment is an effective method of erosion protection.

In AMD treatment plants, settling units vary in size, configuration, and method of solids removal. Excavated settling ponds in the ground are most popular for economic reasons. Mechanical clarifiers or thickeners may be preferred as they enable the operator to exercise control over the treatment process and improve sludge densities.

Before sizing and designing settling ponds or clarifiers, tests should be conducted to determine the behaviour and characteristics of the sludge and quality of the incoming fluid. The tests determine sludge-settling velocity, optimum pH, best neutralizing agent, dosage rate, sludge density, and volume.

Earthen ponds are commonly used as settling basins. Earthen basins should not be located in swamps, marshes, on floodplains, or over abandoned wells or mine workings. A layer of impervious material about 0.6- to 0.9-m thick should be placed at the bottom of the pond to prevent seepage.

The performance of the separator is related to other variables:

- Flow turbulence in the basin
- Velocity distribution throughout the basin
- Particle interaction
- Particle resuspension

Expressed empirically:

$$\frac{\text{Depth}}{\text{Settling velocity}} = \text{Detention time} \tag{4.24}$$

$$\frac{D}{V_s} = \frac{L \times W \times D}{f} \tag{4.25}$$

or

$$V_s = \frac{F}{L \times W} = \frac{\text{Flow rate}}{\text{Area}} \tag{4.26}$$

where
    Vs = settling velocity (m/min)
    D = depth (m)
    L = length of pond (m)
    W = width of pond (m)
    F = flow rate (m³/min)

Therefore, all particles with settling velocity higher than or equal to hydraulic surface loading (flow/area) are removed.

Settling velocity is influenced by many other factors as well:

- Inlet and outlet devices
- Wind-induced turbulence
- Nonquiescent flow current
- Flocculation of particles

Settling performance is dependent on hydraulic surface loading, which can be calculated by dividing the design flow (L/d) by the surface area of the pond or clarifier (1 m²), with the resulting hydraulic surface loading of L/d/m². Common values lie in the range of 175–350 L/d/m². The AMD sludge exhibits three types of settling behaviour. The type 1 sludge settles rapidly with a clear liquid-solid interface. Figure 4.10 illustrates this settling phenomenon. In the first part of the figure, the precipitated sludge forms a homogeneous mixture. In the second part, the stratification begins and the particles begin settling onto already settled particles. Water trapped inside forms a gelatinous mass. The adjacent upper layer (zone C) is a transition zone characterized by a suspended solids concentration lower than zone D, but greater than zone B. The supernatant zone A develops as the liquid-solids separation is completed. Settling continues in part III, and zones A and D increase in depth while B and C decrease. In IV, only zones A and D remain, with most of the solids present in zone D. At this point, zone D begins to compress. Compaction forces of the sludge displace the trapped water.

Type 2 settling involves sludge generated from mine drainage containing low concentrations of iron and aluminium with a pH of 6.5–7.5 and little acidity. Lime addition is not always required. Because no natural nuclei exist for flocculation, the sludge has poor settling rates and most remain in the solution. The settled particles are light and fluffy in character and produce sludges with only 0.5% solids. Satisfactory liquid-solids separation does not occur if the treatment does not involve neutralization. The volume of sludge is determined by iron content, suspended solids, and other precipitable elements present in the water. Polymer or flocculent addition can be useful to enhance the settling rate.

Type 3 settling behaviour does not depend on the solid content and pollutional loadings. It is produced by limestone or sodium carbonate neutralization. A two-phase separation system is observed. Most of the solids settle rapidly, with a distinct liquid-solids interface and a turbid, cloudy supernatant.

Sludge production increases as the iron and aluminium concentrations increase. The sludge formed in the pH range of 6.4–7.2 possesses the best settling properties, but the iron oxidation rate is low. The ideal pH for iron oxidation is around 8.5.

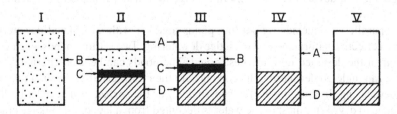

FIGURE 4.10 Type 1 settling.[7]

Sludges produced from limestone treatment have a higher density than those generated by other lime treatment processes. This results in a smaller volume required for sludge disposal. Sludges produced from highly acidic waters neutralized with caustic soda (NaOH) have very low densities with long detention times. Sludges produced from soda ash (Na_2CO_3) treatment possess densities somewhere between those generated by lime and limestone.

Polyelectrolytes may be added to improve sludge-settling rates. These are water-soluble, high molecular weight, organic polymers that may be cationic or anionic. These changed polymers adhere to sludge particles and improve settling. Determinations of polymer dosages and whether cationic or anionic treatment should be used are made from treatable tests.

Sludge generated in the AMD treatment process is generally composed of hydrated ferrous or ferric oxide, gypsum, hydrated aluminium oxide, unused lime, sulphates, calcium carbonates, bicarbonates, and trace amounts of silica, phosphate, manganese, copper, titanium, and zinc. The characteristics of the sludge vary with mine drainage quality and neutralization method. Important characteristics include settleability, density, dewaterability, particle surface properties, and viscosity. Sludge settleability is determined by settling rate and final sludge volume. Sludge density is reported as per cent solid by weight. Sludge dewaterability is determined by the ability of sludge to be concentrated into a more manageable and less voluminous form by centrifuging, filtering, or lagoon. The electrostatic charge or particle surface property influences flocculation characteristics of the particles. The viscosity of sludge measures flowability when pumping sludges.

Neutralization of mine drainage with lime produces hydrated sludges which are light, gelatinous, and very voluminous. The ferric hydroxide sludges from AMD have high iron concentrations and are generally a fluffy mass with very low solids.

The chemical characteristics of AMD sludge (called yellow boy) vary with raw drainage and method of treatment. Sludge is generally composed of hydrated ferrous or ferric oxides, gypsum, hydrated aluminium oxide, varying amounts of sulphates, calcium carbonates, bicarbonates, and trace quantities of silica, phosphate, manganese, titanium, copper, and zinc.

Table 4.2 presents the primary composition of some typical mine drainage sludges generated from different mine waters and neutralization processes. The chemical quality of mine water, the type of neutralization process, and the chemical agents applied influence sludge composition and its characteristics.

The important physical properties that influence sludge handling and treatment processes include its density, settleability, viscosity, particle size and surface properties, and dewaterability.

Figure 4.11 illustrates the relationship of settling time to sludge volume for various neutralizing agents.

Main sludge disposal methods include lagoon, deep mines, underground disposal, filtration, bed drying, and centrifugation.

When land is available, lagoon offers one of the cheapest methods of sludge storage and dewatering. Mine drainage treatment plants are usually located in remote or isolated areas where land is readily available. Lagoons function as settling ponds or impoundments for collecting and dewatering the sludge, and as permanent storage and disposal facilities.

In mine drainage treatment, settling ponds are used in series in which the first pond serves as the primary settling unit and the second serves as a polishing pond. The system isolates the majority of the solids in the first pond, but little sludge dewatering occurs. The primary pond is equipped with a sludge removal device that transfers the sludge to a disposal lagoon in which atmospheric dewatering can occur.

A dual or parallel arrangement of settling ponds together with an isolated dewatering lagoon offers the most effective natural method of sludge dewatering. Two settling ponds, each with volume sufficient to treat the designed flow, are best suited. The ponds can be used alternately so that the inactive pond can undergo dewatering. Its contents are then transferred to the final disposal lagoon for further dewatering. In both ponds, the supernatant, or the surface water, must be decanted to allow the sludge to dry. After most of the water is decanted from a lagoon, shrinkage cracks appear

TABLE 4.2
Chemical Analyses of Sludges

				Weight, % (Dry Basis 105°C/24 hr)				
Alkali Used:	Hydrated Lime-Air Oxidation			Hydrated Lime Bio-Oxidation	Dolomite		Calcined Dolomite	
Mine Water:	Bennett's Branch	Proctor 1	Proctor 2	Proctor 2	Proctor 2	Tyler	Proctor 1	Proctor 2
Component								
Al	3.8	4.7	3.1	8.0	2.8	5.5	4.5	4.8
Fe	19.5	17.7	23.1	24.3	13.0	7.4	13.5	23.2
Ca	6.9	5.8	5.2	4.8	17.2	10.7	6.7	5.2
Mg	6.6	4.3	5.1	1.0	3.8	11.8	9.8	5.8
SO_4	5.7	6.8	5.8	11.5	4.4	1.6	2.3	5.5
H_2O at 180°C	12.5	15.8	14.8	—	10.2	8.7	11.7	14.7
Compound Composition								
$Al(OH)_3$	11.1	13.7	8.9	23.1	8.2	16.0	13.2	13.9
$Fe(OH)_3$	37.6	34.0	44.3	46.6	24.9	14.2	25.9	44.7
$CaCO_3$	11.4	7.5	7.0	0.0	38.3	25.0	14.4	7.2
$MgCO_3$	—	—	—	—	—	21.0	12.0	6.1
$3MgCO_3 \cdot Mg(OH)_2 \cdot 3H_2O$	25.2	16.4	19.4	5.1	14.4	22.4	24.6	15.6
$CaSO_4 \cdot 2H_2O$	10.7	12.7	10.9	20.7	8.4	3.1	4.4	10.4
Total	96.0	84.3	90.5	95.5	94.2	101.7	94.5	97.9

Source: From Penn Environmental Consultants. Design Manual, Neutralization of Acid Mine Drainage, U.S. EPA Rep.-600/2-83-001 (1983).

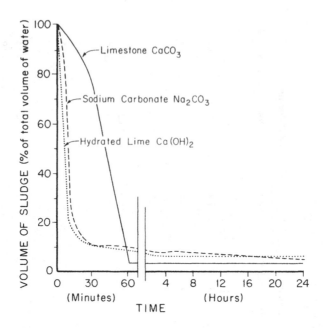

FIGURE 4.11   Treatability test settling curves.[7]

over the entire pond. Sludge lagoons are difficult to cover with soil. Sludge exhibits thixotropic tendencies, i.e., it will tend to liquefy upon vibration and then exhibit low compressive strength, which presents difficult problems for the weight and movement of the machinery employed in covering operations using soil.

Spread burial of the sludge utilizing the benefits for soil conditioning (i.e., alkalinity, minerals) can be economical if land area is available. Controlled amounts of sludge are applied to an area, and the area is then covered by soil in layers. The heavy metal content and other toxic constituents of the sludge should be evaluated for their long-term effect on vegetation.

Disposal of mine drainage sludge in abandoned deep mines through boreholes has been practised. Sometimes this means of disposal offers the most economical and simplest method for sludge disposal. However, all legal and environmental ramifications must be considered. For underground disposal, the sludge must have a pH of 7.0 or above, and all of the iron must be in the ferric form. For underground disposal, the following environmental factors are considered:

- Mine hydrology—presence of water in the mine, water level, location, and quantity of discharge from the mine
- Connection to nearby workings and other mines
- Quality of mine water, pH, acidity, and iron
- Sludge input—volume of sludge, mine volume, leachate quality from the sludge, sludge concentration
- Effect of sludge on discharge from the mine
- Geology—structure of the coal beds, strike, dip, and presence of faults

Vacuum filtration can be applied to mine drainage sludge for dewatering. Operational variables include the amount of vacuum, drum speed, drum submergence, filter media, and sludge conditioning before filtration.

Pressure filtration is an accelerated process of vacuum filtration. This process has not often been used on mine drainage sludge because of high capital, labour, and maintenance costs.

Porous bed drying of mine sludge has been employed. A variety of filter media such as sand, crushed limestone, coal, red dog, and gravel has been used. Water is removed by decanting

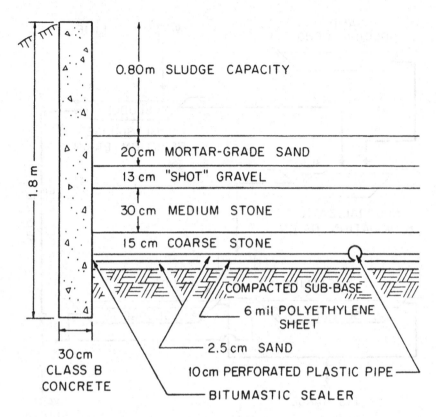

FIGURE 4.12   Cross-sectional view of the drying bed construction.[7]

the surface water in ponds, by percolation through the bottom of the bed, and by evaporation. Figure 4.12 illustrates a typical drying bed. The method was found to be impractical in the eastern United States, which experiences freezing weather conditions in some seasons.

The high-density sludge process developed by Bethlehem Steel Corporation is illustrated in Figure 4.13. This process achieves settled sludge concentrations between 15% and 40% compared to a maximum of 15% from the conventional lime neutralization process. The resulting sludge storage or disposal volume is reduced significantly.

4.7   REVERSE OSMOSIS

Reverse osmosis (RO) can be highly effective in removing most of the dissolved solids in AMD. The product water is low in dissolved solids, usually <100 mg/L, but may contain chemical or bacterial constituents that exceed drinking water standards. The product water can be used for other industrial purposes.

Osmosis occurs if two solutions of different concentrations in the same solvent are separated from one another by a membrane. If the membrane is semipermeable (i.e., permeable to the solvent and not to the solute), then the solution will flow from the more dilute solution to the more concentrated solution until equal concentration results. In RO, the direction of solvent flow is reversed by the application of pressure to the more concentrated solution. As a result, the concentrated solution termed "brine" becomes more concentrated. The solvent, termed "the permeate," is the product of the process.

Tubular, hollow-fibre, and spiral-wound membrane types have been tested for use in treating AMD. The spiral-wound configuration with a formamid-modified cellulose acetate membrane may be slightly superior to others with respect to the average flux (permeate flow rate), long-term flux stability, and dissolved solids rejection.

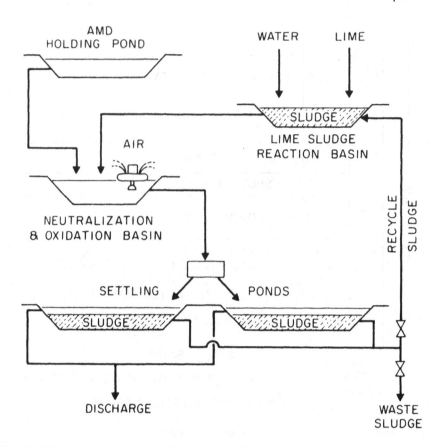

FIGURE 4.13   High-density sludge neutralization process.[7]

Problems with membrane fouling can occur as the concentrations of various compounds increase during the process. Most important is the possibility of iron fouling and calcium sulphate formation. Iron fouling can be minimized by lowering the feed water pH or flushing the RO membrane by operating at lower pressures for short periods. When the raw AMD contains high concentrations of sulphates, gypsum ($CaSO_4$) can form if its solubility is exceeded. In this case, this process may not be applicable.

Bag filters or cartridges should be used to minimize fouling by suspended solids with feed water. Membrane life can be increased, and rejection levels improved. The filters should have the capability to remove particles larger than 20 µm. The filters are placed in the suction side of the RO feed pumps.

In AMD treatment, the pH of the feed should be maintained between 2.8 and 3.0. At a pH lower than 3.0, ferric iron remains dissolved. When pH values exceed 3.0, ferric hydroxide may begin to precipitate on the membrane surface.

Although a low pH is necessary to improve operating conditions when treating AMD, it is lower than the optimum range of 5.0–6.5 for the cellulose acetate membranes. The life of this type of membrane is reduced. New pH-resistant membranes are now available.

Disinfection can be used to reduce iron fouling problems and to inhibit microbial activity in the raw mine drainage feed. Ultraviolet (UV) light, proven to be an effective bactericide, can be used to prevent accumulations of iron-oxidizing bacteria on the membrane surface.

One of the design parameters critical to a successful RO application is an accurate permeation rate (flux rate) over the membrane life. This is essential to estimate the quantity of installed surface area, cleaning cycles, and membrane replacement. With a constant feed rate, a decline of permeate flow will occur due to membrane fouling. Even without fouling, a slight flux decline will occur because of the membrane compaction.

The system is designed to produce a constant permeate output based on the daily design flow. Use of pressure-compensating flow controls that automatically adjust the operating pressure for flow variations can accomplish a constant permeate output.

The system is designed to operate at 281 kg/cm^2 because, at this operating pressure, minimum membrane compaction is experienced, while maintaining adequate flux ratios to assure high effluent quality.

As the flux declines, the pressure control system compensates for the decreased flow by increased operating pressure, thereby maintaining a constant product flow. Due to compaction, a loss of flux is typically offset by appropriate sizing and start-up at a reduced pressure with a gradual increase in operating pressure over the life of the modules. To reduce fouling, each module should operate with a 10:1 brine:product flow ratio. This brine velocity prevents "boundary layer" development, a stagnant layer of water against the membrane surface.

Module configuration is an important design item. The type of pressure vessel manifolding arrangement is determined by the desired recovery level and the need to maintain an adequate brine: product flow ratio. Pressure vessel arrangements and modules are designed so that the raw feed enters a parallel bank of pressure vessels, and the concentrate from this bank enters as the feed for the next parallel arrangement of vessels.

In high-recovery continuous flow systems, only a small number of modules should process the most concentrated portion of the feed stream so as to confine any fouling due to chemical precipitation to a minimum number of modules.

Two controlling factors limit this overall recovery of water from the treatment process. The first factor is the precipitation of calcium sulphate ($CaSO_4$). Acid mine water typically contains high concentrations of calcium and sulphate ions. With the progress of the RO process, calcium and sulphate concentrations increase. Once the solubility is exceeded, calcium sulphate precipitates. The empirical limit for RO recovery levels with AMD is given by:

$$R = 100 - 0.55(Ca \times SO_4)^{-1/2} \tag{4.27}$$

where Ca and SO_4 concentrations (mg/L) are those used in the raw feed. Gypsum ($CaSO_4$) is only slightly soluble. When concentrated, it precipitates and forms a hard, tenacious scale on tanks, piping, and membranes.

The second factor influencing recovery rates is the desired quality of the permeate. As the drainage is processed, the concentration of fatal dissolved solids in the permeate increases linearly with the total dissolved solids. Increasing recovery increases the concentration of pollutants in the waste (reject) stream and in the product water. The maximum recovery of the process is determined by the final use of the permeate.

In most cases the spiral-wound cellulose acetate membrane rejects 99% or more of the dissolved salts in the raw AMD feed. Table 4.3 shows expected permeate water quality.

Membrane life is affected by pH level, temperature, and operating pressure. The useful membrane life is the time taken to lose 40% of its initial flux.

The RO process removes dissolved solids from the AMD. The process generates a highly concentrated waste stream that requires treatment before disposal. The exact volume and salt content of the concentrate stream depends on the influent quality as well as the recovery rate. The possible treatment and disposal methods include:

- Lime neutralization
- Evaporation—mechanical or atmospheric
- Contract disposal

TABLE 4.3

Anticipated Permeate Water Quality

Parameter[a]	Raw Water Quality	Product Water Quality
pH (units)	3.4	4.3
Specific conductance (μmhos)	1,020	32
Acidity	210	32
Calcium	150	1.2
Magnesium	115	1.4
Iron, total	110	1.2
Iron, ferrous	71	0.8
Aluminium	15	0.8
Manganese	43	0.4

Source: From Penn Environmental Consultants. Design Manual, Neutralization of Acid Mine Drainage, U.S. EPA Rep.-600/2-83-001 (1983).

[a] All values expressed as milligrams per litre unless otherwise noted.

Lime neutralization of the waste stream is a possible treatment and disposal method.

Evaporation techniques include a mechanical, wiped-filmed unit, which can reduce the volume by 75% or more.

Contract hauling and disposal by a waste disposal firm can be another alternative.

4.8 ION EXCHANGE

The ion exchange process can be used to remove unwanted dissolved ions in AMD to produce water of excellent quality for many industrial uses. Ion exchange can produce drinking-quality water, but it will additionally require the use of filtration and disinfection so as to comply with public health regulations.

Ion exchange involves the reversible interchange of ions between a solid medium and the aqueous solution. To be effective, the solid ion exchange medium must contain ions of its own, be insoluble in water, and have a porous structure for the free passage of the water molecules. Within the solution and the ion exchange medium, the number of charges (charge balance) must stay constant. Ion exchange materials show an affinity for multivalent ions; therefore, they tend to exchange their monovalent ions. This reaction can be reversed by increasing the concentration of monovalent ions. Thus, the ion exchange material can be regenerated once its capacity to exchange ions has been depleted.

Ion exchange is commonly used for softening of hard water. The ion exchange material is charged with monovalent cations, usually sodium (sodium chloride). The hard water is passed through a bed of ion exchange material, and the divalent calcium and magnesium cations are exchanged for sodium ions as follows:

$$Ca^{2+} + 2Na^+ (resin) \rightarrow Ca^{2+} (resin) + 2Na^+ \qquad (4.28)$$

Ion exchange processes which have been developed for the treatment of AMD include the sul-bisul process, the modified desal process, and the two-resin process.

4.9 SUL-BISUL PROCESS

The sul-bisul process (Figure 4.14) employs a two- or three-bed system, depending upon the mine drainage quality. Cations are removed by a strong-acid resin in the hydrogen form, or by

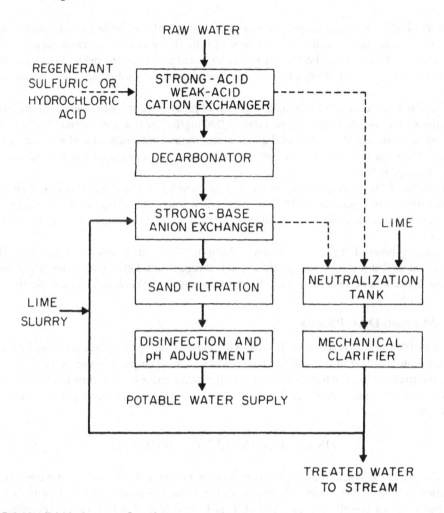

RAW WATER

REGENERANT
SULFURIC OR
HYDROCHLORIC
ACID

STRONG-ACID
WEAK-ACID
CATION EXCHANGER

DECARBONATOR

STRONG-BASE
ANION EXCHANGER

LIME

SAND FILTRATION

NEUTRALIZATION
TANK

LIME
SLURRY

DISINFECTION AND
pH ADJUSTMENT

MECHANICAL
CLARIFIER

POTABLE WATER SUPPLY

TREATED WATER
TO STREAM

FIGURE 4.14   Sul-bisul process flow sheet.

a combination of weak and strong-acid resins. The AMD feed is first passed through the cation exchanger which removes the metal cations and exchanges these for hydrogen protons, or (H⁺) ions. This reaction is expressed by the chemical equation:

$$Fe^{2+}SO_4 + 2HR \rightarrow MR_2 + H_2SO_4 \tag{4.29}$$

where R represents the strong-acid exchange groups on the resin, and M represents a divalent metal cation, such as iron (ferric), calcium, or manganese.

The water produced from the first stage contains additional sulphuric acid. The water is decarbonated to remove carbon dioxide formed during the cation exchange process. Then, a strong base anion resin (R′) operating in the sulphate-to-bisulphate cycle removes both the sulphate and hydrogen ions during this exchange reaction:

$$R'SO_4 + H_2SO_4 \rightarrow R'\left(HSO_4\right)_2 \tag{4.30}$$

The sulphate ions in the solution and on the resin are converted to the bisulphate form by the high acid content in the feed. This conversion of bivalent sulphate to monovalent bisulphate provides for twice the amount of sulphate to be stored in the resin. Removal of sulphur produces good quality water. Regeneration of the cation exchange bed is accomplished via either hydrochloric or sulphuric acid. In the regeneration process of the anion bed bisulphate ions are converted back to the sulphate form by the feed water. Addition of lime slurry to the regenerant speeds this reaction. The product water must be filtered and chlorinated according to public health regulations before use as potable water. Wastes from the regeneration process would have to be treated before discharge.

The sul-bisul process can be used to demineralize brackish water containing predominantly sulphate anions with dissolved solids content up to 3,000 mg/L. The raw water should have an alkalinity content of about 10% that of the total anion concentration and a sulphate:chloride ion ratio of at least 10:1. The process is especially suitable for alkaline waters containing calcium sulphate, such as mine drainage waters.

Limitations of the process include low exchange capacity of the anion exchange resin and its inefficient method of regeneration. The expended anion resin can be regenerated by the raw water itself, requiring a considerable volume of water over a significant length of time if the sulphate content is low.

The large volume of regenerants requires disposal. This water must be sufficiently alkaline and abundant so that it can be used as the anion bed regenerant. Otherwise, other alkalis must be employed. When this becomes necessary, the process may not be economically competitive.

4.9.1 Modified Desal Process

The modified desal process (Figure 4.15) is another ion exchange process that has been investigated for purification of AMD in order to produce potable water. The process uses a weak base anion resin in the freebase form, which is converted to the bicarbonate form to treat the raw AMD. The weak base resin exchanges sulphates for bicarbonates, allowing the cations to pass through the bed according to

$$MSO_4 + 2R''HCO_3 \rightarrow R''_2SO_4 + M(HCO_3)_2 \qquad (4.31)$$

where R'' is the weak base exchange group on the resin matrix and M is a divalent metal ion.

Aeration of the solution of metal bicarbonates is required to oxidize ferrous iron to ferric iron and to purge the carbon dioxide gas. The effluent is then treated with lime to precipitate metal hydroxides, settled to remove suspended solids, and then filtered and chlorinated before use as potable water.

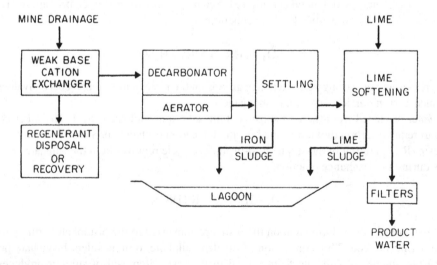

FIGURE 4.15 Modified desal process.[7]

Ammonia is used as the alkaline regenerant for displacing sulphate from the exhausted resin. Lime is used to precipitate the ammonia regenerant for reuse. Thus, ammonia is recycled. Lime and carbon dioxide used in the process can be recovered by roasting the lime sludge wastes in a kiln. All the principal chemicals used in the process can be recycled so that the net discharge of the process would approach zero. The products of the process include potable water, iron hydroxide, and calcium sulphate.

Use of the modified desal process is not limited by total dissolved solids or pH levels, but large quantities of carbon dioxide are needed to achieve good resin utilization for high total dissolved solids or alkaline feed waters. Application of this process is limited to waters containing <2,200 mg/L of sulphate. Mine waters containing iron in the ferric form may cause fouling of the anion bed due to precipitation of ferric hydroxide.

4.9.2   TWO-RESIN PROCESS

The two-resin process (Figure 4.16) involves the use of a strong-acid cation exchanger in the acid form (H+) followed by a weak base anion exchanger in the freebase (DH-) form. In the cation column, proton (H^+ ions) are exchanged for the metal ions in the AMD. Following cation exchange, the anion column feed is rich in sulphuric acid.

The removal of total metal cation greatly increases the regenerant dosage, and the operating cost of the system. The concentration of the residual metals in the cation exchanger effluent can be optimized by controlling the dosage of the regenerant. The anion exchange is fed predominantly by sulphuric acid, which is totally absorbed by the resin. A weak base anion exchange resin absorbs only acids. It cannot split neutral salts. The anion exchange effluent is alkaline, and some precipitation of residual iron and aluminium ions can be expected.

Sulphuric or hydrochloric acids may be used for regenerating the cation exchanger. Because of its lower cost, sulphuric acid is usually preferred; however, gypsum may be formed. Treatment for disposal of both the regenerant streams is required.

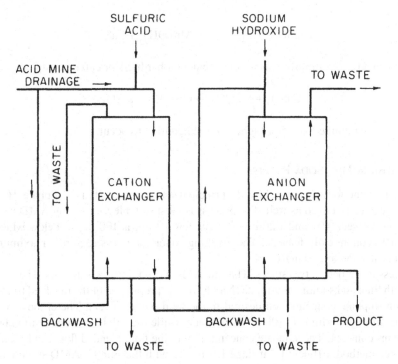

FIGURE 4.16   Two-resin ion exchange system.[7]

4.10 CHEMICAL SOFTENING

The chemical softening process can be used for the treatment of AMD when the effluent water is to be used as potable water or for drinking purposes. Two processes that merit application to AMD treatment are lime-soda and alumina-lime-soda.

4.10.1 LIME-SODA PROCESS

The lime-soda process takes advantage of the low solubilities of calcium and magnesium components and removes these precipitates by sedimentation. Calcium is precipitated as calcium carbonate, and magnesium is precipitated as magnesium hydroxide. Lime and soda ash are the chemicals most often used to bring about these chemical reactions.

The total dissolved solids in the water are not affected by this process. Calcium and magnesium ions are replaced by sodium ions, while the sulphate concentration remains constant. In the lime-soda process, the first four-unit processes (equalization, neutralization, iron oxidation, and solids removal) are the same as for the conventional lime neutralization process. Then, the effluent from the solids removal unit enters a flash in X tank for chemical addition, the first softening process. The next step is the softening reaction (flocculation) tank, settling basins, a re-carbonation chamber, filters, and chlorination.

The neutralization and iron oxidation processes are similar in any typical AMD plant: pH adjustment and iron and manganese removal. At this stage, lime is added for the neutralization of the mine drainage ability and the precipitation of iron, manganese, and aluminium as hydroxides.

Lime is again added to the sedimentation basin effluent as the first step in the 4.16 process. Lime is also required for manganese and magnesium removal, both of which are precipitated as hydroxides according to the equations:

$$MgSO_4 + Ca(OH)_2 \rightarrow Mg(OH)_2 + CaSO_4 \tag{4.32}$$

$$MnSO_4 + Ca(OH)_2 \rightarrow Mn(OH)_2 + CaSO_4 \tag{4.33}$$

Soda ash is then added to remove noncarbonate calcium hardness or calcium sulphate:

$$CaSO_4 + Na_2CO_3 \rightarrow CaCO_3 + Na_2SO_4 \tag{4.34}$$

The pH must be maintained at 9.5 or higher for precipitation to occur.

4.10.2 ALUMINA-LIME-SODA PROCESS

This process is suited for waters in which the principal source of salinity is a sulphate. Heavy metals and hardness can be removed as well. The process is most suitable for treating AMD with sulphate concentrations between 400 and 1,200 mg/L. The lower limit is 100 mg/L, below which sulphate removal is not economically feasible. The drinking water standards require a maximum limit on sulphate concentration to 250 mg/L.

The process is applied in two stages. The raw AMD is split into two streams. The larger stream is treated with lime and sodium aluminate ($NaAlO_2$). The proportion of the raw feed treated in stage 1 depends on sulphate concentration desired in the total plant effluent. The effluent from stage 1 typically contains approximately 100 mg/L sulphate, while the sulphate concentration bypassed to stage 2 remains constant. The sulphate concentration in the final blended flow can be calculated by a mass balance method. Effluent from stage 1 is mixed with the smaller AMD stream, and carbon dioxide is added for pH control. In both stages, the produced solids are removed by filtration.

The key reactions take place in stage 1. The sodium aluminate and lime neutralize the raw acidity, precipitate the heavy metals and magnesium, and remove calcium sulphate. Sulphate is removed by the sodium aluminate, and to a lesser degree, by the iron and aluminium present in the raw AMD. The reaction between AMD, the sodium aluminate, and lime produces insoluble calcium sulpha-aluminates which dominate the sludge produced.

After the stage 1 reaction, precipitating solids are removed in a mechanical settling unit. The resulting sludge is further dewatered by filtration.

The water produced in stage 1 contains excess lime which is removed in stage 2 by mixing with the smaller stream of raw AMD. Carbon dioxide is added to lower the pH to 10.3, the minimum solubility of calcium carbonate. Calcium carbonate is formed by the reaction:

$$2OH^- + Ca^{2+} + CO_2 \rightarrow CaCO_3 + H_2O \tag{4.35}$$

Excess carbon dioxide redissolves calcium to form calcium bicarbonate. Precipitates of calcium carbonate and metal hydroxide precipitates are removed by sand filtration. The resulting filtrate will have a pH of 10.3 and will contain about 35 mg/L of dissolved calcium carbonate at its minimum solubility level. Additional carbonation can reduce the pH to an acceptable level for potable water.

4.11   BACTERICIDES IN AMD CONTROL

Inhibiting or destroying thiobacilli can significantly slow the rate of acid production. The controlled release of bactericides over a long period of time has been developed. The ProMac system, developed by BF Goodrich, is an example. A case study using the ProMac system is discussed as follows.[8]

The huge pile of refuse that accumulated during decades of coal mining operations spewed its nasty toxic acids into the nearby West Fork River at Dawmont, West Virginia.

The abandoned coal tipple and preparation plant site a few miles northwest of Clarksburg proved to be an eyesore, in addition to causing a detrimental effect on the West Fork, which contributes to the water supply of Fairmont to the north.

That was the situation at Dawmont in 1986.

Today, the site is covered with a lush growth of birds-foot trefoil, a legume vegetation that is high in nitrogen and has a strong root system that grips itself firmly to the soil. And the acids that once found their way into the West Fork have been brought under control.

The 36-acre reclamation project experienced such a quick and strong comeback that it earned the Grafton Coal Company a 1988 Reclamation Award from the West Virginia Mining and Reclamation Association and West Virginia Department of Energy.

Grafton Coal Company, the contractor on the reclamation, was recognized for the

timely, efficient, and successful restoration of 36 acres known as the Dawmont Project. The company's innovative techniques in dealing with highly acidic soil and water conditions transformed an eyesore and environmental liability into an aesthetically pleasing and potentially commercial asset, thus representing the highest potential of the Abandoned Mine Lands programme.

The West Virginia DOE decided to use the BF Goodrich Company's ProMac system in the reclamation at Dawmont.

A technology developed in the 1980s, the ProMac System involves controlled-release bactericides designed to substantially reduce acid drainage problems, improve water quality, and reduce associated treatment costs in active mining, at coal preparation plants, in active coal piles, in active refuse piles, and in mine tailings.

For use in mine land reclamation, the ProMac system requires a one-time application that sets up a recovery cycle to help nature permanently restore a site with healthy vegetation and to provide assurance against postreclamation acid discharge.

Reclamation of the Dawmont site began in September 1986 and was completed 1 year later by Grafton Coal. It was necessary to move about 250,000 yd of coal refuse or gob as it is called in the coal industry and spread it over the 36-acre site.

The refuse was the waste material that came from the underground mines around Dawmont, after separation of the usable coal. The gob pile built up from the day the mines began operation in the 1920s until they were closed in the 1960s.

Since the mine closings, the gob pile continued to pour out its toxic acids into the nearby West Fork River and even occasionally experienced minor fire flare-ups. It remained an eyesore.

After regrading the pile to stable slopes, treatment with the ProMac System was begun during the summer of 1987 and finished in September. Application of the chemicals was accomplished with a hydroseeder.

Included in the treatment was a powder that reacts immediately and three different time-release pellets. The time-release bactericides are encased in plastic pellets, with one formula releasing beneficial ingredients for the first 2 years, a second formula doing its job up to 5 years, and the third pellet continuing for 7 years or longer.

By the time the chemicals have released over the 7 years, the natural life cycle will have taken over to maintain vegetative growth and eliminate acid generation.

Following application of the powder and pellets at Dawmont, a small quantity of lime was spread over the site, soil cover of 12–18 in. was provided, and the reclamation completed by seeding, fertilizing, and covering with straw.

With the ProMac system still being considered to be "under the microscope" by the West Virginia DOE, 1 acre at the Dawmont site was left untreated as a control area. The treated and untreated areas are tested on a monthly basis.

In the 2 years since the site was finished, the Dawmont Project has experienced a rapid turnaround. The refuse pH was 3 or less before reclamation and currently is close to 5 on the treated portion. In the 1-acre control area, the pH has never gone above 3.

Reduction of the refuse acidity has shown significant improvement, with the control area still measuring about 1,000 ppm as compared to <50 in the treated area. The generation of iron and manganese also provides evidence that the treatment is achieving results greater than those in the control area.

4.12 DETERMINATION OF ACID-GENERATING POTENTIAL

Acid potential (AP) is defined as the amount of acid that would be produced if all pyrite in the sample reacted according to the following stoichiometry:

$$FeS_2 + \frac{15}{4}O_2 + \frac{1}{2}H_2O = Fe(OH)_{3(s)} + 2SO_4^{2-} + 4H^+ \tag{4.36}$$

Pyrite, 1 mol, containing 64 g sulphur, produces 4 mol of H^+ acidity, equivalent to 200 g $CaCO_3$ acidity. Thus, the AP of pyritic sulphur is 3.125 g acidity (as $CaCO_3$) per gram of sulphur; 1% pyritic sulphur has the potential to produce 31.25/1,000 tons acidity. Total sulphur concentrations can be measured with an automated Fisher sulphur analyzer, consisting of a furnace and SO_2 titrator, according to the instrument instructions. Sulphur speciation can be analyzed for samples containing >0.5% sulphur by hot extraction with hydrochloric and nitric acid to remove sulphate and pyritic sulphur, respectively.

Organic sulphur is defined as the sulphur extraction remaining after acid extraction. AP can be calculated from both total sulphur and, when analyses are available, pyritic sulphur. Total sulphur is often used in mine permit applications to minimize analytical costs.[9]

Neutralization potential (NP), defined as the ability of the stratum to neutralize strong acid, can be determined by treating a 2-g sample with 20–80 mL of 0.1 M HCl, heating nearly to boiling, and swirling periodically until no gas evolution is observed. The samples are made up to 125 mL with distilled water, boiled for 1 min, and cooled to room temperature. The treated sample is then titrated with standard NaOH (0.1 or 0.5 M) to pH 7. NP is calculated as the amount of HCl consumed by the sample and converted to the units of tons of $CaCO_3/1,000$ ton material:

$$NP; tons / 1000 tons = \frac{g\,HCl\,consumed}{g\,sample} \times \frac{50\,g\,CaCO_3}{36\,g\,HCl} \times 1000 \qquad (4.37)$$

Net neutralization potential (NETNP) is calculated for the stratum by subtracting AP from NP. A positive NETNP indicates an excess of neutralizes, while a negative NETNP indicates a deficiency of neutralizers in the stratum.

Geochemically, a number of fundamental concerns are found in the theoretical basis on which the acid-base or NAPP technique is based. The major concerns are:

1 Assumes all sulphur occurs as pyritic sulphide and is acid forming
2 Assumes all ANC is available to neutralize acid
3 Does not consider the kinetics of either the acid-forming or acid-neutralizing processes

The first two concerns may be considered to balance each other; however, failure to consider the site-specific mechanisms and kinetics when assessing the likely field geochemistry of a particular waste can result in significant error.

The NAPP procedure is an essential tool in waste characterization, but the interpretation and identification of the implications for waste management require a more detailed determination of waste geochemistry. Once the broad principles of the site-specific processes have been identified, a monitoring and management programme can normally be developed which is based on the NAPP procedure.

BIBLIOGRAPHY

1. Kleinmann, R. L. P. Biogeochemistry of acid mine drainage and a method to control acid formation, *Min. Eng. 33*(3):300–305 (1981).
2. Singer, P. C. and W. Stumm. Acid mine drainage: The rate-determining step, *Science 167*:1121–1123 (1970).
3. Nordstrom, D. K. *Hydrogeochemical and Microbiological Factors Affecting the Heavy Metal Chemistry of an Acid Mine Drainage System*, Ph.D. thesis, Stanford University, Stanford, CA (1977).
4. Dugan, P. R. Bacterial ecology in strip mine areas and the relationship to the production of acidic mine drainage, *Ohio J. Sci. 75*(6):266–279 (1975).
5. Kleinmann, R. L. P. *Thiobacillus ferroxidans* and the formation of acidity in simulated coal mine environments, *Geomicrobiol. J. 1*(4):373–388 (1985).
6. Walsh, F. and R. Mitchel. A pH dependent succession of iron bacteria, *Environ. Sci. Technol. 6*(9):809–812 (1972).
7. Penn Environmental Consultants. Design Manual, Neutralization of Acid Mine Drainage, U.S. EPA Rep.-600/2-83-001 (1983), pp. 1–183.
8. Sobek, A. A. *Successful Reclamation Using Controlled Release Bactericides*, in American Society for Surface Mining and Reclamation Conference, Charleston, WV (1990).
9. Erickson, P. M. and R. S. Hedin. Evaluation of Overburden Analytical Methods as a Means to Predict Post Mining Drainage Quality, U.S. Bureau of Mines Information Circular 9183 (1988), pp. 11–19.

5 Environmental Impacts of Metal Ore Mining and Processing

5.1 INTRODUCTION

Metal ore mining and its processing including copper, lead-zinc, gold, iron, etc., create adverse environmental effects. Mining ore mining involves the removal of large quantities of rock from the ground. Ore deposits in shallow depths are mined by surface mining methods. Surface mines produce large quantities of waste rock than underground mines. The environmental concern is primarily related to the damage of landscape and acid drainage. Mines are classified based on their major product. However, they may produce large quantities of other metals as co-products. For example, zinc mines in the United States produce 72% of all zinc, but 100% of cadmium, germanium, and indium, and 3.1%, 4.1%, and 6.1% of all gold, silver, and lead in the United States. As a result, metal ore mining and processing creates environmental impacts due to multi-elemental contamination of the environment.

In general, mineral ores occur as metallic sulphides which causes sulphur dioxide in the form of acid deposition. Despite the recent reduction in sulphur-dioxide emission and particulate matter, smelting is still a significant source of gaseous and dust contaminants. Copper smelting is regarded as one of the major man-made sources of sulphur dioxide. For example, the Sudbury Ni, Cu smelters in Ontario, Canada, produced more sulphur-dioxide than all volcanoes in history. Consequently, soil near three Sudbury smelters had a pH value from 3 to 4 and lake water pH from 4 to 5. Topsoil in the vicinity of smelters contains elevated levels of trace elements. The abundance and diversity of soil microorganisms are generally reduced by smelter emissions and acidic soil conditions. This decrease lowers soil fertility. Primary metal smelters contribute to damages of vegetation. Trace element uptake from contaminated soils and direct deposition of contaminants from the atmosphere onto plant surfaces can lead to plant contamination by trace elements. Consequently, plant toxicity and the potential for transfer of contaminating elements along the food chain exists. Soils barren of vegetation are particularly to erosion, which leads to further damage to the environment in the vicinity of smelters.

When the rate of mining of a given element exceeds the rate of its recycling the natural rate of its recycling by a factor of 10 or more, the metal should be considered as a potential contaminant. For this reason the potentially hazardous materials to the environment are Ag, Cd, Cr, Cu, Hg, Mn, Pb, Sb, Sn, and Zn. Estimates of reserve and resources rely on geological and statistical data.

According to the known information, the present level of reserves of Zn and Pb may last only for 10–25 years. Cu may last for another 25 years. Supplies of Ni will be adequate for 50–100 years. The environmental aspects should be made an integral part of mineral production. For example, the eight major Cu primary smelters in the United States currently recover about 98% of their sulphur-dioxide emissions. A continuous flush system installed at the Cu smelter in Bingham Canyon, Utah, will recover 99.9% of sulphur dioxide. The metals produced in major quantities are referred to as major metals such as Fe, Al, Cu, Pb, Mn, and Zn. The remaining metals are recovered in smaller quantities and are called minor metals. Among the major metals, the production of Cu, Pb, and Zn causes the greatest environmental concerns. Among the minor metal, As and Cd are the most hazardous because of their geochemical and toxicological properties.

5.1.1 COPPER

Copper is mined worldwide in various countries. The biggest deposits are located in the United States, Chile, Canada, Zambia, Peru, etc. Mining operations generate huge quantities of wastes and tailings, each day more than mining any other metals. They are disposed of in large piles or valley fills. Open-pit copper mines are filled after mining because they are so large and their ore bodies are not lying flat or near the surface. Modern mines are regraded and revegetated when they are abandoned as they are tailings and waste dumps. In the old mines this was not true, which are the source of metal-rich acid water and sediments. Environmental concerns are centred on emissions of sulphur dioxide and easily vapourized trace elements such as copper, cadmium, and mercury. Copper smelting is responsible for 65% of atmospheric emission from smelting.

5.1.2 LEAD AND ZINC

Lead and zinc often occur together in metallic deposits, but they have different industrial applications and biologic health effects. Zinc is an essential physiological element, whereas lead has no biological function and creates serious environmental and health effects. Emissions from lead smelters and refineries are closely regulated. This increases the cost of producing the lead. As a result, primary zinc production has been increasing, but lead production has been decreasing. Zinc commercially occurs as sphalerite, and lead is mined as galena as its primary ore. Lead is mined in 47 countries, and zinc is mined in 50 countries. Zinc is the most widespread metal in terms of primary production. Secondary lead smelters are located in 43 countries, but only 21 countries have secondary zinc smelters. The principal use of lead is in batteries, which accounts for 80% of the United States lead consumption. Zinc is used to make galvanized steel and various zinc alloys.

Lead has been used that has been so widely used that anthropogenic emissions have greatly outweighed the natural releases of this metal. The primary smelting of sulphide ores is the second most important source of the anthropogenic lead after transportation. In response to numerous environmental concerns, the US Congress has considered legislation to place a direct tax on lead production. If a tax of $1.65 per kg on primary production is imposed, it will effectively stop domestic lead production within 2 years. Because of the close relationship between lead and other metals, such a tax would cut the US production of zinc and other metals such as cadmium and bismuth by greater than 80% and gold production and gold production by 15%.

5.1.3 ARSENIC

Arsenic is well known for its toxicity. Large doses of arsenic, above 100 mg per person, induce acute arsenic poisoning, resulting in death. Arsenic is used largely in wood preservatives, herbicides, and insecticides. Arsenic is recovered largely as a by-product of other metals. Around the Sudbury Cu-Ni smelters in Ontario, arsenic concentrations in lake sediments deposited during the last few decades are several times higher than those from times before smelting began.

5.1.4 CADMIUM

Rechargeable batteries with cadmium, nickel, gold, or mercury, if used in power electric cars, could power consumption, but cadmium is potentially one of the most toxic elements in the environment. Cadmium is particularly hazardous because of its easy uptake by plants, its tendency to accumulate in food chain crops, and its persistent nature once it is in the environment. However, cadmium production is not decreasing and consequently as much as 60% of the total input of cadmium into the air comes from smelting and refining. The metal is recovered from flue dust during roasting and sintering of the mineral Sphalerite and in sludge from the electrolyte refining of zinc. With the production of 1 mg of zinc, 3 kg of cadmium is produced. Therefore, environmental problems associated with zinc production are partly attributed to cadmium release to the environment.

5.2 ENVIRONMENTAL IMPACTS

In the metal mining industry, the valuable part of the crude ore is which has to a small part of the total material mined. In the copper mining industry, the ratio of material handled to the units of valuable material is 40:1. The total portion of the material handled to the units of valuable metal is 420:1. The material handled consists largely of waste material distinguishing the mining industries from other industries. Mining and beneficiation process generates four kinds of waste materials: mine waste (overburden, and barren rock), tailings, dump heap leach, and mine water. The major part of nonfuel minerals is mined surface mining techniques. Surface mines produce more waste than underground mining. In most of the surface mines, the amount of ore mined to waste produced is about 2–10 times compared to underground mining the same ratio 9–27%. Tailings produced are produced from beneficiation plants. The crushed ore is concentrated to release ore particles from the matrix-less valuable rock. Dump leaching, heap leaching, and leaching are the processes usually used to extract metals from low-grade ores. For copper extraction, sulphuric acid is used. As the liquid percolates through the rock, it leaches out the metal. Dump leach piles often cover 100 ha, with usual height to 60 m, and contains tens of millions of Mg of low-grade ores, which becomes waste after leaching. Heap leaching operations are usually much smaller than dump leaching and last over a period of months rather than years. The mine water, which infiltrates a mine, must be removed to continue mining operations. The mine wastes are generally deposited on site in the United States.

Smelting and refining plants produce gaseous emissions and particulate matters, wastewater, and solid wastes. These are emitted in the atmosphere and to the water system or disposed on the land. The most noticeable form of contaminants from the metal industry discharge is from smokestacks. Tall stacks discharge pollutants into the atmosphere. Their discharge is dispersed into the atmosphere. But the discharge is at great heights. The contaminants are sufficiently diluted when dispersed into the lower atmosphere so that they meet the air quality requirements. The major solid wastes are generated by smelting and refining, which may constitute hazardous waste, process waste, and residuals from air pollution control and waste treatment systems.

5.2.1 SOIL DEGRADATION

In mining and smelting areas, soil is affected by the disposal of mine tailings, acid mine drainage and aerial deposition of contaminants from smelters. The exposed soils become acidic and contaminated with trace elements. This contamination and acidification lower soil fertility. The negative impacts of mining on surrounding land are often related to the disposal of large quantities of tailings. The area of land disturbed by mining in the United States is estimated at 924,000 ha per year. In 1976 the total accumulated mineral waste, overburden, submarginal ores, mining wastes, and strip-mining spoils was estimated at >23 billion mg and covers 2 million ha of land.

Abandoned mine tailings have diverse physical, chemical, and ecological properties. However, acid drainage causes elevated levels of trace elements, which are usually the common characteristics of most tailings produced by metal mining. Additionally, wind and water erosion and the associated environmental degradation are common problems related to the tailing material. Leaching of acidifying compounds from tailings causes acidification of adjacent soils.

5.2.2 ACIDIFICATION POTENTIAL

Acidification caused by sulfide minerals in the ore and the surrounding rock is one of the major sources of water pollution from mining operations. Sulfide minerals are the main potentials of acid generation in mining operations. Pyrite is the principal source of acid generation. Other sulfide minerals present in the parent ore are also potent sources of acid pollution from mining operations. Acid minerals on contact with ground water and air produce acid which then flows into ground water. Such acidified water is the main source of water pollution.

One of the remedies of this problem is to apply neutralizing chemicals such as lime into the water from mining. The lime reacts with the sulfuric acid and neutralizes the acid contained in the discharged mine water.

5.2.3 SOIL DEGRADATION BY SMELTING

Soils are affected by the disposal of mill tailings, acid mine drainage, and contaminants from the aerial disposition of smelter. The exposed soils become acidified and contaminated with trace elements. Elemental contamination and acidification lower soil fertility and reduce the variety and the proportion of biological species in the soil. The adverse impact has been largely related to huge amounts produced and disposed of by mining. The area of land disturbed by mining in the United States at the time was estimated at 924,000 ha. It was estimated that the total accumulated mineral waste, including overburden, submarginal ores, and mining wastes, strip-mining spoils to be greater 23 billion Mg covering 2 million ha of land. Abandoned mill tailings have diverse physical, chemical, and ecological properties. Acid drainage due to sulphide oxidation and elevated levels of trace elements are usually common characteristics of most mill tailings, produced by metal mining. Additionally, wind and water erosion and the associated environmental degradation are widespread problems associated with mill tailings. Leaching of acidifying compounds from tailings leads to acidification of adjacent soils.

5.2.4 ACIDIFICATION

Strong acidification takes place in soil areas where metallic ore is mined and processed. Acidifying compounds emitted from smokestacks and draining acid from mine waters and tailings cause acidification. Metal processing does not remove all pyritic minerals such as pyrite, pyrrotite, and chalcopyrite. The oxidation of pyrite and pyrrotite produces acid. The oxidation of ferrous to ferric cations is accelerated by the presence of a bacterium, as explained in Chapter 4. The amount of acidity produced is a function of many variables, such as temperature, oxygen supply, the concentration of sulphides, initial pH of the surroundings, total concentrations of Fe, and the presence of bacteria. Formation of sulphuric acid decreases the pH of the tailings and adjacent soils, which usually results in increased mobility of metals present there.

Additionally, the sulphur-dioxide emissions from smelters contribute to the soil acidification. The amount of sulphur measurement released varies depending on the origin of the ore. Sulphur emissions from copper smelters in Canada were about 0.7 Mg per Mg of copper produced vs. 11 in the United States, and 0.03 in western Europe. In Sudbury basin Cu-Ni mining and smelting basin soils are strongly acidified (pH values ranging 2.0–7.5) as a result of sulphur-di-oxide fumigation from two active copper-nickel smelters. Metals present in the acidic soils become easily mobile and available for plants, therefore their leaching, plant uptake, and runoff from soils increase. Sudbury soils had the concentrations of mobile Al as high as 100 mg per kilogram, which is apparently one of the factors of plant toxicity of these soils. Also high concentrations of exchangeable Cu and Ni were detected in these soils.

5.2.5 SOIL BIOLOGY

The compounds of soil biota are diverse. Included are bacteria, fungi, algae, and soil fauna. Soil microorganisms contribute to the changes in forms of plant nutrients through the mediation process of various biogeochemical cycles involved in nutrient recycling. These processes can alleviate adverse properties in the soils, thus improving conditions of growth of higher plants. The abundance and diversity of soil microorganisms are generally reduced by smelter emissions and acidic conditions. The soil flora of contaminated sites in Sudbury area shoed was characterized by a low diversity of chalcopyrite, absence of cyanobacteria. Similar results have been observed for soils subjected to emissions from various metallurgical plants in Russia. In these areas cyanobacteria were absent from soils within a radius of 20 km. The absence of cyanobacteria in Sudbury soils is apparently a result of strong soil acidification.

5.2.6 TRACE METALS

There is much evidence that lead and zinc production results in most severe soil contamination, not only by lead and zinc but also by associated silver, arsenic cadmium, copper, and nickel. Soil contamination by metals from mining and smelting was comprehensively investigated in Japan, the UK, and Poland. Also, the contaminants from copper-nickel smelters in Sudbury in northern Ontario, Canada into the surrounding environment are well documented.

Although currently the UK no longer produces lead and zinc, several case studies show soil contamination by these metals. The most extensive source of metal concentration in the UK operating has been the metalliferous mining which commenced in Roman times and had been operating until the end of the 19th century. It has been estimated that >4,000 square kilometres of agricultural land in England and Wales is contaminated by one or metals because of the historical operations. It has been observed that extremely high concentrations of cadmium, lead, and zinc detected in the soils reclaimed from an old mine at Sipham in southwestern England, little of these metals have been transferred into pasture plants. This observation is good evidence of the effect of metal forms on plant life. High metal concentrations in Shipham soils result from mineralization, not the emission of contaminants. The metals are strongly fixed to primary soil minerals. The data information on Sudbury area has long history of metal mining. However, it seems that probably two factors are involved: (i) leaching and runoff metals into water systems (ii) removal of fine soil particles, enriched with metals, by erosion. However, it should be pointed out that erosion does not change metal concentration in soils. Erosion results in the translocation of the metal-contaminated soil over a bigger area and its dispersion. Leaching and washing of the metals soils are apparently facilitated by the low soil pH. An acidic reaction is the most important factor in increasing soil metal mobility. Erosion can be of importance for the reduction of local metal concentration, especially in barren and semi-barren areas.

Upper Silesia (southern Poland) is a heavily industrialized and agricultural area. The industries present there include coal mines, metal ore mines, lead-zinc smelters, iron smelters, coal-fired power plants, and many others. Metal pollution is mainly caused by two active lead-zinc smelters. The arable land covers about 321,000 ha of predominantly medium-quality soils, which accounts for about 48% of the total area. At some areas in upper Silesia the deposition of cadmium and lead exceeds more than ten times the proposed limit, which was 2.75 kg per km^2 cadmium and 182.5 kg per km for lead. The arable soils are contaminated with cadmium, lead, and zinc; also occasionally high concentrations of copper have been detected. The most serious concentrations exist at the vicinity of the two active lead-zinc smelters. However, other areas are also affected by the long-range transport of contaminants. It is estimated that only 10% of the arable land has natural concentrations of heavy metals, 30% the soil is slightly contaminated, and 60% is medium to heavily contaminated.

Japan is well known for environmental contamination, including soil contamination from mining and smelting. There are several reasons for severe contaminations occurring in Japan. Japan has a high population density, a large number of metal mines and rapid economic development after the Second World War. The abrupt increase in metal production spread metals from the mines, smelters, and metal processing factories. Mining and smelting industries are the main sources of cadmium contamination of Japanese soils. There are also areas with soils contaminated by arsenic, bismuth, zinc, chromium, and lead from mining/smelting in Japan.

5.2.7 Trace Metal Contamination of Terrestrial Plants

Plants uptake from contaminated soils and deposit contaminants from the atmosphere onto the plant surfaces. They are the sources of elevated levels of trace elements in terrestrial plants growing in mining/smelting areas. The crops in upper Silesia in Southern Poland had elevated concentrations of cadmium, lead, and zinc apparently as a result of emissions from the two lead-zinc smelters in the area. However, the metal levels were not extremely high in agricultural products. In Sudbury soils in Ontario, Canada, areas remain barren or semi-barren of vegetation as a result of severe environmental conditions. Following the environmental improvements in the Sudbury area, several plant species developed colonized barren sites. However, recolonization was confined to relatively favourable sites by species that evolved metal tolerance.

5.2.8 Health Risk

It is a complex process to evaluate the health impacts of mining and smelting operations. The complex factors have to be submitted for assessment. There is no uniform methodology available. Different approaches may lead to conflicting results. For example, in Upper Silesia in Poland is a region with numerous diverse industries, ferrous and nonferrous mining and smelting, hard coal mines, and power plants. Upper Silesia frequently experiences peak ambient sulphur-dioxide concentrations of >300 $\mu m/m^3$ and average values of more than 60 $\mu m/m^3$, which is the level by which the health can be endangered. In addition, ambient concentrations of aerosols in many sites in Upper Silesia have ambient concentrations of acid aerosols that are high enough to cause a significant deterioration in lung function. Additionally, it has been estimated that because of a heavy load of air pollutants and stagnant air masses, the oxygen content of the air can decrease by as much as 20%. This can induce health hazard to heart patients asthmatics. The hazardous conditions prevailing in Upper Silesia are believed to be responsible for 15% higher circulatory problems, 30% more cancer cases, and 47% greater respiratory problems compared to other areas of Poland. The incidence of these diseases can be related to an overall deterioration in environmental quality in Upper Silesia caused by emissions from various industries, which includes metal ore mining and processing. However, it is difficult to quantify the relative contributions of particular sources.

BIBLIOGRAPHY

1. Dudka, S. and D. C. Adriano. Environmental impacts of metal ore mining and processing: A review, *J. Environ. Qual.* 26:590–602 (1997).
2. Adamo, P. S., M. J., Dudka, M. J. Wilson, and W. J. Hardy. Chemical and Mineralogical forms of Cu and Ni in contaminated soils from the Sudbury mining and smelting region, *Canada Environ. Pollut.* 91:11–19 (1996).
3. Amiro, B. D. and G. M. Courtin. Patterns of vegetation in the vicinity of an industrially disturbed ecosystem, Sudbury, Ontario, *Can. J. Bot.* 59:1623–1639 (1981).
4. Davies, B. E. and H. M. White. Environmental pollution by wind blown lead mine waste: A case study in Wales, *U.K., Sci. Total Environ.* 20:57–74 (1981).

6 Acid Rock Drainage and Metal Migration

6.1 INTRODUCTION

The sources of ARD from mining operations include:

- Drainage from underground workings
- Runoff from open-pit workings
- Waste rock dumps from mining activities
- Mill tailings
- Ore stockpiles
- Spent ore piles from heap leach operations

The ARD from underground workings has been known since earlier times because it generally occurs as a point of discharge of substantial flows of low pH water. Many of the ARD sources have been drainage tunnels or mine adits.

The environmental effects of ARD from open-pit mines have been of more recent concern because many of the ARD producers or potential producers are still in operation and the ARD is not being treated. Many of these mines have been developed without plans to deal with ARD. The large areas of exposed rock in open pits can produce large volumes of ARD. Slope deterioration in the long term can cause continuous exposure of fresh rock to the natural elements, producing ARD.[1]

Waste rocks produced from mining operations are exposed to precipitation, runoff, and possibly seepage. Waste rocks containing sulphides are potentially large sources of ARD. The chemical and physical properties of these waste rocks significantly affect the quality of ARD and the change in its quality over time.

Sulphide-rich tailings are well known as potential sources for acid generation. The low permeability of the mill tailings and the flooding occurring in both operating and abandoned tailings impoundments limit the role of ARD generation and release. Consequently, the full potential effect of very large tailings dumps of more recent origin has not yet developed. Many of these tailings dumps are in active use, and their potential impacts with ARD are controlled. The ARD may be produced after abandonment.

Low-grade ore stockpiles are often of particular concern as they can become concentrated sources of ARD. Similarly, the spent heap-leach piles can be sources of ARD, particularly those associated with low pH leachates.

6.2 THE ACID-GENERATION PROCESS

Acid rock drainage is produced by the exposure of certain sulphide minerals, most commonly pyrite, to air and water, resulting in the production of acidity and elevated concentrations of metals and sulphate. The sulphur in the mineral is oxidized to a higher oxidation state, and aqueous iron, if present, is precipitated as ferric iron. Sulphide minerals are often found in rocks that lie below a mantle of soil beneath the water table. Under natural conditions, the overlying soil and groundwater allow very little contact with oxygen; therefore, acid-generation proceeds at a slow rate and its effect on groundwater quality is negligible. When the rock is exposed to air and water by the mining process, the rate of acid generation is accelerated.

The ability of a particular rock formation to generate acid is determined by the relative content of acid-generating and acid-containing minerals. The process by which acid is consumed is known as "neutralization." Acid waters produced by the oxidation of sulphides in a rock may be neutralized upon contact with acid-consuming minerals. As a result, the water flowing from the rock may have a neutral pH and negligible acidity. However, if the acid-consuming minerals are dissolved, washed out, or coated by other minerals through encapsulation then, as acid generation continues, acid water will eventually drain from the rock. Whenever the acid-generation capability exceeds the acid-consuming capacity of the rock, it is generally expected that the water draining from the rock will not be of neutral pH.

It is important to consider the scale of examination when addressing acid drainage. A rock that produces pH-neutral conditions in water passing over it may experience acid generation in microenvironments around sulphide grains. The resulting acidic water may be neutralized by the remainder of the sample as it leaves the microenvironment. If an acid-generating rock has no neutralization potential, then acidic water will be present on the scale of both the sulphide grains as well as the rocks as a whole. A more complex situation occurs when pH-neutral water flowing over a rock invades and flushes the microenvironments of the sulphide grains, resulting in a slower rate of acid generation and a slower consumption of neutralization potential. To clarify terms, "acid-generation" generally refers to the reaction in the microenvironment around a sulphide grain, whereas "acid mine drainage" refers to the chemical composition of water emanating from a rock or waste rock pile on a larger scale.

The time interval between the initial disturbance of the rock and the peak rate of acid generation may range from days to years, depending on a number of environmental factors and the neutralization potential of the rock. In addition, the rate at which acid generation occurs through time will vary depending on several environmental factors and geochemical characteristics of the sulphide minerals. As a result, acid generation is not a simple process; rather, it is a complex set of chemical reactions changing through time, which are currently the topic of a large amount of scientific research.

6.3   SULPHIDE MINERALS

Crystalline substances that contain sulphur combined with a metal (e.g., iron) or semi-metal (e.g., arsenic) but no oxygen are called sulphide minerals (Table 6.1). If a metal or semi-metal are both present in a mineral (e.g., arsenopyrite, FeAsS), the semi-metal substitutes for sulphur in the crystal structure. These minerals form in strongly anoxic (i.e., chemically reducing) environments, as indicated by sulphur, which is present in its lowest natural oxidation state. In oxygenated environments, sulphur exists in higher oxidation states, such as $S_2O_3^{2-}$, SO_3^{2-}, and SO_4^{2-} (sulphate) and forms minerals with oxygen (e.g., gypsum, $CaSO_4-2H_2O$).

Under certain geological conditions, most notably near-surface, low-temperature deposits (bogs and swamps, etc.) and/or rapid deposition (e.g., mid-oceanic ridge sulphide deposits), sulphide may be precipitated in amorphous (noncrystalline) or poorly crystalline forms. For iron-sulphide minerals, amorphous FeS or greigite may form initially and then alter to pyrite via sulphurization. This process may lead to the formation of raspberry-like balls or "framboids" of fine-grained pyrite crystals. This framboidal pyrite has a significantly higher rate of acid generation when exposed to an oxidizing environment than coarsely grained, euhedral pyrite.

Marcasite (Table 6.1) is a low-temperature iron-sulphide mineral that may form instead of pyrite and which reportedly has a higher rate of acid generation under oxidizing conditions than crystalline pyrite. Marcasite may also be found in higher temperature paleoenvironments where it is metastable with respect to pyrite at temperatures >157°C.

At elevated temperatures, sulphide may be mobile, leading to recrystallization as a massive sulphide sometimes found at metal mines. The rate of acid generation from massive sulphide may be

TABLE 6.1

Summary of Common Sulphide Minerals and Their Oxidation Products

Minerals	Composition	Aqueous End Products of Complete Oxidation[a]	Possible Secondary Minerals Formed at Neutral pH after Complete Oxidation and Neutralization[b]
Pyrite	FeS_2	Fe^{3+}, SO_4^{2-}, H^+	Ferric hydroxides and sulphates; gypsum
Marcasite	FeS_2	Fe^{3+}, SO_4^{2-}, H^+	Ferric hydroxides and sulphates; gypsum
Pyrrhotite	$Fe_{1-x}S$	Fe^{3+}, SO_4^{2-}, H^+	Ferric hydroxides and sulphates; gypsum
Smythite, greigite	Fe_3S_4	Fe^{3+}, SO_4^{2-}, H^+	Ferric hydroxides and sulphates; gypsum
Mackinawite	FeS	Fe^{3+}, SO_4^{2-}, H^+	Ferric hydroxides and sulphates; gypsum
Amorphous	FeS	Fe^{3+}, SO_4^{2-}, H^+	Ferric hydroxides and sulphates; gypsum
Chalcopyrite	$CuFeS_2$	Cu^{2+}, Fe^{3+}, SO_4^{2-}, H^+	Ferric hydroxides and sulphates; copper hydroxides and carbonates; gypsum
Chalcocite	Cu_2S	Cu^{2+}, SO_4^{2-}, H^+	Copper hydroxides and carbonates; gypsum
Bornite	Cu_5FeS_4	Cu^{2+}, Fe^{3+}, SO_4^{2-}, H^+	Ferric hydroxides and sulphates; copper hydroxides and carbonates; gypsum
Arsenopyrite	FeAsS	Fe^{3+}, AsO_4^{3-}, SO_4^{2-}, H^+	Ferric hydroxides and sulphates; ferric and calcium arsenates; gypsum
Realgar	AsS	AsO_4^{3-}, SO_4^{2-}, H^+	Ferric and calcium arsenates; gypsum
Orpiment	As_2S_3	AsO_4^{3-}, SO_4^{2-}, H^+	Ferric and calcium arsenates; gypsum
Tetrahed rite and	$Cu_{12}(Sb,As)_4S_{13}$	Cu^{2+}, SbO^{3-}, AsO_4^{3-},	Copper hydroxides and carbonates; calcium and ferric arsenates; antimony
tennenite		SO_4^{2-}, H^+	materials; gypsum
Molybdenite	MoS_2	MoO_4^{2-}, SO_4^{2-}, H^+	Ferric hydroxides; sulphates; molybdates; molybdenum oxides; gypsum
Sphalerite	ZnS	Zn^{2+}, SO_4^{2-}, H^+	Zinc hydroxides and carbonates; gypsum
Galena	PbS	Pb^{2+}, SO_4^{2-}, H^+	Lead hydroxides, carbonates, and sulphates; gypsum
Cinnabar	HgS	Hg^{2+}, SO_4^{2-}, H^+	Mercuric hydroxide; gypsum
Cobaltite	CoAsS	Co^{2+}, AsO_4^{2-}, SO_4^{2-}, H^+	Cobalt hydroxides and carbonates; ferric and calcium arsenates; gypsum
Niccolite	NiAs	Ni^{2+}, AsO_4^{3-}, SO_4^{2-}, H^+	Nickel hydroxides and carbonates; ferric, nickel, and calcium arsenates; gypsum
Pentlandite	$(Fe, Ni)_9S_8$	Fe^{3+}, Ni^{2+}, SO_4^{2-}, H^+	Ferric and nickel hydroxides; gypsum

[a] Intermediate species such as ferrous iron (Fe^{2+}) and $S_2O_3^{2-}$ may be important.

[b] Depending on overall water chemistry, other minerals may form with, or instead of, the minerals listed here.

Source: From *Draft Acid Rock Drainage Technical Guide*, Vol. 1 (Vancouver, Canada: BiTech Publishers, 1989). With permission.

relatively slow, but the rate may be accelerated during the mining process through blasting and grinding. In general, the relative rates of oxidation for sulphide minerals under typical environmental conditions are unclear and detailed experimentation is recommended.

Oxidation of these minerals may lead to the formation of secondary minerals after some degree of pH neutralization or when pH is maintained near neutral during oxidation. Some of these minerals are listed in Table 6.1; other minerals may form in addition to, or instead of, these minerals depending on water chemistry, the extent of oxidation, and the presence of other compounds such as aluminosilicates. These secondary minerals may encapsulate the sulphide mineral and/or any neutralizing mineral, slowing the reaction rate.

6.4   CHEMICAL AND BIOLOGICAL REACTIONS
RELATED TO ACID GENERATION

Acid generation, as well as acid consumption, is the result of a number of interrelated chemical reactions. The primary ingredients for acid generation are:

- Sulphide minerals
- Water or a humid atmosphere
- An oxidant, particularly oxygen, from the atmosphere or from chemical sources

The total exclusion of moisture or oxidant will stop acid generation. In most cases, bacteria play a major role in accelerating the rate of acid generation, and the inhibition of bacterial activity in these cases will lessen the rate of acid generation.

The reactions of acid generation are best illustrated by examining the oxidation of pyrite (FeS2), which is one of the most common sulphide minerals. The first important reaction is the oxidation of the sulphide mineral into dissolved iron, sulphate, and hydrogen (H^+):

$$FeS_2 + 7/2O_2 + H_2O \rightarrow Fe^{2+} + 2SO_4^{2-} + 2H^+ \tag{6.1}$$

The dissolved Fe^{2+}, SO_4^{2-}, and H^+ represent an increase in the total dissolved solids and acidity of the water, and, unless neutralized, the increasing acidity is often associated with a decrease in pH. If the surrounding environment is sufficiently oxidizing, much of the ferrous iron will oxidize to ferric iron:

$$Fe^{2+} + 1/4O_2 + H^+ \leftrightarrow Fe^{3+} + 1/2H_2O \tag{6.2}$$

At pH values above 2.3–3.5, the ferric iron will precipitate as $Fe(OH)_3$, leaving little Fe^{3+} in solution while lowering pH at the same time:

$$Fe^{3+} + 3H_2O \leftrightarrow Fe(OH)_3 \, (\text{solid}) + 3H^+ \tag{6.3}$$

Any Fe^{3+} that does not precipitate from the solution may be used to oxidize additional pyrite:

$$FeS_2 + 14Fe^{3+} + 8H_2O \rightarrow 15Fe^{2+} + 2SO_4^{2-} + 16H^+ \tag{6.4}$$

Based on these simplified basic reactions, acid generation that produces iron, which eventually precipitates as $Fe(OH)_3$, may be represented by a combination of reactions:

$$FeS_2 + 15/4O_2 + 7/2H_2O \rightarrow Fe(OH)_3 + 2SO_4^{2-} + 4H^+ \tag{6.5}$$

On the other hand, the overall reaction for stable ferric iron that is used to oxidize more pyrite is

$$FeS_2 + 15/8O_2 + 13/2Fe^{3+} + 17/4H_2O \rightarrow 15/2Fe^{2+} + 2SO_4^{2-} + 17/2H^+ \tag{6.6}$$

Equations 6.3 to 6.6 assume the oxidized mineral is pyrite and the oxidant is oxygen. However, other sulphide minerals such as pyrrhotite (FeS) and chalcocite (Cu_2S) have other ratios of metal: sulphide and metals other than iron. Other oxidants and sulphide minerals also have different reaction pathways, stoichiometries, and rates, but research on these variations is limited.

The primary chemical factors that determine the rate of acid generation are:

- pH
- Temperature

- Oxygen content of the gas phase, if saturation is <100%
- Oxygen concentration in the water phase
- Degree of saturation with water
- Chemical activity of Fe^{3+}
- Surface area of exposed metal sulphide
- Chemical activation energy required to initiate acid generation

Certain bacteria may accelerate or decelerate the rate at which some of the above reactions proceed, thereby increasing or decreasing the rate of acid generation. *Thiobacillus ferrooxidans*, in particular, is known to accelerate these reactions through its enhancement of the rate of ferrous-iron oxidation. *T. ferrooxidans* may also accelerate the reaction through its enhancement of the rate of reduced sulphur oxidation. Experimental testing of the many other bacterial species capable of oxidizing iron and sulphur is generally limited. In addition, several species are known to reduce sulphur and iron, potentially counteracting acid generation (Table 6.2).

Most testing of *T. ferrooxidans* has involved oxidation of pyrite (FeS_2); however, the bacterium may accelerate the oxidation of sulphides of antimony, gallium, molybdenum, arsenic, copper, cadmium, cobalt, nickel, lead, and zinc.

For bacteria to thrive, environmental conditions must be favourable. *T. ferrooxidans*, for example, is most active in waters with a pH around 3.2. If conditions are not favourable, the bacterial influence on acid generation will be minimal. This apparent importance of environmental conditions explains the contradiction in reported experimentation that shows bacterial influence ranges from major to negligible. Experimental laboratory-based and in-field tests with bactericides indicate

TABLE 6.2
Bacterial Species That Influence Rate of Sulphur and Iron Oxidation

Bacterial Species	Type	Optimal Growth of Chemical Environment
Thiobacillus ferrooxidans	Sulphur oxidizing	pH = 2.5–3.5
	Iron oxidizing	
T. novellas	Sulphur oxidizing	pH = neutral to alkaline
T. thioporus	Sulphur oxidizing	
T. denitiricans	Sulphur oxidizing	pH = neutral to alkaline
		Nitrate supply for reduction to N_2
Arthrobacter sp.	Sulphur oxidizing	—
Bacillus sp.	Sulphur oxidizing	—
Flavobacterium sp.	Sulphur oxidizing	—
Pseudomonas sp.	Sulphur oxidizing	—
Desulfavibrio sp.	Sulphur reducing	—
Desulfotomaculum sp.	Sulphur reducing	—
Salmonelis sp.	Sulphur reducing	—
Proteus sp.	Sulphur reducing	—
Suffolosu sp.	Sulphur reducing	—
Metalloginium sp.	Iron oxidizing	—
Sidenocapsa sp.	Iron oxidizing	—
Leptothrax sp.	Iron oxidizing	—
Gallionella sp.	Iron oxidizing	—
Vibrio sp., *Bacillus* sp.	Iron oxidizing	—
Aerobacter aerogonus	Iron oxidizing	—

Source: From *Draft Acid Rock Drainage Technical Guide*, Vol. 1 (Vancouver, Canada: BiTech Publishers, 1989). With permission.

bacterial activity enhances the rate of acid generation (as indicated by sulphate and acidity) by a factor of up to 5, with one extreme measurement of a factor of 20.

In situations in which bacterial acceleration is significant, additional factors determine the bacterial activity and the associated rate of acid generation:

- Biological activation energy
- Population density of bacteria
- Rate of population growth
- Nitrate concentration
- Ammonia concentration
- Phosphorus concentration
- Carbon dioxide content
- Concentrations of any bacterial inhibitors

The bacterial contribution to sulphide and iron oxidation can be complex through contributions from many species. However, research indicates that *T. ferrooxidans* often plays a major role in organic-enhanced oxidation in natural environments. This justifies the attention paid to this bacterium but does not always justify the exclusion of other bacteria from consideration.

Following the oxidation of a sulphide mineral, the resulting acid products may either be immediately flushed away by water moving over the rock or, if no water movement occurs, may accumulate in the rock while remaining readily available for flushing. If the acid products are flushed away from the sulphide mineral, they may eventually encounter an acid-consuming mineral; the resulting neutralization will remove a portion of the acidity and iron from the solution and will neutralize the pH. Sulphate concentrations are not usually affected by neutralization unless mineral saturation with respect to gypsum is attained. Consequently, sulphate sometimes may be used as an overall indicator of the extent of acid generation, even after neutralization by acid-consuming minerals has occurred.

The most common acid-consuming mineral is calcite ($CaCO3$), which consumes acidity through the creation of HCO_{3-} or $H_2CO_3{}^0$:

$$CaCO_3 + H^+ \rightarrow Ca^{2+} + HCO_3^- \tag{6.7}$$

and

$$CaCO_3 + 2H^+ \rightarrow Ca^{2+} + H_2CO_3{}^0 \tag{6.8}$$

There are also other acid-consuming minerals such as $Al(OH_3)$:

$$Al(OH)_3 + 3H^+ \rightarrow Al^{3+} + 3H_2O \tag{6.9}$$

It is not unusual for a rock to contain both sulphide minerals and acid-consuming minerals. The balance between the two types will determine whether the rock will eventually produce acid conditions in the water passing over and through it, and this balance forms the basis of the experimental procedure used in static tests.

6.5 METAL LEACHING AND MIGRATION PROCESSES

Acid-generation processes in rocks, which produce low pH water, are capable of dissolving heavy metals contained within the rocks. This water then migrates from the generation site and enters the receiving environment. High-metal loadings in the water are the most harmful to the environment.

A series of reactions occurs along the path as the low pH water migrates from the source to the receiving environment. The resulting quality of the water is determined by the following factors:

- Nature of the sulphides
- Availability and type of soluble constituents
- Nature of alkaline reactants
- Physical properties of the waste

Some examples of the quality of ARD are given in Table 6.3.

Several naturally occurring physical, chemical, and biological properties of mine waste affect metal solubility and contaminant migration. The mobilization of metals is mainly controlled by chemical factors, while the processes that occur along the migration route are controlled by physical and chemical factors.

Physical properties that influence metal solubility include waste particle size and shape, and temperature and pressure of pore gases. However, chemical factors are more predominant than physical properties in the metal mobilization process. Physical properties are important in the migration rate of ARD and in the reactions that occur along the migration path. Important characteristics include:

- Climatic conditions
- Waste permeability
- Availability of pore water
- Pore water pressure
- Movement mechanism, whether by stream flow or diffusion

These factors control the rate of movement of contaminant fronts, the amount of dilution, and the degree of mixing that occurs as the ARD moves from the source to the environment. The physical properties of the subsurface are different than those of waste, so a number of contaminant fronts

TABLE 6.3
Examples of Acid Rock Drainage Quality

Parameter[a]	Seepage from Abandoned Uranium Mine Tailings Pond in Ontario	Waste Rock Dump Seepage from Active Silver Mine in British Columbia	Mine Water from Underground Copper Mine in British Columbia
pH	2.0	2.8	3.5
Sulphate	7,440	7,650	1,500
Acidity	14,600	43,000	—
Iron	3,200	1,190	10.6
Manganese	5.6	78.3	6.4
Copper	3.6	89.8	16.5
Aluminium	588	359	—
Lead	0.67	2	0.1
Cadmium	0.05	0.5	0.143
Zinc	11.4	53.2	28.5
Arsenic	0.74	25	0.05
Nickel	3.2	8.0	0.06

[a] Units are mg/L except pH.

Source: From *Draft Acid Rock Drainage Technical Guide*, Vol. 1 (Vancouver, Canada: BiTech Publishers, 1989). With permission.

may develop, all moving at different rates. Surface water yields tend to occur before groundwater yields in hard-rock waste dumps because of lower retardation and the resulting rapid migration through the waste rock. The quality of the yields, whether surface or groundwater, is a function of the dilution and buffering reactions that occur en route.

The solubility of the metals is generally determined by the pH of the leachate. Other chemical factors include the specific metal being dissolved, Eh, adsorption characteristics, and the chemical composition of the leachate. As the ARD moves away from the sulphide source through the waste material, more acid-generating material may be encountered, causing a reduction in pH. The drainage over alkaline material may also cause complete or partial neutralization.

With the gradual lowering of the pH level the dissolved metal load generally increases. However, a combination of chemical conditions could cause increasing mobilization of metals even at neutral or alkaline conditions. During the neutralization process of the drainage, precipitation of many of the soluble metals may occur, and the resultant drainage will contain the residual metals.[2]

An interesting phenomenon that has been observed in copper and massive sulphide ores is elevated zinc loadings in neutral drainage. Dissolved copper precipitates out as the pH is raised; the zinc remains at relatively high concentrations until the pH is raised to values above 9.5. As the ARD front moves through the waste or subsurface strata, the chemical composition of the front undergoes continuous change.[3,4]

Biological activities along the route may influence metal dissolution. Metal leaching occurs where iron-oxidizing bacteria are present with iron and metal sulphides. Biological species can also attenuate the mobility of metals by absorption and precipitation.

6.6   PREDICTION OF ACID DRAINAGE

The prediction process for determining potential acid generation from metal mining operations includes:

- Comparison with similar and neighbouring mines
- A systematic sampling programme to collect representative samples
- Static tests on the samples
- Kinetic tests using anticipated on-site conditions using potentially acid-generating samples
- Modelling

In the exploration stage, samples of the ore and waste rock should be collected for acid-base accounting analyses. These results early in the mine planning stages would indicate whether acid drainage might be a concern.

The geological units of ore and waste rocks should be identified based on lithology, mineralogy, and continuity of units. A comparison should be made with neighbouring mines and similar geological and paleoenvironmental areas for obtaining an initial indication of potential acid generation. A sampling programme for each geologic unit should be implemented. These samples should be subjected to acid tests to determine the potential for net acid production.

If the static test results give an uncertain indication of acid generation, then kinetic tests could be of value in determining acid production potential. If the potential for net acidity is identified for any geological unit, the mine plan should be revised. Mathematical models should be utilized along with kinetic tests to predict acid generation over a longer time period.

A simple approach to assess acid-generation potential involves geological comparison with nearby mines. This approach assumes that all factors influencing the acid-generation process are identical for the mines. This is rarely the case in vein deposits as the host rocks, alterations, and mineralogies are often dissimilar. On a larger scale, comparison over a wider geographical region is likely to be unreliable as the nongeological factors that affect acid generation, like climate and physiography, will vary.

A basic approach for the assessment of acid-generation potential is to compare paleoenvironmental and geological characteristics. In this process, it is necessary to classify both deposits from the standpoint of acid generation. Some of the existing mineral deposit classification models can be useful in the prediction of ARD.

The geological factors controlling the generation of ARD include:

- Oxidation state of minerals
- Sulphide mineral compositions
- Texture and crystal development in sulphides
- Presence of acid-consuming minerals
- Presence of rock structures that increase the permeability

Any available database to refine a geological classification in terms of ARD potential should also be helpful.

In the prediction process, a reliable sampling programme is the initial step. Beginning a sampling programme is complicated because static tests on samples from a defined geologic unit may indicate significant variability in acid-generation potential. This variability may indicate that the geologic unit actually consists of two or more units from the standpoint of acid generation. The sampling plan should be revised to define the additional units. Such an iterative sampling programme may be necessary to clearly define acid-generating units.

The sampling programme should be directly based on the mine plan. The samples should be taken from different areas in the mine plan. This approach will help predict the timing of acid generation as mining progresses.

The design of a sampling programme is initiated at the exploration stage when geological units are identified. In the next stage, a more detailed sampling programme is required to define more reliably the potential for net acid generation. The sampling programme should also respond to any change in the mine plan. A minimum number of samples to characterize each geological unit in terms of its potential to generate net acidity will be needed.

The potential sources of samples are outlined in Table 6.4.

TABLE 6.4
Potential Sources of Samples for Acid-Generation Prediction

Mine Component	Existing Mines	Proposed Mines
Pit walls	Drill core	Drill core
	Pit walls	Underground exploration passages
		Trenches
Underground	Drill core	Drill core
workings	Walls	Underground exploration
	Excavated rock	passages
Waste rock/	Waste rock piles	Drill core
overburden piles	Drill core	Underground exploration passages
Tailings	Tailings	Pilot plant for mill process
	Impoundments	
Ore stockpiles	Ore stockpiles	Drill core
		Underground exploration passages
Spent ore	Heap leach	Pilot plant for heap leach

Source: From *Draft Acid Rock Drainage Technical Guide*, Vol. 1 (Vancouver, Canada: BiTech Publishers, 1989). With permission.

For large-scale kinetic tests such as on-site rock piles, large amounts of a specific unit are required. For proposed mines such large volumes of specific units may not be available.

6.6.1 Static Tests

A static test defines the balance between potentially acid-generating minerals (potential acidity) and acid-neutralizing minerals (neutralization potential) in a sample. In particular, acid-generating compounds include reactive sulphide minerals and acid-neutralizing compounds include carbonate minerals. A sample will theoretically generate net acidity at some point only if the potential acidity exceeds the neutralization potential (NP); otherwise the sample will not produce net acidity as long as the NP is not dissolved more quickly than the generation of acidity.

Despite the theoretical simplicity, static tests cannot be used to predict the quality of drainage emanating from waste materials at any future time. Acid-generation processes and therefore drainage quality are time dependent and functions of a large number of complex factors such as mineralogy, rock structure, and climate. For this reason, static tests should be treated as a qualitative predictive method; i.e., they can only indicate whether potential exists for the generation of net acidity at some unknown time.

Several types of static tests are available such as acid-base accounting and APP: sulphur ratio. However, all of these tests are simply variations on a basic procedure and all require variations of the same basic analyses for determining the balance between potential acidity (AP) and NP. Consequently, the basic, common procedure will be presented and the names of the variations will be deemphasized.

The initial step in defining the acid-generating/acid-neutralizing balance in a sample begins with a measurement of total sulphur in a sample, commonly performed with a Leco furnace/analyzer. The measurement of total sulphur allows the calculation of "maximum potential acidity," which may overestimate the potential for acid generation if all sulphur in a sample is not acid generating. Therefore, additional analyses may be performed to refine the potential acidity. The analyses, which have not yet proven to be as reliable as total sulphur, are:

- Sulphur species, which defines short-term leachable sulphate and leachable sulphide using acid extractions
- Reactive sulphur, which defines short-term oxidizable sulphide using hydrogen peroxide

The unproven nature of these additional analyses makes them options in a static test.

Following the delineation of AP, the next parameter, NP, is defined. The measurement of NP provides a gross value for neutralization; however, this value may overestimate the capacity of the sample to neutralize the pH to an environmentally acceptable level above 6. An analysis of carbonate content will provide a more meaningful measure of NP from the perspective of pH neutralization. The carbonate analysis is recommended as an optional portion of static tests.

Paste pH is measured in a paste formed by water and the ground sample. The pH value will indicate the immediate reactivity of neutralizing minerals in the sample and will indicate whether significant acid generation occurred prior to the measurement.

Following these analyses, the potential for net acidity is calculated by subtracting AP from NP with a negative value indicating the potential for net acidity. Alternatively, a ratio of NP to AP can be used (APP: sulphur ratio), but the subtraction method (acid-base accounting) is adopted here.

The subtraction of maximum AP (based on total sulphur) from the gross NP yields the "net neutralization potential" (NETNP). Theoretically, a sample can be expected to generate net acidity at some point if the NETNP is <0. However, based on general experience, values of NETNP in the range of −20 and +20 tons of CaCO3/1,000 tons of sample (−2 to +2% $CaCO_3$) may be considered to have the ability to generate net acidity. This range of uncertainty is attributed to the sources of error in:

- Obtaining the objective of defining true potential acidity and neutralization
- Converting total sulphur to acidity using a restricted conversion factor
- Analysis

The subtraction of AP (based on reactive sulphide) from carbonate content yields the "net neutralization potential from species" (NETNP(S)). This value will presumably reflect the actual NETNP due to the narrower range of uncertainty and, thus, provide more reliable predictions, although no database exists to confirm these conclusions. The primary sources of error are similar to those for the NETNP (above), except that estimating long-term reactive sulphide from a short-term test may result in some uncertainty.

In the event the samples from a geologic unit indicate the unit has or may have the potential for net acid generation, kinetic tests should be conducted.

6.6.2 KINETIC TESTS

Static tests identify the geologic units at a site that may have the potential to generate net acidity. Geochemical kinetic tests involve weathering (under laboratory-controlled or on-site conditions) samples of these units in order to confirm the potential to generate net acidity, determining the rates of acid generation, sulphide oxidation, neutralization, and metal depletion and test control/treatment techniques. This information is critical because, for example, the rate of acid generation may be negligible or, in extremely rare cases, may be severe for only a short period of time so that long-term control or treatment techniques may not be necessary. Based on the results of kinetic tests, the optimization of treatment and control techniques to address the specific severity and duration of acid drainage from a geologic unit will minimize the overall costs of acid-generation abatement.

Whereas static tests provide some information on overall potential acid generation independent of time, kinetic tests explicitly define reaction rates through time under specific conditions. As a result, kinetic tests are significantly more expensive and continue for months or years. Laboratory kinetic tests conducted in the short term only provide semiquantitative information on drainage water quality because they do not reproduce site conditions. In order to provide quantitative data on water quality at the site, waste material test pads can be monitored for several years. Ultimately, true prediction of long-term drainage quality will only be possible through quantitative mathematical models that can reliably extrapolate results beyond the time of the tests.

The initial step in a kinetic test is the definition of material characteristics in addition to those measured in static tests, specifically surface area, mineralogy, and total metals. These characteristics are important to the interpretation of the results from kinetic tests as they can affect the acid-generation process or overall water quality.

The particle size of a material can affect acid production and acid-consumption results. Smaller grain-sized materials have a greater surface area per unit weight and a greater density of broken crystal bonds.

The mineralogy of a sample may also be directly related to reaction rates. Both the chemistry and crystal form of the minerals in a sample control the rates of acid generation and neutralization. For example, poorly crystalline minerals react faster than their crystalline counterparts, and some sulphide minerals oxidize faster than others. Additionally, the mineralogy of a sample may determine the metals that could be leached during acid generation and the extent to which pH may be neutralized by the sample.

Total metal analysis assists in the evaluation of the water quality from the tests. First, total metal analysis indicates any metals present in high levels that may warrant attention. Second, the leaching rates of a metal when compared to the total metal content will suggest when a metal may be depleted within the sample, resulting in negligible leach concentrations, even though it is difficult to extrapolate laboratory test concentrations to field leaching conditions.

Once the material characteristics have been determined, which include the sulphur and carbonate content determined during static tests, the overall programme objectives must be defined before the selection of a kinetic test. The programme objectives should be based on the mine plan and the proposed handling of acid-generating rock. Programme objectives could include one or more of the following:

- Selection or confirmation of disposal options
- Determination of the overall water quality impact
- Determination of the effect of the flushing rates through a sample on water quality
- Determination of the influence of bacteria on the acid-generation sample

Kinetic tests are selected for each acid-generating component based on the information required to meet programme objectives.

Both small-scale controlled tests (e.g., humidity cells) and large-scale on-site weathering trials have been used in assessing acid-generation reactions. The controlled tests have the advantage of simulating specific climatic and weathering conditions. On-site tests may be considered more representative than controlled tests because of the natural conditions under which the tests are conducted; however, because results vary as climatic conditions change, the interpretation and extrapolation of the test results are more complicated.

The data from kinetic tests are evaluated to define the rate and temporal variation of acid generation and water quality of a sample or a treatment/control technique. The results are assessed to determine if they are environmentally acceptable with respect to the proposed mine plan. For example, if the proposal is made to mix waste rock with limestone and the tests indicated that acidic drainage occurred in a kinetic test, then the results would not be environmentally acceptable.

If the results are not environmentally acceptable, then the mine plan and the programme objectives must be redefined. The mine plan must be redefined to ensure that the appropriate acid-generation control and treatment techniques are used. The programme objectives may have to be redefined to incorporate the changes in the mine plan and to test for additional appropriate acid-generation control and treatment techniques.

Additional tests would not be necessary if the existing data (through extrapolation) were sufficient to evaluate environmental acceptability concerning the new mine plan. If the existing data are not sufficient, then additional kinetic tests should be conducted to meet the new objectives.

When the results are environmentally acceptable, experimental results can be extrapolated to other conditions or into the future using mathematical models.

6.7 CONTROL OF ACID GENERATION

The ARD control can be divided into three stages:

- Control of acid-generation process
- Control of acid-generation migration
- Collection and treatment of ARD

Controlling ARD by preventing or inhibiting acid generation is the most preferable stage of control. The objective of acid-generation control is to prevent or reduce the rate of acid formation at the source by inhibiting sulphide oxidation. This can be achieved by excluding one or more of the principal ingredients or by controlling the environment around the sulphides.

The primary components in the acid-generation process are

- Wastes containing reactive sulphide
- Oxygen
- Water

Factors that influence acid generation include:

- Bacterial activity
- Temperature
- pH

Acid generation can be controlled by eliminating or reducing one or more of the essential components or by controlling the environmental factors at the source in order to retard the rate of acid generation. This control can be achieved in one of the following ways.

1. Sulphide removal or isolation: If sulphide minerals in waste rock and tailings are removed, reduced, or isolated by coating or through some other means, then the sulphide-oxidation-producing acid will not occur. High sulphide content in the waste can be concentrated and separated from the bulk of mine waste. The procedures used to concentrate, remove, or isolate the sulphides are termed "conditioning" of tailings and waste rock.

2. Exclusion of water: Total exclusion of water to prevent acid generation may not be practical. Water includes surface water, infiltration due to precipitation, and groundwater seepage. The main source of water depends on the type and location of the waste facility. In underground mines, groundwater seepage is prominent, while for waste rock dumps and tailings deposits, surface water and infiltration are important. Water can be excluded with impermeable barriers such as a synthetic membrane cover, but in the long term, degradation would result from water penetrating the barrier to facilitate acid generation.

3. Exclusion of oxygen: The elimination of oxygen from waste rocks would prevent the oxidation of sulphide minerals or reduce the rate of contaminant production. Although it is possible that acid can be generated under anaerobic conditions, this has not been a significant factor in mining wastes. Significant reduction in oxygen level can be achieved through the placement of a cover with an extremely low oxygen diffusion characteristic. Appropriate cover materials include soil, water, and synthetic materials.

4. pH control: If the pH of the water can be maintained within the alkaline range, acid generation can be inhibited. The pH may be controlled by the addition of alkaline materials to potentially acid-generating wastes. Blending of acid-consuming waste with acid-producing waste to achieve a net acid-consuming mixture can be a successful approach. Adding and mixing imported alkaline material such as ground limestone can be an efficient procedure.

5. Control of bacterial action: When the pH within a reactive waste pile drops below 4, the rate of acid generation increases fivefold or more by the presence of the bacteria *T. ferrooxidans*. The use of bacterial control compounds such as anionic surfactants (sodium lauryl sulphate), organic acids, and food preservatives can control bacterial action.

6.8   AVAILABLE CONTROL MEASURES

The principal objective in the selection of ARD control measures is to achieve the necessary environmental control in the most cost-effective way. The effectiveness of any control measure is determined by several site-specific factors:

- Degree of the acid-producing potential of the mine waste, including the nature, quantity, and reactivity of sulphide minerals present, neutralizing potential of the rock, etc.
- Physical characteristics of the waste
- Climate, topography, and surface and groundwater hydrology
- The expected time period over which the measure will be effective
- The sensitivity of the receiving environment to AMD

6.9   CONDITIONING OF TAILINGS/WASTE ROCK

The generation of ARD may be reduced by placing tailings and rock dumps in a condition that is favourable for ARD prevention. The sulphide content of tailings may be reduced by means of bulk sulphide flotation prior to placement. This process will produce a sulphide concentrate and flotation tailings. The latter will contain residual sulphide, and hence may still be potentially acid generating; however, control of acid generation will be easier to achieve for this, the bulk of the waste. The disposal of sulphide concentrate remains a consideration. An option that has been identified is pressure leaching of the concentrate to produce acid, filtering the acid off, and disposing of the remainder of the concentrate. The cost of flotation and disposal of sulphide concentration will influence the feasibility of this method. Another approach to tailings disposal may be to utilize the "dry" tailings disposal technique combined with an additive such as cement or bentonite, for example. These procedures could be used to produce a compacted soil cement with the intention of reducing oxygen and water access to the sulphide minerals. No records are available of these methods having been used to control acid generation.

The potential for isolation of pyrites by developing a coating of some form has been evaluated; however, these methods are still experimental and do not yet indicate adequate, economical control of ARD.

Placing tailings in a systematic managed manner in order to achieve a uniform deposit with maximum density and minimum segregation results in the minimum permeability to both air and water. Layered tailings placement, with minimized pool areas and maximized discharge densities, is a placement method often adopted. This technique is often referred to as "subaerial" or "semidry" placement. While this technique may have an advantage under certain conditions, abatement of ARD does not necessarily occur. If the tailings remain in a saturated state, reductions in acid generation due to oxygen exclusion and reduced infiltration (due to reduced surface permeability) are noticeable but still comparatively small. However, once the tailings are allowed to dry (which is the case in this method), shrinkage cracks extending from the surface into the tailings deposit may cause a dramatic increase in permeability. Evidence indicates that this secondary permeability permits both oxygen and water entry into the tailings and continued acid generation. While underdrainage is maintained, this may increase the rate of both oxygen entry and ARD. Thus, the direct beneficial effect of layered tailings on ARD abatement is small and, in some instances, may be detrimental. Of greater importance is the improved consolidation characteristics and surface trafficability, which permits easier cover placement.

The relatively poor control of ARD provided by layered or subaerial deposition is demonstrated by the experience with South African gold tailings, in which layered tailings deposition is practised extensively. Oxidation and acid generation have penetrated many metres, in some cases tens of metres, into these tailings.

For acid-generating waste rock, merit may be found in segregating and isolating the high sulphide wastes during mining. This may serve to concentrate the high sulphide wastes in one location. While processes that concentrate high sulphide wastes may have definite potential in terms of waste management aspects, some form of ARD control is still required.

6.10   WASTE SEGREGATION AND BLENDING

Waste segregation involves the careful removal and separate handling of various geologic units at a mine site. Mines with acid-generating geologic units may also have other geologic units with excess acid-consuming capability. As a result, the segregation and separate handling of each unit provides two primary benefits. First, the volume of rock that may generate acidity and require treatment or control is minimized. Second, if acid-consuming units contain carbonate, which readily and reliably reacts to acidic pH conditions, these units can be blended with the acid-generating units in experimentally defined proportions for pH control. This is practised effectively at some coal mines in the eastern United States.

The blending of acid-generating and acid-consuming rock units is similar to the alternative control technique of adding limestone or other neutralizing additives to the acid-generating waste. Consequently, successful blending is primarily dependent on the same factors as limestone addition:

- The movement of water through the system
- The nature of the contact of acidic waste/water with the acid-consuming rock/water
- The proportion of excess acid-consuming rock
- The type and reactivity of the acid-consuming minerals

These factors determine the required procedure for blending.

Because acid-consuming rock units rarely contain $CaCO_3$ and other highly reactive carbonates in high proportions, overall costs/benefits of transporting and adding low-volume, highly neutralizing additives may be less than blending with higher volume, less reactive acid-consuming rock. The difference in the volume of the mixture, associated catchment area, and monitoring requirements are also factors in determining the overall cost/benefit analysis of the alternatives. However, situations occur in which a mine plan may require similar handling, transportation, and disposal for both acid-generating and acid-consuming rock and, if blending is experimentally demonstrated to be successful and reliable, the sole cost of the blending of the rock units may be more economical.

The costs of segregation and blending are site specific and dependent on the mine plan, the handling and transportation of the material, and the technique of blending.

6.11 BACTERICIDES

The rate of sulphide oxidation and acid generation is enhanced in some environments by microbiologic activity, particularly that of *T. ferrooxidans*. This bacterial activity can accelerate the oxidation both of ferrous iron to ferric iron in water and of reduced sulphur in the sulphides to a higher oxidation state. The purpose of bactericides is to create a toxic environment for bacteria so that the inorganic rate of acid generation cannot be enhanced. This does not imply that acid generation will cease.

The most popular bactericides for acid-generating materials include benzoate compounds, sorbate compounds, anionic surfactants such as sodium lauryl sulphate, and phosphate compounds. Laboratory and field experiments indicate that bactericides reduce the rate of acid generation (as indicated by sulphate and acidity) as well as concentrations of certain metals generally by factors of up to 5, with one reported case of a 20-fold reduction. The overall effectiveness of each bactericide compound appears to be similar, generally between 50% and 95% effective in the short term. It should be noted that the period during which bactericides remain effective is limited due to the fact that they degrade and are removed by infiltrating and percolating water. The results quoted in Table 6.5 do not take into account the degradation and depletion of bactericides.

Available studies do not discuss the impact of bactericides on pH, presumably because little effect on pH is seen, or effects are unpredictable. A 10-fold decrease in acidity theoretically changes the pH by only 1 pH unit (e.g., from pH 2 to pH 3), although aqueous buffering by sulphate and metals would further limit the pH change. The one detailed set of pH data in the referenced studies indicated the successful inhibition of *T. ferrooxidans* resulted in a maximum pH increase of only 0.3 units from pH 2.9 to pH 3.2.

Based on the laboratory and field data, bactericides may reduce the rate of acid generation but will not eliminate AMD. Consequently, bactericides must be used in conjunction with other control techniques for proper environmental protection. Additionally, some concern has been expressed over environmental toxicity of the bactericides that must be applied in strong concentrations to eliminate bacteria.

Bactericides are applied to surfaces of piles and fields using sprayers or hydroseeders. The cost is often several thousands of dollars for each hectare of surface. Because bactericides degrade and are removed by infiltrating water, occasional reapplication is necessary. The timed release of bactericide from rubber pellets is reported to extend the lifetime of one application.

TABLE 6.5
Effectiveness of Bactericidal Methods

Method	Results
Spray and controlled-release pellets	80% reduction in acidity, sulphates, irons, manganese, and aluminium
Sodium lauryl sulphate potassium benzoate,	Complete inhibition of *T. ferrooxidans*; 92% to 84%
potassium sorbate on silver mine waste rock	lower acidity
As above on coal mine waste	Short-term reduction in acidity
Sodium lauryl sulphate controlled release from rubber pellets	50% to 95% reduction in acidity
Sodium lauryl sulphate	60% to 95% reduction in acid production
BF Goodrich ProMac System Co.	58% to 72% reduction in acidity, 58–68% reduction in sulphate

Source: From *Draft Acid Rock Drainage Technical Guide*, Vol. 1 (Vancouver, Canada: BiTech Publishers, 1989). With permission.

6.12   BASE ADDITIVES

In mining environments with sulphide-rich rock, the potential for acid drainage is based on the relative proportions of acid-producing and acid-consuming materials. Acid-consuming minerals are also known as "alkaline," "basic," or "neutralizing" material. If the potential for acid drainage exists through excess acid-producing material, one potential control technique is the addition of excess neutralizing material, particularly carbonate and hydroxide compounds, which produce a neutral to alkaline pH in the associated water.

The common additives are limestone ($CaCO_3$), lime (CaO or $Ca(OH)_2$), and sodium hydroxide ($NaOH$). These additives are usually used in solid rather than dissolved form because the liquid represents a less concentrated source through solubility constraints. For example, 1 m^3 of $CaCO_3$-saturated water provides approximately 1 kg of $CaCO_3$, whereas 1 m^3 of high-purity limestone provides around 3,000 kg of $CaCO_3$.

The success of base additives to control acid drainage depends primarily on

- The movement of water through the system
- The nature of contact of acidic rock or water with neutralizing additives or water
- The proportion of excess neutral material
- The type and purity of neutralizing additive

The movement of water can affect the success of this technique in several ways. For example, the movement of water can affect the rate of acid generation, particularly in a saturated system in which the sole source of oxygen is dissolved in and carried by the water. If background groundwater with little oxygen moves upward into a saturated acid-generating rock pile, the rate of acid generation can be expected to decrease as soon as the oxygen added by disturbance and transport is depleted. Such a situation would lessen the severity of the acid drainage and minimize the necessary quantity of additive for pH control.

An occasionally overlooked complication related to water movement is the consumption of neutralizing additives by pH-neutral water. For example, rainfall passing through a surficial layer of additive and upwelling groundwater passing through a basal layer of additive can usually dissolve some of the additive in excess of the quantity needed to control acid drainage. The rate of water flow will determine the amount of required excess additive required to avoid early depletion of the additive.

The rate and direction of water movement provide the primary connection between acid generation and acid neutralization because these processes cease in the absence of water. This is closely related to the second factor determining successful control, the nature of contact of acidic rock or water with neutralizing additives or water. The nature of the contact can be divided into three basic scenarios:

- The additive lies above or "upstream" of the sources of acid generation
- The additive is mixed with the acid source
- The additive lies below or "downstream" of the acid source

The first scenario of acid/neutralizing contact involves the dissolution of additive into ambient water, followed by the movement of this water into acid-generating material, such as with rainfall moving downward through a surficial layer of additive into an acid-generating rock pile. Unless the rate of acid generation is low, this scenario is not effective because sufficient acidity can often be released into the water to overcome the alkalinity of the water contributed by the additive.

The second scenario of acid/neutralizing contact involves the mixing of additive with the acid-generating material. This provides a continual dissolution of additive in response to the acid generation, preventing the development of acid drainage outside the microenvironment around sulphide grains. In fact, laboratory experiments have demonstrated that the rate of acid generation can be slowed significantly, leading to decreased additive consumption, if alkaline water invades the sulphide microenvironment. As the thoroughness of mixing of additives and acid-generating material decreases, the potential success decreases as the first scenario comes into play and "hot spots" of unhindered acid generation arise. This accounts for the recognized poor success of layers of additive within acid-generating material.

The third scenario of acid-neutralizing contact has the additive downstream of the acid source, such as in a basal layer or a collector trench. In this case, the optimum use of additive requires the flow of acid water through the additive rather than over the top of the layer, which may not occur in a collector trench lined with a base additive. The elimination of hydraulic short circuits, which would allow water to flow around the additive, is critical to this scenario. Unlike the second scenario, no opportunity exists for in situ control of the rate of acid generation.

The third factor determining the potential success of neutralizing additives is the amount of excess additive. The deficit of natural neutralization potential as defined by static and kinetic tests indicates the minimum required quantity of additive as $CaCO_3$ equivalent. This quantity must then be increased to account for the dissolution by ambient precipitation and groundwater, and for the encapsulation of additive by precipitates. As acidic water comes into contact with additive and is neutralized, metal compounds such as hydroxides precipitate from the water and may encapsulate the additive, slowing or preventing further neutralization. This is a recognized problem with base additives that has not yet been solved, requiring a significantly higher quantity of additive than would otherwise be required. The alternative of forcing alkaline water into the microenvironments around sulphide grains, thereby encapsulating the sulphide minerals, has not been addressed experimentally in detail, but may warrant attention.

The fourth factor determining the potential success of neutralizing additives is the type and purity of additive. The common additives of limestone, lime, and sodium hydroxide differ in their solubilities and, thus, differ in the pH they create upon dissolution. Limestone, a common natural mineral, often raises aqueous pH to around 7.0–8.5. However, dissolution of limestone into water is restricted by high calcium concentrations, such as are found in gypsum-saturated waters (like most acid drainage), through the "common-ion effect." The common-ion effect may limit pH neutralization to pH 5–7. Lime and sodium hydroxide, which are not found in surficial natural environments because of their highly alkaline character, create aqueous pH values approaching 10 and above upon dissolution. Like limestone, their dissolution can also be restricted by the common-ion effect. Still, lime and sodium hydroxide provide greater neutralization than limestone per unit weight, but the greater unit cost and the environmentally unacceptable alkaline pH detract from the benefit.

The purity of the additive is important for successful control in that the lower the purity, the greater the quantity needed for equivalent neutralization. Lime and sodium hydroxide are manufactured and can be obtained in essentially pure form. On the other hand, limestone is quarried and its purity is often <100% as $CaCO_3$. The impurities are other carbonates and other minerals that may not contribute to neutralization. For any choice of additive, a neutralization-potential test identical to the tests carried out on acid-generating material should be carried out so that consistency is maintained.

After the additive is adjusted to a proper grain size to maximize reactivity and geotechnical stability, standard earth-moving equipment is required for the application of base additives as layers at locations upstream or downstream of the acid-generating material. For mixing of additive and acid-generating material rather than layering, the additive must be brought to the proper grain size to maximize reactivity and ease of application, as well as to minimize the potential for additive migration after application. Viable application procedures include slurry spraying, mechanical mixing during disposal, and slurry injection into boreholes following disposal.

The purchase, transport, and application of base additives may cost tens to hundreds of dollars for each cubic metre. These costs will be relatively low if a mine site is located near a limestone deposit that can be easily quarried and transported.

6.13   COVERS AND SEALS TO CONTROL ACID GENERATION

Covers and seals offer the ability to restrict the access of oxygen and water to reactive wastes. The restriction of water can serve to limit both the formation of acid and the subsequent transportation of the oxidation products into the environment. The exclusion of oxygen is more practical than the exclusion of water for the purpose of acid-generation control.

To limit oxygen and water entry, the cover must itself have a low permeability to either air or water, and it must not have holes or imperfections where entry can occur. If holes or cracks occur in, for example, a cover on a waste rock dump, then oxygen entry takes place as a result of a convective flow of air into and out of the dump in response to natural barometric pressure changes and thermal currents through the dump. Cracking of a dump surface, as a result of the large settlements to which dumps are prone over the long term, may result in an inflow of surface runoff. The resistance of the cover to cracking, the burrowing effects of roots and animals, and erosion and degradation due to weathering and frost action determine the long-term effectiveness of the cover.

A variety of materials may be used to provide surface covers depending on local availability and site conditions. These include different types of soils, synthetic membranes, water, and a combination of soil and water, which result in saturated soil or bog conditions, and various other materials such as concrete, asphalt, etc. Alternative cover materials (other than water) and their permeability to water are shown in Table 6.6.

The most effective means of excluding oxygen is by means of a water cover. The other cover materials are generally more effective as inhibitors of infiltration in the control of ARD migration.

6.14   SOIL COVERS

Soil covers show promise as oxygen inhibitors as can be seen in the published data from the Rum Jungle site in Australia. The effectiveness of soil covers as oxygen barriers is influenced by the moisture content maintained in the cover. A cover that can be maintained in a saturated condition will be more effective, primarily due to the low diffusivity of oxygen in water, and to the absence of desiccation cracking. In this regard, composite covers with layers of different soil types to prevent desiccation of clay or till have been suggested as potentially beneficial. The reduction in oxygen entry due to soil covers is greatest for very coarse waste rock dumps where oxygen entry by convection is large. Convective transport of air into coarse waste rock is driven by changes in both temperature and barometric pressure and can be large through small holes or cracks in a cover. The

TABLE 6.6

Alternative Cover Materials

Cover Material	Permeability to Water (m/sec)	Advantages/Disadvantages
Compacted clay	10^{-9}–10^{-11}	Availability of large quantities problematic in many areas; subject to erosion, cracking, and root penetration; good sealing if protected and maintained
Compacted till	10^{-7}–10^{-9}	As above, but generally more permeable
Compacted topsoil	10^{-5}–10^{-8}	As above, but less robust, more permeable; questionable longevity
Peatland bog	10^{-5}–10^{-6}	Need to maintain in saturated condition; normally impractical for elevated waste dumps and sideslopes
Concrete	10^{-10}–10^{-12}	Subject to cracking, frost, and mechanical damage
Asphalt	10^{-20}	As above
HDPE synthetic	Impermeable	Requires proper bedding and protective cover; highly impermeable; life span unlikely to exceed 100 years; subject to root and mechanical penetration

Source: From *Draft Acid Rock Drainage Technical Guide*, Vol. 1 (Vancouver, Canada: BiTech Publishers, 1989). With permission.

long-term permanence of soil covers in resisting disruptive forces, such as erosion, cracking, frost action, root action, and burrowing animals has yet to be proven in the field.

While soil covers inhibit the access of oxygen to the waste and may control acid generation, they are generally more effective in controlling infiltration and the migration of ARD.

6.15 SYNTHETIC MEMBRANE COVERS

Synthetic membranes, such as polyvinyl chloride (PVC), high-density polyethelene, etc., have the potential to provide covers with an extremely low conductivity to air and good oxygen exclusion. These covers may be installed to provide good medium-term control; however, long-term degradation, loss of plasticity, and ultimate cracking limit the long-term effectiveness of such membranes.

Because of their vulnerability to puncture, membrane liners must be installed with adequate bedding preparation and protective surface covers. They are of low permeability and offer the potential of acting both as oxygen and infiltration barriers. Thick (2 mm) high-density polyethelene (HDPE) membranes are less susceptible to the disruptive forces affecting soil liners, except for the likelihood of tearing under the differential settlement and long-term weathering. To allow for long-term degradation it is probably necessary to provide for liner replacement in 50–100 years.

Although the present value of the replacement cost of a synthetic cover on a 100-year basis may be quite small, the monitoring of the condition of the cover, administration of the fund, and actual implementation of the replacement are impediments to this approach, despite the fact that it can provide a relatively positive seal to moisture and oxygen penetration.

Other synthetic materials include geopolymers, asphalts, and cements such as high-volume polypropylene fibre-reinforced, sulphate-resistant shotcrete. The cost of such synthetic covers is often prohibitively high, and these often suffer from cracking and disruption over a long period of years.

As in the case of soil covers, synthetic membrane covers are more effective in the control of ARD migration.

6.16 WATER COVER

Water cover is currently the most promising oxygen-inhibiting technique, and hence the most promising acid-generation control measure. The solubility of oxygen in water and the diffusion rate of oxygen through water are both very low. Thus, in the absence of convective transport, the rate of oxygen transport through water is sufficiently low to be of no concern in terms of acid generation. Evidence is steadily accumulating that underwater disposal of potentially reactive wastes reduces acid generation to negligible levels. Although oxidation of sulphide and resultant acid generation may not be halted entirely by placing wastes underwater, the rate of acid generation is generally reduced sufficiently to make the impact negligible. Care must be taken when considering placing old wastes that have previously generated acid below water because the solution of acid products contained in the waste may occur. The availability of water and the cost of maintaining this cover over the long term are obvious and important site-specific criteria that would influence its use as a water cover. Water cover may be achieved by the disposal of waste into natural waters, into manmade impoundments, or into flooded underground mine workings and open pits. However, it may not be feasible to achieve water cover on existing deposits and some types of waste facilities.

6.17   SATURATED SOIL OR BOG

The effectiveness of a saturated soil layer for the exclusion of oxygen has been demonstrated and may be suitable for some categories of waste. Bog conditions can be achieved by the combination of a shallow soil cover with a shallow water cover provided by a water-retaining structure. Under these circumstances, the waste will be effectively underwater. The soil helps to prevent total loss of coverage when the water depth reduces during dry periods, and it also prevents convective currents and wave action. Vegetative accumulation is also believed to have a marginal but beneficial effect on ARD abatement.

6.18   SUBAQUEOUS DEPOSITION

It has already been identified that the disposal of acid-generating waste underwater is currently the most promising abatement measure. Field evidence is growing that the disposal of reactive mine wastes underwater curtails oxidation to negligible levels. This is due to the very low diffusivity of oxygen through water (approximately 2×10^{-6} cm^2/sec). This concept can include disposal of waste into natural waters, flooded mine workings, or manmade reservoirs. A number of factors and possible limitations associated with underwater disposal are to be considered, however.

Current legislation and the politically controversial nature of lake and marine disposal of mine tailings (acid generating or otherwise) suggest that gaining approval for an application for this method of disposal for tailings will be difficult to achieve. For this reason it is probably beneficial to fully investigate any "on-land" means to achieve water cover. Lake and marine disposal of tailings should generally only be considered when all on-land options have been exhausted. This may not necessarily be the case for other waste types, for example, waste rock.

6.19   DISPOSAL INTO MANMADE IMPOUNDMENTS

Because available evidence indicates that water cover provides the most secure method of acid-generation control, consideration should be given to the construction of a water retention facility. The practicality and cost of a manmade reservoir relative to alternative measures are clearly dependent on site-specific criteria, for example, topography and volume of waste to be stored. The cost of flooding existing waste facilities is likely to be very high.

The design of a facility to provide water cover for combined tailings and waste rock may prove beneficial and cost-effective for proposed developments. Combined tailings and waste rock disposal may have definite advantages in terms of acid-generation control, particularly if the tailings are not acid generating and are discharged at elevated pH (>7). If intimate mixing of tailings and waste rock can be achieved, the permeability of the coarse waste rock would be significantly less than if the rock were placed alone. This has the advantage of reducing potential water movement through the waste rock.

There are, however, limitations and design considerations that place manmade water cover facilities at a disadvantage, including:

- Water retention dams that require the detailed design of embankment and spillway facilities, careful construction control, and maintenance over the long term. Depending on the site-specific conditions, this may not prove economical.
- Reliable water sources must be available to provide a continuous water cover of sufficient depth to avoid exposure of the waste and erosion due to wave action or water flow.
- Minimum water cover needs to be maintained in low precipitation and drought periods.
- Water reservoirs may induce unacceptable seepage. If other soluble deleterious products are found in the wastes, these may result in increased contaminant loading of the environment. Whether these are significant for the specific project and site conditions needs to be determined.

6.20   DISPOSAL INTO FLOODED MINE WORKINGS

Flooding underground mine workings and open pits is a means of controlling acid generation from their exposed rock faces. This method also provides a potential disposal area for acid-generating waste.

Flooding of worked-out coal mines has been successful in the control of AMD in several instances, with acidity reductions of 45% to 99% being reported. The potential benefits from flooding underground mine workings have been reported for several anthracite coal mines in eastern Pennsylvania. Field investigations at the mines, which were allowed to flood some 14–20 years ago, revealed that the mine waters, which were formerly highly acidic, are now slightly alkaline. Sulphate reductions of approximately 54% and 74% in mine waters were seen in comparison to 1960s data. In addition, marked decreases in the iron, aluminium, manganese, calcium, and magnesium levels were observed.

The disadvantages associated with the storage of waste in flooded mine workings are:

- At single pit operations it is necessary to store reactive mine waste rock for the life of the mine and to then incur rehandling costs in moving the material back to the pit when the operation is producing no revenue.
- If all waste removed from the pit is reactive, the swell factor (usually about 30%) will generally result in an excess volume of reactive waste to available underwater storage, particularly as the pit will likely flood to a less than full point.
- Any sulphides in the pit walls above the final water elevation will oxidize, causing a deterioration in water quality unless preventative measures can be applied.
- Backfilling and flooding preclude future underground development that might be associated with the ore body.

6.21   LAKE DISPOSAL

Lake disposal is a subaqueous method of the disposing of mine waste into an existing natural system. Lakes have been used in the past for the disposal of both acid-generating and non-acid-generating tailings. Available evidence indicates that the deposition of acid-generating rock underwater, such as into a lake, will effectively control the rate of acid generation. However, lake disposal of tailings also presents other problems related to turbidity and metal mobilization, which may affect

TABLE 6.7

Important Characteristics in the Assessment of Lake and Marine Disposal Sites

Site characteristics	Proximity of site to mine
	Route to lake from mine for tailings transport
	Regional climatic conditions
	Water/recreation use
Physical characteristics	Bathymetry
	Thermal stratification
	Hydrology — turnover and flushing events
	Hydrogeology — recharge/discharge characteristics
Chemical characteristics	General water quality including pH, buffering capacity, metal concentrations, and alkalinity (seasonal variations); suspended solids loading
Biological characteristics	Identification of resident fish
	Identification of salmonids
	Identification of benthos communities
	Identification of salmon, crab, and shrimp fisheries (marine)
	Productivity
	Unique systems associations

Source: From *Draft Acid Rock Drainage Technical Guide*, Vol. 1 (Vancouver, Canada: BiTech Publishers, 1989). With permission.

the biological communities in the lake. Consequently, the environmental concerns related to lake disposal include:

- Toxicity of reagents and heavy metals from the mill process
- Excessive nutrient additions from the use of explosives
- Increased turbidity due to suspended solids causing a reduction in light penetration
- Direct physical impact from the placement of waste on the habitat

Therefore, in selecting a lake for the disposal of tailings or waste rock, the relationships between the various site, physical, chemical, and biological characteristics should be investigated. As an example, the overall climatic conditions and depth (from bathymetry) will determine the development of density and temperature stratification, which in turn may lead to an annual lake turnover. Mixing and stratification control the chemical characteristics (baseline conditions) and therefore, the ability of the lake to assimilate the loading of acidity from the waste material. Finally, all three nonbiological characteristics combine to influence the lake ecosystem and its productivity and the tolerance of the ecosystem to mine materials disposal. Table 6.7 outlines some of the characteristics that should be determined to define the suitability of a lake location.

The characteristics of the tailings or waste rock that should be assessed are acid-generation potential, leaching characteristics, and settling properties of tailings. These characteristics will determine the physical and chemical impact of the material on the lake environment. For example, the leaching characteristics will indicate which metals, if any, could dissolve in lake water.

6.22   MARINE DISPOSAL

Marine disposal is another subaqueous method for disposing of mine waste into an existing natural system, thereby achieving water cover. This option is available for mines situated in proximity to a marine water body and is popular in countries such as Norway. As with lake disposal, very limited research has been carried out at the sites related to acid-generation control, and most of the impact observed is due to suspended solids and metal leachings from tailings solids.[5]

Many of the site, physical, chemical, and biological characteristics of importance in lake disposal must also be evaluated at marine disposal sites; in addition, several special factors should be considered. In particular, seawater has a greater buffering capacity and higher alkalinity than freshwater, which will affect the chemical interaction between water and tailings. Further, strong tides and currents will disturb the tailings, carrying them to other sites and affecting fisheries resources over a wider area.

Some of the waste material's characteristics that should be assessed are acid-generation potential, settling properties of tailings, and metal leaching characteristics. In particular, metal leaching in a saline environment should be considered.

Because many nearshore coastal environments frequently have fisheries value, reliable evidence for minimal environmental impact of marine disposal would be difficult to obtain. However, one may find local inlets and other coastal areas in which fisheries value is minimal and waste disposal may be environmentally acceptable. Otherwise, this option may only be acceptable if long pipelines or barges carry the waste far offshore, but the associated costs would be relatively high.

6.23 MIGRATION CONTROL OF ARD

Where acid generation is not prevented, the next level of control is to prevent or reduce the migration of ARD to the environment. Because water is the transport medium for contaminants, the control technology relies on the prevention of water entry to the ARD source. Control of water exit is of little value because in the long term all water entering the ARD source must exit, long-term storage being negligible. Water entry may be controlled by:

- Diversion of all surface-water flowing towards the ARD source
- Prevention of groundwater flow into the ARD source
- Prevention of infiltration of precipitation into the ARD source
- Controlled placement of acid-generating waste

Diversion facilities usually consist of ditches. Diversion of surface flows, while easily implemented, is often difficult to maintain over many years. The best long-term solution to such surface flows is to select a disposal site that minimizes the need for diversion. Site selection is generally not an option in the case of open pits and underground operations, while factors other than ARD control may take precedence in site selection for other facilities.

If the ARD source is located over a groundwater discharge area, interception and isolation of the groundwater are very difficult to achieve and maintain over a long time. While measures such as underdrains and sealing layers can be employed, their long-term performance is questionable. The most effective solution is to select a site that is not located in a groundwater discharge area. Site selection may not be a practical solution.

The long-term secure prevention of infiltration is the most difficult to achieve. Covers of different types may be considered.

Controlled placement of acid-generating wastes includes cellular construction of dumps and tailings deposition methods that increase density and reduce permeability.

6.24 DIVERSION OF SURFACE WATER

Surface water can be prevented from entering acid-generating facilities by: (1) construction of drainage ditches and berms, and/or (2) site selection to avoid high runoff. The construction of ditches and berms is generally considered to be a short-term control because of high maintenance costs for long-term usage. However, maintenance requirements can be reduced for long-term structures by designing for extreme flow conditions with consideration for debris accumulation and providing appropriate erosion protection. While maintenance requirements can be substantially reduced, some form of periodic inspection and maintenance will nevertheless be required. Design flows for diversions can be reduced for facilities such as stockpiles, waste and spoil sites, and tailings deposits by locating these facilities near catchment watersheds. Site selection to reduce runoff, however, may have unfavourable implications to construction and operating costs.

6.25 UNDERGROUND MINES

Surface water inflow is generally not the main source of water in underground mines. Surface water may flow into underground mines through portals, ventilation shafts, or possibly through cracks in the rock that develop by mining-induced settlement. At the Balaklala, Mammoth, and Walker mines in California, subsidence has created caved areas that channel surface water into underground workings. Diversion can be achieved by ditches and berms in the short term and concrete plugs in the long term.

6.26 OPEN PITS

Diversion of surface water around open pits and strip mines can be achieved with ditches and berms. Creeks and streams may have to be rerouted around these operations. These structures require periodic inspection and maintenance in the long term. The period between inspections depends on the design criteria and level of design adopted. Long-term facilities should be designed for severe flows, and possibly with allowance for flow blockage. Erosion protection, such as drop structures or riprap, should also be provided. Design flows for short-term structures are generally less severe due to the shorter return period.

Diversion of surface water may also be conducted within the pH limits in which one portion of the pit is not an ARD source. This can be achieved with in-pit ditches or berms and by sloping of the pit floor. This is a short-term control and may be expensive to operate because of practical problems and restrictions on mining activities.

6.27 WASTE ROCK DUMPS AND SPOIL PILES

Diversion of surface water around waste rock dumps and spoil piles is achieved with ditches or berms and rerouting of creeks. Requirements for diversion can be reduced by site selection. However, other factors such as haul distance to the dump location influence site selection. Favourable sites are at the crest of slopes, or on small plateaus and near the upstream end of a watershed; however, unfavourable cost implications may be found with these locations. Long-term facilities will require more stringent design than short-term facilities and will require some form of maintenance.

Diversion of surface water around acid-generating dumps cannot be avoided by the use of underflow through base drains. Even if the drain is composed of non-acid-generating waste, the water flowing through the drain will become contaminated by water that has infiltrated through the overlying waste. Separation of these waters with barriers in the dump is not considered practical. Differential settlement within the dump would damage or destroy such a barrier.

6.28 TAILINGS DEPOSITS

The approach and methods for the diversion of surface water around tailings deposits are the same as those described for waste rock dumps. However, natural topography is important in site selection for tailings dams. The most suitable location for a tailings dam, in terms of minimizing construction materials, is most often one that requires a substantial diversion of surface water. Construction cost and the cost of long-term maintenance of diversion facilities need to be optimized in site selection. Other forms of ARD control also need to be considered. For example, the control of acid generation by means of water cover eliminates the need for surface water diversion. In the case of acid heap-leach operations, the inherent low pH of the material and pore water is an important consideration in the design of control measures.

6.29   STOCKPILES AND SPENT HEAP-LEACH PILES

The methods for the diversion of surface water around stockpiles and spent heap-leach piles are the same as those described for waste rock dumps. Proximity to the mill site will usually be a principal factor in site selection for stockpiles. The presence of chemicals from the leachate used for mineral extraction in heap-leach operations may require a more stringent design than is required for stockpiles.

6.30   GROUNDWATER INTERCEPTION

When groundwater enters or comes into contact with acid-generating waste, this provides a transport medium for the contaminants. The entry of groundwater into waste facilities may be prevented by interception or isolation of groundwater before it enters the waste, or site selection to avoid groundwater discharge into the waste. The objective is to minimize groundwater contact with acid-generating material or other water that has become contaminated by acid generation.

All collection and interception methods are prone to failure over the long term. Therefore, selection of a site that avoids groundwater discharge is the best method of control. Site selection is not an option for underground mines and open pits, but should be considered for waste dumps, spoil piles, tailings deposits, stockpiles, and spent ore piles. Any potential storage or waste disposal site will be located in either a groundwater recharge area, a groundwater discharge area, or an area with sufficient seasonal variation in groundwater level to be both a recharge and discharge area, depending on the time of the year. The latter case may be the most severe in terms of environmental impact. The performance and cost of different groundwater interception and isolation methods vary over a wide range, depending on hydrogeological and other site-specific parameters.

6.31   UNDERGROUND MINES

Underground interception or control can be achieved by various methods for underground mines, depending on the site geohydrology. Regional dewatering around underground openings can be achieved with wells from surface or perimeter drainage galleries. Localized control of groundwater in conductive aquifers encountered by mining can be achieved by grouting the aquifer or dewatering with drainholes. In any method in which the water is collected within the mine, effort will be required to prevent it from contacting acid-generating material. Exploration drill holes can be a large source of groundwater flows into underground workings. Inflow through boreholes can be controlled by properly grouting and sealing the holes after drilling.

6.32   OPEN PITS

Dewatering around open pits is a common practice for improving slope stability. Groundwater interception and removal can be achieved with perimeter wells, drainage adits, and horizontal drainholes. However, an important consideration is that ARD could develop in wells and adits if these pass through acid-generating rock and oxygen is present. A means of preventing this acid generation is to keep the adit, well, or drain flooded, hence, minimizing oxygen access.

Horizontal drainholes are not well suited to groundwater interception for the prevention of ARD if these drainholes flow into the pit where contact with acid-generating rock is likely to occur. A pipe collection system can be installed to prevent ARD contamination. However, a low level of performance should be anticipated because of damage caused by mining activities, and bench and pit slope failures. Groundwater isolation by grouting may be effective for areas of high inflow.

All of these methods are short-term controls, although drainage adits that drain away from the pit could provide long-term isolation.

6.33   WASTE ROCK DUMPS AND SPOIL PILES

Groundwater interception or isolation under waste rock dumps and spoil piles will generally be difficult to implement and maintain. Site selection that avoids groundwater discharge is the best method of preventing groundwater from flowing into dumps. If a waste facility is located in a groundwater recharge area, and acid generation and migration are not controlled, there is a risk of groundwater contamination occurring. In some cases it may be necessary to situate the dump on a groundwater discharge area and provide measures for acid-generation control or facilities for collection and treatment.

6.34   TAILINGS DEPOSITS

Groundwater interception or isolation at tailings impoundments is generally not practical. Avoiding groundwater discharge by site selection is difficult because the best disposal sites are usually valleys, which are often also groundwater discharge areas.

6.35   STOCKPILES AND SPENT HEAP-LEACH PILES

Groundwater isolation under stockpiles and heap-leach piles can be achieved with engineered barriers, though these are not likely to provide long-term control. Isolation can be achieved by placing the stockpile on an impermeable layer overlying a drainage layer to remove groundwater inflow. Integrity of the impermeable layer is critical for successful isolation. If damaged, the impermeable layer is very difficult to repair without removing the stockpile.

6.36   COVERS AND SEALS TO CONTROL INFILTRATION

The transport medium for contaminants is water, and the principal source of this water is the infiltration of precipitation. The control of infiltration is therefore important in controlling ARD migration. The most practical way of controlling the infiltration of precipitation is by means of low-permeability covers or seals. Soil and synthetic materials are commonly used to construct covers. These can be applied to rockfaces in open pits or underground mines, waste rock dumps, spoil piles, tailings deposits, stockpiles, and heap-leach piles. The length of time during which control is required is an important consideration in selecting the most appropriate cover material or combination of materials.

6.37   SOIL COVERS

Soil covers show promise as oxygen inhibitors, but they are generally more effective in controlling infiltration of precipitation. The effectiveness of soil covers as inhibitors of infiltration depends on factors such as climate, cover design, and construction.

   For minimizing cost, a simple, single-layer soil cover is preferred. A fine-textured soil, such as clay or silt, is required to limit infiltration. To effectively limit oxygen transport, it is necessary to maintain the layer at a high moisture content. A single soil layer, however, is limited in its effectiveness for the following reasons:

- Without capillary barriers, a simple soil cover is prone to large seasonal variations in moisture content. This could result in desiccation cracking and an increase in permeability. In addition, decreasing the moisture content of the soil increases the rate of oxygen diffusion. These seasonal variations are greatest near the surface, and their effect is, therefore, greatest on thin covers. For single-layer soil covers to be effective, they need to be relatively thick to maintain a saturated zone during the dry season. The cover thickness required is probably a function of the climate.

- The fine-grained soils required to limit infiltration may be susceptible to frost. Ice segregation may result in degradation of the cover and increased permeability. Frost heave may also cause the surface of the cover to become irregular, allowing ponding and increased infiltration.
- A simple soil cover does not have the ability to prevent moisture from being sucked up from underlying tailings by capillary action. Likewise, it does not limit the migration of salts from the tailings to the surface due to surface evaporation and transpiration.
- A simple, single-layer fine-grained soil cover may not be able to adequately withstand wind and water erosion or burrowing and root action. Some form of erosion protection, such as vegetation or riprap, is normally required.

These limitations on the effectiveness of a single soil layer can be overcome by using complex covers.

The effectiveness of a soil cover is greatly improved by adopting a complex cover design consisting of several layers, each performing specific functions to improve water and oxygen exclusion and long-term stability. These layers and their specific functions are described below.

Erosion protection can be provided by vegetation or by a layer of coarse gravel or riprap. The establishment of vegetation on the waste dumps is desirable for aesthetic and land use reasons. Therefore, revegetation is usually the most desirable method of providing erosion control. However, where revegetation is not practical or will not sufficiently control erosion, coarse gravel or riprap may be required.

Studies for uranium tailings deposits in Canada indicate that forest cover would adequately control sheet and rill erosion and wind erosion, but no methods of analysis are available to assess the effectiveness of vegetation on gully erosion.

A special study on vegetative covers was recently carried out as part of the Uranium Mill Tailings Remedial Action Project (UMTRAP) in the United States (U.S. Department of Energy, 1988). This study investigated the use of vegetation to stabilize uranium tailings and specifically includes the use of vegetation to intercept infiltration. The principal finding of the study is that properly developed plant communities on complex soil covers can be effective in stabilizing covers and controlling infiltration on topslopes of waste piles. The study showed that the appropriate vegetative cover will adapt to climatic change, will repair itself after severe disturbances such as fires and droughts, and will persist indefinitely with little or no maintenance. The plants were found to protect topslopes against sheetwash erosion; however, resistance to gully erosion depends more on the overall pile configuration than on the vegetation and soil.

Certain physical, chemical, and vegetative stabilization methods have been evaluated for purposes of mine reclamation by the U.S. Bureau of Mines. This study incorporated field testing of these different methods and costs for the various stabilization procedures.

The purpose of the moisture retention zone is to provide a zone for moisture retention to limit the desiccation of underlying layers. It also provides a growth medium to support vegetation. Moisture retention is therefore desirable for two reasons:

- It helps to keep the infiltration/oxygen barrier moist. This helps prevent desiccation cracking and reduces oxygen diffusion.
- By retaining moisture after a precipitation event, it supports vegetation and allows time for evapotranspiration to occur, thus reducing infiltration.

The soil used to construct the moisture retention zone would generally be a loam soil with a substantial sand fraction.

The upper drainage/suction break layer serves two primary purposes:

- To drain water laterally from the surface of the infiltration barrier, preventing ponding
- To prevent moisture loss from the infiltration barrier due to upward capillary suction

Prevention of ponding reduces infiltration. Keeping the infiltration barrier moist helps to reduce oxygen diffusion and prevents desiccation cracking. This layer can also be designed to prevent intrusion by burrowing animals if it incorporates large gravel. For drainage to be effective it must be constructed with a cross fall of 1% or greater.

The effectiveness of this layer would be expected to decrease with time as it becomes clogged with roots and organic debris and fines and as the drainage slope is modified by the long-term settlement of the underlying tailings or rock waste.

The infiltration barrier is a low-permeability layer consisting of fine-grained soil or synthetic materials (or a combination of both). Its purpose is to prevent the downward infiltration of moisture and the diffusion of oxygen into the waste. The lower the permeability of this material, the more effective it is as a barrier to infiltration. The objective of this layer is to provide a sufficient barrier to enable the overlying coarse-grained layer to drain infiltration and prevent ponding.

Capillary barriers are used beneath the infiltration barrier to reduce infiltration. The principle is that if negative pore water pressure is maintained in the low-permeability material at the interface with the underlying coarse-grained capillary barrier, infiltration into the lower layer would be prevented. It was found that this would only be effective if ponding on the low-permeability layer does not occur which, in practice, would be difficult to achieve. However, for soil covers over fine-grained waste deposits such as tailings, a capillary barrier beneath the infiltration barrier may be useful in preventing suction of contaminated pore water from tailings up into the cover during dry periods.

The long-term performance of a complex soil cover could be greatly reduced if fine-grained materials are allowed to migrate into the coarse-grained layers. Filter layers should be considered.

A basic layer could be incorporated into the design to reduce the pH of infiltrating water and therefore acid-generation rates. Alkaline materials such as limestone could be spread over the surface of the waste before placing the cover or mixed into the cover layers.

Limestone is commonly mixed with a waste rock during placement at coal mines with great success, and research is being done on the addition of phosphate rock. The potential for acid-generation control by surface applications of alkaline materials is less attractive than mixing them with the waste. Limestone has low solubility in near-neutral water, and the resulting alkaline charge is therefore small and may be insufficient to control ARD. Surface inflows tend to be concentrated at isolated locations, such as depressions, cracks, permeable zones, etc. At these locations, the available alkaline materials are quickly exhausted. The addition of a basic layer would not significantly reduce AMD where unsaturated conditions predominate, such as in waste piles. It would be more beneficial in saturated tailings and might be usefully employed in tailings impoundment covers.

Information on the relative effectiveness of soil covers in controlling AMD may be obtained from the results of mathematical model simulations of covers and from the results of monitoring a limited number of actual covers. The results of the infiltration modelling runs are discussed below and illustrated in Figure 6.1a and b.

Bare tailings can be expected to have high runoff rates, modest evapotranspiration losses, and substantial net infiltration or seepage. With an unvegetated surface, the runoff can be expected to be quite high. In the example model runs, runoff, evaporation, and seepage rates account for 19%, 63.5%, and 17.5% of the annual precipitation, respectively. When the tailings permeability is increased by a factor of 3, runoff rates decline by 3% and seepage rates increase by 3%.

Vegetation has a marked effect on the water balance at a tailings site. With the growth of a "moderate" vegetation cover, runoff rates decrease from 19% to 9% of the annual precipitation, while evapotranspiration increases from 63.5% to 73.6%. The major finding is that seepage rates are not changed. With a good vegetative cover, runoff rates are again reduced further, to 3.3% of the annual precipitation. Although evapotranspiration rates are increased, this may not offset the reduced runoff. One should not conclude from this that this phenomenon is universally applicable to all sites.

Direct application of soil to the tailings area surface may have mixed effects. If the soil retains its low permeability, runoff will increase substantially and seepage rates will be greatly reduced.

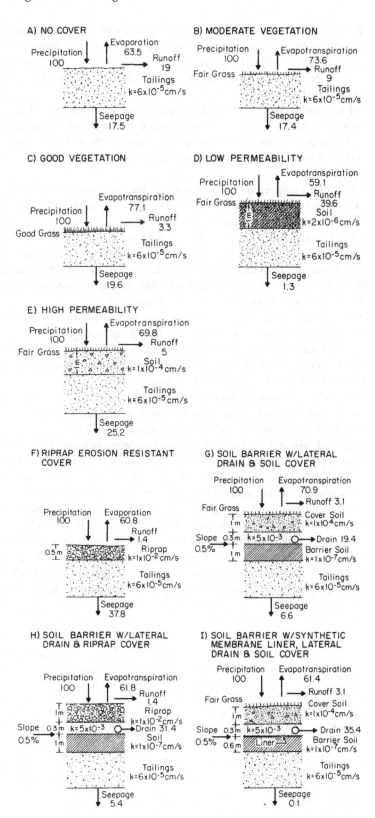

FIGURE 6.1 (A-I) Effect of cover types on infiltration rate. (From *Draft Acid Rock Drainage Technical Guide*, Vol. 1 (Vancouver, Canada: BiTech Publishers, 1989.)

A compacted till cover with a permeability of 2×10^{-8} m/sec will reduce seepage rates to <2% of the annual rainfall. If this cover cracks and weathers (as is expected), infiltration rates increase substantially. The example indicates that if the effective permeability of the cover increases to 1×10^{-4} m/sec, seepage rates exceed those for bare tailings. The increased permeability results in a major reduction in the rate of surface runoff.

Rock or gravel is often applied to stabilize the surface of a tailings area. This pervious layer effectively eliminates runoff and therefore can substantially increase infiltration rates. For the modelled case, the rock/gravel cover increases seepage rates from 17.5% to 37.8%. This is more than a factor of 2 and further demonstrates how the permeability of the surface layer can affect the overall amount of seepage produced.

A properly constructed engineered cap can greatly reduce infiltration rates. The example modelled includes a cap with 1 m of soil for frost protection and vegetation, 0.3 m of lateral drainage layer, and 1 m of a low-permeability seepage barrier. This cap reduces surface runoff to 3.1% of the annual rainfall. The lateral drain intercepts 19.4%, while evapotranspiration accounts for 70.9%, leaving 6.6% as seepage. This is a 62% reduction in the total seepage as compared to bare or vegetated tailings. The major finding is that these layers are effective but not 100% efficient in limiting seepage. At 6.6% infiltration this represents approximately 60 mm of precipitation or 60,000 m^3/ year from one 100-ha disposal site at the Elliot Lake project.

Rock/gravel surface layers have a major effect, increasing the infiltration. With a pervious surface zone, the lateral drains become more efficient, reducing the seepage rates.

An engineered cap with a synthetic membrane liner is by far the most effective infiltration barrier. The seepage rates predicted for an engineered cap with a liner that was 99% efficient are 0.1% of the annual rainfall. The life of the liner, however, needs to be considered.

Although theoretical simulations are useful in comparing alternative cover types, the true effectiveness of covers in controlling AMD can be determined only by monitoring the performance of actual covers in the field. Unfortunately, monitoring results are limited.

The best-documented case of a soil cover in use on an actual mine waste dump is that of the Rum Jungle uranium and copper mine in Australia. Composite covers were placed on three acid-generating overburden heaps. The top surface covers consisted of a 225-mm compacted clay liner, overlain by a 250-mm sandy clay loam retention zone layer, which was overlain by a 150-mm gravelly sand erosion layer. Rehabilitation of the heaps also included reshaping their surfaces and providing surface drainage systems. A typical cross-section of the rehabilitated heaps is shown in Figure 6.2.

Measurements of oxygen concentrations in the pore gas in the heaps show a marked reduction in oxygen concentrations after the installation of the compacted clay cover. Although measurements indicate that the transition rate of gas through the seal has increased since its initial placement, due to desiccation cracking in the dry season, the oxygen concentrations in the heaps are still much less

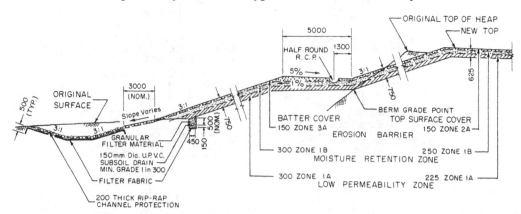

FIGURE 6.2   Cross-section of rehabilitation heap showing cover details. (From *Draft Acid Rock Drainage Technical Guide*, Vol. 1 (Vancouver, Canada: BiTech Publishers, 1989.)

than they were before rehabilitation. The effect of this reduction of oxygen concentration on oxidation rates has not been quantified.

Pre- and postrehabilitation measurements on and near the heaps indicate that the cover has provided some reduction in infiltration. However, the amount of that reduction remains questionable. Estimates based on lysimeter measurements indicate reductions >90%, while others based on groundwater estimates indicate only a 50% reduction.

The final measure of the effectiveness of the covers, however, is the reduction of metals loads in the local river system. Precipitation, flow, and metal load data are summarized in Tables 6.8 and 6.9. The four wet seasons during and after rehabilitation (1984–1985) have all experienced below-average rainfall and well below-average runoff. The reductions in postreclamation metal loads are indicated in Table 6.10. Although samples taken from the East Finniss River show large reductions in metals loading, it is unclear whether these reductions are due to covering the heaps or due to variations in precipitation rates i.e., it is not clear whether the reduction in the loads is due to reduced acid generation and migration resulting from cover placement or merely due to reduced migration resulting from low precipitation in the years following cover placement.

One interpretation of the results of surface water monitoring of the East Finniss River indicates a decline in annual copper loads since the placement of the waste dump covers.

Measures of groundwater quality criteria beneath rehabilitated dumps at Rum Jungle indicate that the groundwater quality has not changed significantly during the 4 years since cover placement. It would appear that there is a store of contaminants in the groundwater and within the dump. The response of downstream surface-water quality to cover placement may well be affected by the

TABLE 6.8

East Branch of the Finniss River Revised Pollution Loading Values

Season	1971/72	1972/73	1973/74
Rainfall (mm)	1,542	1,545	2,000
Total flow $m^3 \times 10^6$	31	22	69
Metal load (t)			
Copper	77	67	106
Manganese	84	77	87
Zinc	24	22	30

Source: From *Draft Acid Rock Drainage Technical Guide*, Vol. 1 (Vancouver, Canada: BiTech Publishers, 1989). With permission.

TABLE 6.9

Summary of Monitoring Results for the East Branch of the Finniss River

Season	1982/83	1983/84	1984/85	1985/86
Rainfall (mm)	1,121	1,704	1,112	910
Total flow $m^3 \times 10^6$	9.5	48	11.7	11.4
Metal load (t)				
Copper	23	28	9	4
Manganese	6	21	7	8
Zinc	5	9	4	3

Source: From Lapakko, K. Prediction of AMD from Duluth complex mining waste in North Eastern Minnesota, in *Proceedings of the Acid Mine Drainage Seminar/Workshop*, Halifax, Nova Scotia (Ottawa: Environment Canada, 1987), pp. 187–221. With permission.

TABLE 6.10

Percentage of Reduction in Pollution of the East Branch of the Finniss River

Season	1983/84 (Stage 1 & Part Stage 2 Complete)	1984/85 (Stage 3 Complete)	1985/86	Target
Metal				
Copper	70%	80%	91%	70%
Manganese	76%	88%	86%	56%
Zinc	67%	73%	79%	70%

Source: From Lapakko, K. Prediction of AMD from Duluth complex mining waste in North Eastern Minnesota, in *Proceedings of the Acid Mine Drainage Seminar/Workshop*, Halifax, Nova Scotia (Ottawa: Environment Canada, 1987), pp. 187–221. With permission.

release of these stored contaminants. Ongoing monitoring is essential until trends in surface and groundwater quality are established.

6.38 SYNTHETIC COVERS

The use of synthetic membranes as liners for tailings impoundments has been investigated.

Flexible membrane liners are commonly referred to as geomembranes. The common types are

- Polyethylene (PE)
- High-density polyethylene (HDPE)
- Chlorinated polyethelene (CPE)
- Chlorosulphonated polyethylene (CSPE, commonly known by the Dupont trademark Hypalon®)
- PVC
- Ethylene propylene diene monomer (EPDM)
- Butyl rubber

Occasionally neoprene and polyurethane are also used. Collectively, synthetic membranes display a number of advantages and disadvantages that may be summarized as follows:

6.38.1 ADVANTAGES

- They can contain a wide variety of fluids with minimum seepage due to low reported permeabilities of typically 1×10^{-1} cm/sec or less.
- They have a relatively high resistance to chemical and bacterial deterioration.
- They are readily installed for many applications.
- They are relatively economical to install and maintain.

6.38.2 DISADVANTAGES

- They are relatively vulnerable to attack from ozone and ultraviolet (UV) light.
- They have a limited ability to withstand stress from heavy machinery.
- They have not been in service long enough to evaluate long-term performance.
- They are comparatively susceptible to laceration, abrasion, and puncture.
- Some materials are prone to cracking and creasing at low temperatures or stretching and distorting at high temperatures.
- Although readily installed, there are often difficulties associated with the material's seams.

Polymeric membranes offer wide-ranging chemical resistance and may be readily inspected. However, they are susceptible to damage during installation, largely due to improper subgrade preparation and vehicular traffic. They require very careful installation, and their performance is dependent on careful and successful field seaming. Field seaming is, in general, a detailed and sensitive operation. Weather, including temperature and precipitation, is generally the governing factor. In this regard, the elastomeric liners, namely butyl, polychloroprene, and EPDM, would appear to present the most problems in field seaming. Of the remaining liner types considered, successful field seaming has been demonstrated with HDPE, PE, CSPE, CPE, and PVC. It is noted, however, that serious concerns have been expressed about the long-term weatherability of PVC and PE.

Proper subgrade preparation and construction are crucial for a successful liner installation and would typically consist of the excavation of compressible materials; sterilization of the subgrade; removal of all roots, sticks, stones, and debris; grading and proof-rolling; and installation of the sand cushion, liner, and soil cover. Installation of the liner and field seaming should be carried out by approved installers meticulously following liner supplier instructions. Soil cover is desirable but will require liner inclinations flatter than about 3 horizontal to 1 vertical. This is a severe limitation when applied to waste dumps.

With the exception of polyurethane, the base polymeric resins and asphalt show promise for long-term resistance to the major anticipated constituents of AMD. However, CPE may be affected by weak sulphuric acid solutions.

Thin, flexible membrane liners are susceptible to overstressing by strains associated with large differential deformations in the subgrade. It may be necessary to subexcavate and replace compressible materials encountered over the subgrade prior to liner installation. Similar concerns exist for liners placed on slopes and where a potential exists for excess hydrostatic or gas pressure buildup beneath the liner.

Seepage through liners occurs primarily through liner defects. A rational approach to evaluating apparent or field liner permeabilities is via detailed monitoring of existing installations. Test results show that the type and thickness of the membrane liner have a relatively small influence on leak rates, while the low-permeability subbase is important in restricting flow through a flawed liner.[1]

Estimates of liner release rates indicate that an asphaltic membrane would reduce seepage to about 50% of an unlined basin for a field liner permeability of 1×10^8 cm/sec. Polymeric liners with an effective permeability of 1×10^{-10} cm/sec would reduce seepage to <10% of an unlined basin.

In West Virginia, a PVC liner was used to cover a 45-acre backfilled site to prevent seepage into acid-producing materials. Results showed substantial decreases in flow and acidity from associated seeps.

Used as the barrier layer in combination with soil material layers in a complex cover, geomembranes should prove to be very effective in limiting oxygen and water transport.

Asphaltic and spray-on surface sealants can be applied to the surface of the waste to form a barrier to infiltration and oxygen diffusion. A number of products are available, including:

- Alkyd
- Asphalt
- Concrete
- Epoxy
- Polyester
- Polysulphide
- Polyurethane
- Silicone
- Synthetic rubber
- Thermoplastic molten sulphur
- Vinyl

These materials have, in general, been developed for applications, such as caulking sealants, soil stabilizers, waterproof barriers, and corrosion protective coatings. To date their application in mine waste covers is limited.

Surface sealants can be formulated to produce either flexible or rigid linings for covers. Surface sealants can be installed with three basic techniques:

1. *In situ chemical cure*. The materials chemically cure or harden after being applied to the surface. These materials usually involve more than one specific chemical.
2. *Heat application*. Materials that are solid in the desired operating temperature range are applied at elevated temperatures to improve ease of application.
3. *Surface drying*. The material is formulated in a water emulsion or diluted in a solvent carrier for application. The carrier evaporates, leaving a solid coating.

Combinations of the above techniques are also feasible in many cases. The object is to prepare the material for ease of application, usually with conventional spraying equipment. The actual technique for application is a function of the specific material.

The primary advantages and disadvantages of surface sealants are:

Advantages
- Either sufficient flexibility to conform with or sufficient strength to support the design load (pedestrian or vehicle traffic, for example)
- Good weatherability and service life
- Compatibility with the stored product
- Immunity to biological attack
- Sufficient puncture and abrasion resistance
- Capability of being placed with minimal defects
- Easily reparable
- Ease of application and production of an integral liner with no joints

Disadvantages
- Relatively difficult to regulate the rate of application and thus the thickness and uniformity of the sealant.
- As a class, these materials are relatively expensive. The high initial cost vs. relative ease of application for the spray-ons should be considered for specific applications.

"Geopolymer" is the term given to a compound of minerals, principally containing silica, phosphate, and oxygen, that bond to form a ceramic-type product. The suitability of using this product as a control measure for acid generation is currently being investigated. It is anticipated that geopolymers may be mixed with tailings to form a solid mass, preventing oxygen and/or water access to sulphides. A possible alternative is to mix the geopolymer with soil or other material and apply this as a cover to the waste. The behaviour of geopolymers when mixed with different waste materials is not fully known at this stage, nor is their resistance to natural processes such as freezing and thawing. Geopolymers are still in the development stage and require extensive research to establish their suitability as a control measure.

Shotcrete is the name given to concrete pneumatically delivered through a hose and applied to a surface at high velocity. Shotcrete may be effective in the control of acid generation when applied as a cover to certain wastes. The advantage of shotcrete is that it can be applied to steep rock slopes or other surfaces that may be difficult to cover using other methods.

The effective use of shotcrete as a cover is dependent on the stability of the underlying material. This method has been used very successfully on rock faces and on compacted materials. However, if the material to which the shotcrete is applied undergoes consolidation or settlement, causing relative displacement at the surface, cracking of unreinforced shotcrete will occur. Once the shotcrete liner has cracked, the effectiveness of the cover is lost. Experience has shown that displacement of uncompacted waste dumps often occurs, and for this reason unreinforced shotcrete is not appropriate as a cover for these materials. High-temperature-induced stresses in a shotcrete cover may also result in cracking.

The resistance of shotcrete to cracking may be increased by providing reinforcement. Conventional steel-mesh reinforcing is expensive, difficult to handle, and subject to corrosion in the long term. Steel-fibre reinforcement is easier to apply; however, it is also vulnerable to corrosion. A method of reinforcement using high-volume polypropylene fibre reinforcement, which is corrosive resistant and relatively flexible, shows promise. A benefit of the fibre-type reinforcement is that it reduces crack widths in the shotcrete. The shotcrete cover is then given the facility to accommodate movements larger than mesh-reinforced shotcrete.

6.39 PLACEMENT OF COVERS

The establishment of covers on mine waste and tailings is complicated by the difficulties of access, trafficability, and stability of the surfaces onto which the cover is to be placed. These difficulties often render a particular cover type impractical or prohibitively expensive. The placement of some cover types requires access by wheeled vehicles working on fairly flat surfaces (asphalt covers). Others require careful bed preparation and moderate slopes (synthetic membranes), while others a firm surface to compact against (clay layers).

Rock waste surfaces are conveniently subdivided into the dump surface and the dump slopes, with different conditions applying to each.

The upper surface of a rock waste dump is usually readily accessible, trafficable, and nearly flat. The placement of any type of surface cover, except a water cover, is usually not difficult.

During dump development, the material on the upper surface of dumps placed by trucking is often broken down and compacted under the wheel traffic of the dump trucks. This results in a fairly compact, lower permeability upper surface. This surface reduces infiltration, and ponding is often experienced on such surfaces. Despite the initial coarse nature of the material in such dumps, it may be necessary to install a suction-breaking layer to prevent downward suction on low-permeability cover layers.

Dumps are subject to long-term consolidation and settlement under their own weight and as the dump rock weathers. These settlements are large (a few per cent of the dump height) and uneven, reflecting the natural variation of the waste rock and dumping procedures. Differential settlements result in disruption of the drainage pattern on the dump surface and cracking of cover materials. Settlement and crack patterns are often such that drainage is towards cracks, resulting in considerably increased infiltration.

Dump slopes are usually placed at their angle of repose. At this angle, slopes are inaccessible, untrafficable, and marginally stable. Cover placement on such steep slopes is essentially impractical. Crest dumping of cover materials has been attempted at some sites, creating uncompacted (permeable), uneven layers of questionable stability.

For dump slopes to be accessible, it is necessary to first reslope. At a slope of 3 horizontal to 1 vertical (3:1), the slopes are trafficable by tracked vehicles and it is possible to place soil-type cover materials. At this slope, it is possible to also place synthetic membrane liners but the stability of cover layers over such membranes is questionable. The cost of resloping large dumps to 3:1 is very high, requiring large expenditures of dozer time unless the resloping had been planned for and the dumps constructed with a staggered dump slope.

Dump slopes are subject to the same concerns regarding different settlement as are the dump surfaces. Erosion on the steep dump slopes is a major long-term concern.

Wet, unconsolidated tailings always present difficult access conditions. Access improves as the tailings are drained and consolidate. Where tailings have been spigotted onto beaches, the sand fraction is deposited near the spigots and drains more freely than the slimes, which accumulate nearby and in the pond. Drained sandy beaches may be trafficable within days of deposition, while pond areas may never achieve this condition. Thus, it is possible to place and compact covers on beach areas with little preparation. In pond areas it may be necessary to apply drainage measures to remove free and near-surface water and to use geofabrics on slimes, followed by thin layers of the cover. These techniques have been successful in placing a cover over slimes that could not support foot traffic at the start of cover placement. Covers may also be placed during winter when freezing conditions allow access, as was done for the cover placement over wet tailings at the Beaverlodge mine.

To prevent capillary suctions in covers, it may be necessary to utilize capillary barriers over tailings fines and slimes.

Prior to placing covers on tailings it is necessary to first develop a tailings surface that has an adequate slope and drainage pattern. Much can be done to achieve such slopes by adopting an appropriate tailings placement and management method. Reshaping of tailings surfaces after mine closure with earth-moving equipment may be difficult and prohibitively costly.

After closure, tailings continue to consolidate and settle as a result of dissipation of pore pressures and thawing of included ice. These settlements can be a substantial portion of the total tailings depth and result in disruption of the drainage pattern, leading to extensive ponding on the tailings surface and cracking of covers.

The placement of covers on steep rock surfaces, such as pit walls, poses a particular problem. Two approaches can be used. The first requires the construction of thick self-supporting covers. The second requires adherence of the cover to the rock face and relies on the rock face to support the cover. The use of gunite or shotcrete methods is appropriate for the second. Both asphalt and concrete materials can be considered. Because of the corrosive nature of ARD upon cement and steel, the use of synthetic fibres and silica fume concrete is appropriate.

6.40   WASTE ROCK AND TAILINGS PLACEMENT METHODS

Control of ARD migration in waste rock dumps can be assisted by engineered placement methods such as cellular dump construction, compacting, mixing with low-permeability material, etc.

Cellular dump construction, when used in conjunction with a cover layer, can significantly reduce the area exposed to precipitation. Cellular construction may utilize a layer construction or lateral cell construction. This method is a short-term control as it applies only to the construction period.

Control of ARD migration by compacting or mixing with low-permeability material are both intended to reduce the bulk permeability of the waste rock dump, which will reduce infiltration. They can be used together or separately.

Compacting of waste rock will require dump construction in thin layers and is only suitable for soft rocks. This approach will also reduce settlement and increase dump stability. The result is long-term integrity and improved cover performance. Compacting will be very expensive.

Mixing low-permeability material into the waste rock will reduce infiltration. A non-acid-generating waste material or, preferably, a net alkaline waste, can be mixed during dumping. Dump stability must be addressed. This approach should provide good long-term reduction of infiltration.

Control of ARD migration in tailings impoundments is limited to methods that reduce the permeability of the tailings with density control. These methods include predisposal thickening, subareal deposition, flocculant addition, and installation of a dewatering system such as wick drains. It may be possible on some sites to reduce tailings permeability with the addition of clay to the tailings line. All of these methods will provide long-term control of ARD migration. Methods that increase the density of tailings may be required for stability in earthquake-prone regions.

6.41   MONITORING

In general terms, environmental monitoring in and around a mine site is intended to define baseline conditions and to identify changes in conditions during and after mining. This information is generally used for decision making regarding mitigation and reclamation strategies.

The environmental conditions monitored typically include physical processes, such as water flow and geotechnical stability, chemical characteristics such as water quality, along with biological response and impacts such as productivity. The major objective of a monitoring programme in the ARD context is to monitor the effectiveness of the prevention/control/treatment techniques and to detect at the earliest point in time whether the techniques are unsuccessful.

In the preoperational phase, baseline monitoring defines existing environmental conditions of the physical, chemical, and biological aspects of the area. This information leads to the identification of areas that are particularly sensitive to changes in environmental conditions and provides essential data to allow the assessment of changes or impacts caused by each component of the mine and mining activities.

In terms of acid generation, test work is conducted in the preoperational phase to determine the potential of each waste material from the various mine components to generate net acidity and acid drainage. Each of the potentially acid-generating materials may be further tested to determine the rate and duration of acid generation and its associated water quality. The design and testing of the required control and treatment techniques may also be conducted in this phase. The mine plan is adjusted in order to reliably implement the control/treatment techniques, to reliably eliminate acid drainage, and to minimize combined costs of environmental protection, mine construction, and operation.

Using the baseline information and the final mine plan, a monitoring programme is established for mine operation. Two types of monitoring stations have been defined for this approach: an effluent discharge point and the receiving environment. An effluent discharge point is generally but not necessarily located on the mine property (on-site), while the receiving-environment stations will generally but not necessarily be located offsite.

In the proposed programme, monitoring stations, including both surface water and groundwater, are established in or near all environmentally sensitive areas potentially affected by the development.

A minimum of one surface water and/or groundwater monitoring station should be selected at a defined discharge point for each component to be an "advance warning" station. This would provide early warning of potential failure of acid prevention/control/treatment techniques. The advance warning stations are located at a point of direct discharge from the mine component into the receiving environment. Each mine component usually has at least one discharge point that can be selected as an advance warning station. These stations should be monitored at least monthly for pH, sulphate, alkalinity, acidity, iron, and electrical conductance. A significant decrease in pH or alkalinity and an increase in sulphate may be an indication of the onset of acid drainage. However, extreme care must be used in separating site-specific trends in water quality, such as seasonal variations in pH, from the onset of acid drainage.

Monitoring stations in the receiving environment in the vicinity of each mine component can be established within, upgradient of, and downgradient of the component for surface and groundwater flows. Downgradient stations, as defined by the movement of surface water or groundwater from the component, should be placed at various distances from the mine component in the receiving surface water and groundwater. Upgradient stations provide data for comparison with downgradient stations to determine the degree and spatial extent of impacts due to each component.

All stations should be monitored at least semiannually for a full set of water quality analyses: pH, sulphate, alkalinity, acidity, iron, electrical conductance, major cations and anions, nutrients, and a suite of metals. Station monitoring may also be used for monitoring water flow rates.

Biological monitoring is not considered to be as reliable, as rapid, or as consistent as water quality measurements and visual observations for the detection of acid drainage. Consequently, biological monitoring is not emphasized here for detecting acid drainage (although certain mines such as those located near important fisheries may be required to monitor productivity, species diversification, or metal levels in fish tissue). Nevertheless, an annual biological survey of the mine site and surrounding offsite region is recommended as contingency monitoring to check for changes in vegetation or fisheries that may indicate the migration of acid drainage undetected by an established monitoring network. A visual inspection to identify changes in colour due to iron staining in seeps and streams in the area is also recommended. Colour changes, for example, could arise during a first-flush event where acid products are released between sampling periods of the monitoring stations.

The monitoring programme implemented during the operational phase of the mine should detect changes in environmental conditions at any station. If these changes are significant, more frequent monitoring should be performed at that station and at other stations to confirm the presence and spatial extent of the change. If an adverse impact is determined, alternative control or treatment techniques should be designed, tested, and implemented. The monitoring programme should be revised to monitor the success of the new techniques.

If no unacceptable impact exists towards the end of the mine life, a long-term monitoring programme for closure would be defined. Each of the steps should be performed because operational conditions at a mine ("baseline" for closure) will be different than preoperational conditions.

The long-term monitoring programme implemented during and after closure would decrease in frequency of sampling as time after closure increases. If significant changes in environmental conditions are detected at any station, additional monitoring should be performed at that station and at other stations to confirm the presence and spatial extent of the change. If the adverse impact is confirmed, alternative control or treatment programmes must be designed, tested, and implemented. The monitoring programme must then be revised to monitor the success of the new techniques. If no unacceptable impacts are detected over an acceptably long period of time, the site can proceed with abandonment.

6.42   SPECIFIC MONITORING PROGRAMMES FOR EACH MINE COMPONENT

Each mine component will require monitoring for the effectiveness of treatment and control technologies and for the detection of failures of these systems at the earliest time. Therefore, the monitoring stations should include at least one advance warning station at a direct discharge point in addition to other stations located at different distances in the receiving environment from the mine component. As discussed above, the advance warning station should be monitored more frequently than other stations.

The objective of the advance warning stations is to detect any significant change from the background quality that indicates the onset of acidic drainage from the mine component. Because the stations are monitored monthly for various parameters, including pH, sulphate, alkalinity, and acidity, trends can be established for background conditions.

The sampling frequency of the advance warning stations or other stations may be revised to incorporate a special sampling frequency. For example, at some sites dry spells may be followed by heavy rains; therefore, sampling of the first flushing of the waste rock stockpile after a storm event can be incorporated into the monitoring programme.

Emphasis of the receiving-environment monitoring programme should be oriented directly downgradient from the mine site as defined by the direction of the surface and groundwater flow from the site. Receiving-environment monitoring defines existing conditions and identifies changes in the conditions during and after mining. Because the advance warning stations are nearest to the area of impact and are monitored at frequent intervals to detect a failure in the treatment/control techniques, the receiving-environment monitoring stations determine the longer term trends in water quality over a wider area.

A proposed monitoring programme for each mine component is described in the following subsections. A summary of the surface and groundwater flows around and through each component and how they are potentially impacted during and following mining operations is outlined. Some of the geotechnical aspects of each component related to geotechnical monitoring are also described. The minimum recommended monitoring programme for each component is given along with a variety of options for monitoring each component.

6.43   ENVIRONMENTAL MONITORING OF OPEN PITS

Potential sources of water entering a pit are precipitation, surface-water drainage, and groundwater discharge. A pit is open to the atmosphere so variations in precipitation affect the daily volume of water in the pit, particularly if precipitation represents a major total of pit flow. Surface-water drainage is frequently diverted away from the pit perimeter in order to lower pumping costs. Groundwater frequently moves towards a pit because an excavation often represents a depression in the water table. Piezometers or monitor wells may be installed around the pit perimeter to intercept groundwater before it enters the pit if water control or geotechnical stability is a concern. All water entering a pit is usually combined and directed to sumps where it is dumped from the pit. The resulting surface flow downgradient of the pit may be directed into a holding pond for monitoring and any required treatment.

From a geotechnical viewpoint, a major concern is the stability of the pit walls. Instability represents a danger to the workforce and a delay in mining. Because water decreases friction through the increase in pore pressures, groundwater can be a major cause of instability in a pit and may be controlled by interceptor wells or underground galleries at the pit perimeter.

During operation and at closure, the location for the advance warning station in the vicinity of the open pit should be at the discharge from the sump and/or the retention pond. The advance warning station is monitored once every month throughout operation and closure.

In addition, the seeps from the pit walls should be sampled once every 6 months to monitor the exposure of new zones over time, and to determine the contribution of each wall to the ARD from the pit. Therefore, it may be possible to separate nonacidic and acidic flows from the pit walls to reduce the amount of impacted water.

Because no groundwater flow from the pit occurs during operation, no advance warning groundwater stations are necessary. After closure, discharge of mine water to the surface is terminated. This will allow the pit to flood and the water level in the pit will recover towards the preoperational water table. As this occurs, the pit will become an integrated part of the groundwater flow system with groundwater flowing into, through, and out of the flooded pit. The surface water within the pit or flowing out of the pit can be established as the advanced warning station that will be monitored monthly at the closure phase.

Receiving-environment surface monitoring stations should be located on surface watercourses that receive direct discharge from the sump and/or retention ponds holding pit water, and at selected stream junctions downstream of the mine site. A minimum recommended sampling programme would be one sampling location in each surface water flow directly receiving discharge from the sump and/or retention pond. The minimum monitoring frequency for the surface water locations should be once every 6 months, with the option for increased frequency depending on results of the advance warning station.

Groundwater monitoring at closure for the receiving environment is dependent on the hydraulic conductivity of the subsurface strata. If aquifers having high permeability are located near the pit, a groundwater monitoring network should be designed that is downgradient and hydraulically connected to the aquifers. The groundwater in an area that has a low permeability moves more slowly, and monitoring stations located nearer the pit should be able to detect any changes. A minimum recommended sampling programme would be one groundwater receiving-environment sampling location in each aquifer in the area. The monitoring frequency for the groundwater locations should be once every 6 months.

During the operational phase of an open-pit mine, surface water is usually diverted away from the perimeter. As a result, any impacts on preexisting surface drainage will be primarily a consequence of diversion. Monitoring of the diversion at established stations is an option but would be considered only if concerns over the ditch bank and bed material were raised.

Precipitation may represent a major portion of pit water in wet climates. Precipitation and evaporation may be monitored at weather stations in or around the perimeter of the pit.

A pit often represents a depression in the local water table, resulting in groundwater flow towards and into the mine from the walls and bottom. This movement can be defined and monitored with piezometers and monitor wells installed around the pit perimeter. The flow rate and quality of groundwater can be monitored on pit walls, and surface water quality can be monitored at the collector ditches within the pit, at the sump, and at the retention pond outside the pit.

6.44   ENVIRONMENTAL MONITORING OF UNDERGROUND WORKINGS

The primary source of water in underground workings is groundwater because the workings often represent a depression in the local groundwater regime. A secondary source of water may be surface water directed into a shaft, adit, or decline intentionally for drilling water or unintentionally. Unlike an open pit, water flow in underground workings does not usually respond as rapidly or as extremely as in open pits because of the buffering process of infiltration to the groundwater system. Mine water flows under gravity to an adit or is pumped from the workings, resulting in a surface-water flow that may be directed to a retention pond.

Geotechnical concerns in underground workings focus on wall and roof stability to minimize dangers to the workforce and delays in mining. Additionally, any collapses in the workings may lead to increases in groundwater flow due to permeability enhancements and to subsidence on upper levels and the land surface.

During operation and at closure, the location for the advance warning station in the vicinity of the underground workings should be the discharge from the sump and/or the retention pond. The advance warning station should be monitored once every month.

In addition, the seeps from the walls could be sampled once every 6 months to monitor the exposure of zones over time and to determine the contribution of each wall or zone to the ARD from the workings. Therefore, it may be possible to separate the nonacidic and acidic flows to reduce the amount of impacted water.

After closure, the discharge of mine water to the surface should be terminated. This will allow the workings to flood and the groundwater levels to recover towards preoperational levels. As this occurs, the workings will become an integrated part of the groundwater flow system with groundwater flowing into, through, and out of the mine. The advance warning station would remain the same as during operation—the discharge point from the sump and/or retention pond.

Receiving-environment surface monitoring stations should be located on surface watercourses that receive direct discharge from the underground workings and at selected stream junctions downstream of the mine site. A minimum recommended sampling programme would be one sampling location in each surface-water flow directly receiving discharge from the sumps and/or retention ponds. The minimum frequency of monitoring the surface water locations should be once every 6 months, with the option for increased frequency depending on the water quality of the advance warning station.

Groundwater monitoring at closure for the receiving environment is dependent on the hydraulic conductivity of the subsurface strata. If aquifers with high permeability are located near the underground workings, a groundwater monitoring network should be designed that is downgradient and hydraulically connected to the aquifers. The groundwater in an area that has a low permeability moves slower, and monitoring closer to the mine component should be capable of detecting any changes. A minimum recommended sampling programme would be one groundwater receiving-environment sampling location in each aquifer in the mine site area. The monitoring frequency for the groundwater locations should be once every 6 months.

During the operational phase of the underground workings, surface water is usually diverted away from any shafts, declines, or adits. As a result, any impacts on preexisting surface drainage will be primarily a consequence of diversion. Monitoring of the diversion at established locations will indicate the extent of the impacts.

Underground workings often represent a local depression in the hydraulic head, resulting in groundwater flow towards and into the mine. This dewatering of the groundwater system can be defined and monitored with piezometers and monitor wells installed from the land surface or from the workings. Additionally, the flow rate and quality of groundwater can be monitored at stations in the workings, in the collector ditches within the workings, or at any sumps and retention ponds outside the workings.

6.45   ENVIRONMENTAL MONITORING OF WASTE ROCK DUMPS, ORE STOCKPILES, AND HEAP-LEACH SITES

Dumps, stockpiles, or heap-leach sites are usually exposed to climatic events, so that precipitation represents a primary source of water moving through the rock. The resulting infiltration moves downward under gravity through the rock to the base of the pile. At the base, infiltration may (1) mix with upwelling groundwater and flow from the basal perimeter of the pile, (2) completely enter the underlying groundwater flow system if hydraulic conductivity is sufficiently high to accept all infiltration and hydraulic gradients have a downward component, and (3) partially enter the ground-water system if hydraulic conductivity is restricted and partially exit at the basal perimeter of the pile. Heap-leach sites may be constructed with low-permeability pads so that negligible amounts of infiltration reach the underlying groundwater system during the active life of the site.

Geotechnical monitoring of dumps, stockpiles, and heap-leach sites addresses the physical integrity of the rock after placement. Consolidation/settlement and slope stability could cause changes in the hydraulic conditions within the structure and could cause migration of rock from the site through physical movement such as slumping and toe collapse.

During operation and at closure (if the piles are not being removed), the location of the advance warning stations should be at the discharge point from a retention pond or drainage ditches located at the base of the piles. If groundwater drainage is found in the area to be primarily due to the hydraulic conductivity, additional advance warning stations should be established that require the installation of a groundwater well network. Both surface and groundwater advance warning stations should be monitored once a month.

Receiving-environment surface monitoring stations should be located on surface watercourses that receive direct discharge from the retention pond or ditches around the pile and at selected stream junctions downstream of the mine site, as well as one upgradient water station. A minimum recommended sampling programme would be one upgradient and one downgradient sampling location in each surface water flow directly receiving discharge from the pile. The minimum monitoring frequency should be once every 6 months for the surface water locations with the option for increased frequency depending on the water quality at the advance warning station.

Groundwater monitoring at closure for the receiving environment is dependent on the hydraulic conductivity of the subsurface strata. If aquifers having high permeability are located near the pile, a groundwater monitoring network should be designed that is downgradient and hydraulically connected to the aquifers, in addition to one upgradient groundwater station. The groundwater in an area that has a low permeability moves more slowly, and monitoring locations nearer the mine component should be able to detect any changes. A minimum recommended sampling programme would be one upgradient and one downgradient groundwater sampling location in each aquifer in the receiving environment. Monitoring should occur every 6 months for the groundwater locations.

Hydraulic monitoring of dumps, stockpiles, and heap-leach sites may include periodic measurements of precipitation at exposed surfaces and with periodic measurements of rates and quality of infiltration into the rock. Within a pile, monitoring would indicate the preferred water pathways and the variation in quality during flow, although monitoring of water in unsaturated, coarse-grained material is difficult and not always feasible. If a water table is positioned within the pile, a piezometer could be installed to monitor the water quality. Additionally, subsurface stations consisting of piezometers or monitor wells may be installed around the perimeter to monitor the direction, flow rate, and quality of groundwater in the area.

Monitoring of the temperature gradients may provide early indications of changes within the pile and can be accomplished using temperature probes within the piles. This method may indicate the initiation of acid generation in localized areas.

6.46   ENVIRONMENTAL MONITORING OF TAILINGS IMPOUNDMENTS

Two primary sources of water entering a tailings impoundment are the mill effluent discharge and precipitation. Water remaining above the tailings surface will flow to low-lying areas and form ponds. From a groundwater perspective, high water levels in an impoundment lead to downward movement through the tailings pile with a lateral component of flow towards the perimeter. Groundwater may leave an impoundment through the base or the retaining walls and enter the local groundwater flow system. This seepage may then enter deeper flow systems or discharge locally to the surface.

During operation and at closure, the location of the advance warning station should be at the direct discharge points from the impoundment. Direct discharge points can include seepage locations through dams and embankments to monitor the groundwater quality, and discharge from the spillways to monitor the surface water quality. These discharge locations should be monitored every month.

When the tailings become consolidated after closure, additional advance warning stations can be located within the tailings impoundment by installing piezometers in the tailings to monitor the quality of the groundwater pore water.

Receiving-environment surface monitoring stations should be located on surface watercourses that receive direct discharge from the tailings impoundment and at selected stream junctions downstream of the impoundment. A minimum recommended sampling programme would be two sampling locations in each surface watercourse which directly receive discharge from the tailings impoundment, one upgradient and one downgradient. The minimum monitoring frequency for the surface water locations should be once every 6 months, with the option to increase frequency depending on water quality at the advance warning station.

Groundwater monitoring at closure for the receiving environment is dependent on the hydraulic conductivity of the subsurface strata. If aquifers with high permeability are located near the tailings impoundment, a groundwater monitoring network should be designed that is downgradient and hydraulically connected to the aquifers. The groundwater in an area that has a low permeability moves more slowly, and monitoring nearer the impoundment should be able to detect any changes. A minimum recommended sampling programme would be one upgradient groundwater station for each hydraulically connected strata impacted by the tailings impoundment, and one groundwater sampling location in each aquifer in the mine site area. The monitoring frequency for the groundwater locations should be once every 6 months.

Hydraulic monitoring within an active impoundment is not always possible due to the unconsolidated nature of fresh tailings, sometimes limiting access only to the perimeter dams and retaining walls. As a result, measurements of mill effluent rate and quality can be made at the mill. Measurements of local precipitation and evaporation can be made at the perimeter. The movement of surface water within the impoundment can be visually defined, and the movement of groundwater can be generally assessed at the perimeter with piezometers.

Geotechnical monitoring of an impoundment at established stations addresses the physical integrity of an impoundment during and after operation. Dam instability, for example, may lead to the migration of tailings solids from an impoundment to a downstream environment. Tailings settlement may affect the rate and direction of water movement, which is a primary pathway for the interaction of tailings with the surrounding environment.

6.47   ENVIRONMENTAL MONITORING OF QUARRIES

Like open pits, potential sources of water entering a quarry are precipitation, surface-water drainage, and groundwater discharge. A quarry is open to the atmosphere and surface-water drainage is not usually diverted around the site, therefore, variable flow rates through the quarry can be expected. Groundwater frequently moves towards a pit because an excavation often represents a depression in the water table; however, a shallow quarry may lower the water table below the base so that groundwater inflow becomes minimal. All water entering a quarry usually combines and is either pumped out or flows under gravity from the site. The resulting surface flow downgradient of the quarry may be directed into a holding pond for monitoring and any required treatment.

The prevention of acid generation from a quarry should primarily be the assessment of its acid-generation potential. The use of a potentially acid-generating quarry is unlikely; therefore, a minimum monitoring programme is used only to confirm that acid generation is not occurring.

From a geotechnical viewpoint, the major concern is the stability of the quarry walls while equipment and workers are in the area. Shallow quarries may be relatively stable without any stability engineering.

The advance warning station can be located at the retention pond or at the surface discharge flow from the quarry and is monitored once a month. No groundwater advance warning stations are necessary as not all quarries will have a groundwater component. Receiving-environment surface monitoring stations are optional.

The water derived from a quarry may be monitored for flow rate and quality. If unacceptable quality is anticipated, this water may be directed to a holding pond for analysis and treatment.

The groundwater monitoring component for receiving-environment monitoring is dependent on the hydraulic conductivity of the subsurface strata. If aquifers with high permeability are near the quarry, a groundwater monitoring network should be designed that is downgradient and hydraulically connected to the aquifers. The groundwater in an area that has a low permeability moves more slowly, and monitoring nearer the quarry should be able to detect any changes.

Receiving-environment surface monitoring stations should be located on surface watercourses that receive direct discharge from the quarry and at selected stream junctions downstream of the quarry. An optional sampling programme could be one sampling location in each surface water flow directly receiving discharge from the quarry.

6.48 ENVIRONMENTAL MONITORING OF HAUL ROADS

For road construction, stability, and maintenance, rock is often taken from mine quarries and mine waste rock. This rock is crushed to an appropriate size and is further crushed by road vehicles. As a result, these roads can be thought of as small-scale waste rock dumps, but their geochemical reactivity is higher because of the finer grain size and continual grinding. Precipitation falling onto roads passes through the rock and either enters the underlying groundwater system or moves as overland flow to low-lying areas. Because the prevention of ARD from a haul road is in the assessment of acid generation prior to the use of the material, the construction of an acid-generating haul road is unlikely. However, due to the use of marginal acid-generating material or material used without knowledge of its acid-generating capability, a minimum monitoring programme is proposed.

Because monitoring of acid generation from haul roads may be very difficult, the minimum recommended programme would be a visual inspection of the area adjacent to the haul roads every 6 months to check for the discolouration of local materials. Stations may be located in sediment traps if they have been constructed and would be monitored every month.

6.49 IMPACT OF AN ABANDONED MINE ON WATER QUALITY

The Argo Tunnel, located 30 mi (55 km) west of Denver, CO, was completed in 1904. The tunnel is 21,968 ft long and intersects 27 mines. The tunnel and the mills are no longer in use, but the tunnel continues to drain the mine workings.[6]

The flow of the drainage was 1.2 cfs. The pH of the drainage averaged 2.8 and the average metal concentrations are given in Table 6.11. The effect of pH on the dissolution of metals in the mine water is given in Figure 6.3.

Water samples were taken from Clear Creek, into which the Argo Tunnel drains. The analysis of the samples is given in Table 6.12.

TABLE 6.11

Composition of the Argo Drainage

Parameter	Conc. (mg/L)	Amount/Day (lb)
Fe	340	2,000
Mn	160	1,000
Zn	75	500
Cu	13	85
As	0.43	3
Cd	0.32	2
Pb	0.12	1
SO4	2,700	18,000
Dissolved solids	4,000	27,000

Source: From Boyles, M. J. Impact of Argo tunnel acid mine
drainage, Clear Creek County, Colorado, in *Water
Resources Problems Related to Mining* (Baltimore,
MD: American Water Resources Association, 1974),
pp. 41–53. With permission.

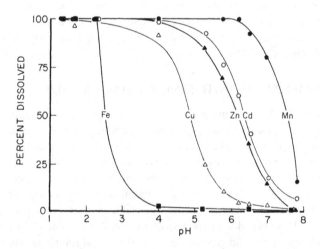

FIGURE 6.3   Heavy metal behaviour as a function of pH in Argo water. (From Boyles, M. J. Impact of Argo tunnel acid mine drainage, Clear Creek County, Colorado, in *Water Resources Problems Related to Mining* (Baltimore, MD: American Water Resources Association, 1974), pp. 41–53. With permission.)

In addition to water analysis, sediments from the bottom of Clear Creek were analyzed and the results are given in Figure 6.4.

6.50   HYDROLOGIC SOLUTION TO ACID MINE DRAINAGE

Hydrologic solutions may be found to solve or reduce AMD problems. The study of water movements in a lead-zinc mine in northern Idaho revealed alternatives to costly treatment of the poor-quality effluent. Delineation of areas of acid production and areas of recharge is important in this process.

TABLE 6.12

Impact of the Argo Drainage on Clear Creek, August 16, 1973

Metal	USPHS Drinking Water Standards (µg/L)	Max. Suggested Cone, for Fish and Other Aquatic Life	State	Conc. (µg/L) on August 16, 1973		
				Clear Creek Upstream from Argo	Argo Drainage	Clear Creek 1:1 Miles Downstream from Argo[a]
Fe	300 SL[b]	300	Dis	30	320,000	80
			Total	350	320,000	1,400
Cu	1,000 SL	10–20	Dis	10	13,000	40
			Total	25	13,000	80
Mn	50 SL	1,000	Dis	200	150,000	860
			Total	240	150,000	860
Zn	5,000 SL	30–70	Dis	240	71,000	560
			Total	250	71,000	610
Cd	10 CR[c]	10	Dis	<1	320	<1
			Total	<1	320	<1
Pb	50 CR	5–10	Dis	<1	110	<1
			Total	10	110	10
sO₄	250,000 SL	—	Dis	15,000	2,700,00	22,000

[a] Past the point of total mixing.

[b] SL = Suggested limit.

[c] CR = Cause for rejection limit.

Source: From Boyles, M. J. Impact of Argo tunnel acid mine drainage, Clear Creek County, Colorado, in *Water Resources Problems Related to Mining* (Baltimore, MD: American Water Resources Association, 1974), pp. 41–53. With permission.

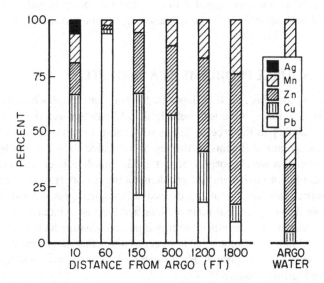

FIGURE 6.4 Composition of bottom sediments compared to Argo water. (From Boyles, M. J. Impact of Argo tunnel acid mine drainage, Clear Creek County, Colorado, in *Water Resources Problems Related to Mining* (Baltimore, MD: American Water Resources Association, 1974), pp. 41–53. With permission.)

The discharge from the Kellog Tunnel, which includes drainage from most of the mine, averages 2,700 gal/min with an average pH of 3.3.

Mining methods used in the mine have included vertical slice, cut, and fill, with square sets and caving. Wastefill from development headings was used for backfilling stopes. Some of the wastefill used in the upper levels contains a high concentration of lead, zinc, and pyrite.

Vertical drainage of water occurs through a series of interconnected stopes between the mine levels. As the water drains down the mine openings, some of it moves through stopes containing ore and wastefill rich in pyrite. The water becomes more acidic as it passes over the waste. Movement of water in the mine provides the transport mechanism for the acid water. The reduction or control of water movement in the mine can be an alternative procedure to reduce the acid drainage problem. Three possible approaches include:

- Block the water movement on a selected level
- Route the water around potential acid-producing areas
- Reduce recharge to the mine

Water discharge data collected from various points in the mine indicate a close correlation with surface runoff events. Three probable areas of recharge to the mine have been identified:

- A subsidence area and related fracturing resulting from caving
- A drift and a stope intersecting the streambed
- Proximity of upper level stopes to the land surface

Three projects for reducing recharge to the mine have been considered. One project includes building an impermeable cutoff to bedrocks at the dams built for the raises. This would provide a more effective diversion around the caving area. The groundwater and surface water recharge to the caving can be effectively eliminated with the repositioning of the raises in the actual bedrock lows and construction of more impermeable cutoffs.

The diversion of water around the areas of losses in the creeks is another project. Dams to bedrocks with collector inlets and pipes capable of carrying the spring runoff past the recharge area could be installed. Elimination of recharge in these areas could reduce most of the flow to upper levels of the mine.

6.51   WATER RESOURCE PROBLEMS IN A LEAD BELT

Southeastern Missouri contains rich lead deposits. Several major mines have been opened in that area. Several creeks flow through the lead belt. Table 6.13 summarizes the physical and chemical parameters of water quality in the creeks. Tables 6.14 and 6.15 summarize metal contents of the water. Studies of the stream sediments in the region indicated only low levels of heavy metals. Generally, the heavy metals were incorporated in the #325 mesh or finer fractions.[7]

One of the evident major environmental problems is the transport of heavy metals under conditions in which wastewater has stimulated excessive biological growth in receiving streams. In some instances, these dense, gelatinous mats of algae and their associated aquatic populations have coated streambeds, blocking photosynthetic energy input, eliminating local stream populations, and causing aesthetic problems. These biological mats act as filters that trap nutrients and sediments that are high in trace metals. They also trap finely ground rock flour, tailings, and minerals that escape from the flotation and tailings reservoirs.

A treatment system was designed to encourage the growth of algae on the mine property and to trap any suspended heavy metals on the algae. A series of broad, shallow, rapidly flowing meanders was built with a sedimentation pond placed at the end of the meanders to prevent heavy metals from

TABLE 6.13

Summary of Physical and Chemical Parameters of Water Quality

	Station No.									
	1	2	3	4	5	6	7	8	9	10
PH	8.10	7.45	7.74	7.73	7.73	7.66	7.58	8.2	8.1	8.0
Temperature	0–25°C	0–25°C	0–25°C	0–25°C	0–25°C	0–25°C	0–25°C	0–25°C	0–25°C	0–25°C
Turbidity (JTU)	3.7	2.9	1.2	1.0	7.7	1.3	0.76	1.0	1.0	1.0
Dissolved oxygen (DO) (ppm)	5.4	4.6	4.7	5.5	5.5	5.6	5.3	5.6	5.2	5.5
Alkalinity (mg/L)	180	200	198	152	150	150	150	144	156	135
Hardness (mg/L)										
Calcium	150	140	210	100	95	100	100	75	80	75
Total	300	280	360	250	235	200	210	155	160	135
Chloride (mg/L)	0	0	30	20	0	0	0	0	0	0
Chem. Ox. Dem. (COD) (mg O_2/L)	30	20	50	40	75	95	35	15	20	30
Phosphorus (mg/L)										
Ortho-	0.1	0.1	0.1	0.1	0.1	0.1	0.1	0.1	0.1	0
Total	0.3	0.3	0.4	0.3	0.2	0.1	0.1	0.2	0.1	0.1
Nitrite (mg/L)	0	0	0	0	0	0	0	0	0	0
Nitrogen (mg/L)	0.1	0.1	0.2	0.1	0.2	0.1	0.1	0.2	0.1	0.1
Ammonia	0	0	0	0	0	0	0	0	0	0
Total organic	26.4	8.4	30.2	25.7	16.4	18.0	12.0	15.6	10.0	22.4
Specific conductance (μmho/cm)	360	200	260	450	100	160	340	300	380	280

Source: From Hardie, M. G. and J. C. Jennett. Water resources problems and solutions associated with the new lead belt of S.E. Missouri, in *Water Resources Problems Related to Mining* (Baltimore, MD: American Water Resources Association, 1974), pp. 109–122. With permission.

TABLE 6.14

Metal Concentration of Streams at Sampling Sites in the New Lead Belt Study Area—Dissolved Metal Content

Element	Station No.									
	1	2	3	4	5	6	7	8	9	10
Pb. max.	0.280	0.022	0.010	0.130	0.061	0.075	0.066	0.012	0.045	0.160
min.	<0.005	<0.005	<0.005	<0.005	<0.005	<0.005	<0.005	<0.005	<0.005	<0.005
mean	0.026	0.009	0.066	0.012	0.014	0.011	0.009	0.006	0.009	0.008
Mass flow (lb/yr)	581	196		135	232	86	6,900	396	606	147
Zn max.	0.460	0.084	0.027	0.250	0.097	0.034	0.036	0.023	0.128	0.088
min.	<0.010	<0.010	<0.010	<0.010	<0.010	0.005	<0.010	<0.010	<0.010	<0.010
mean	0.063	0.018	0.011	0.068	0.028	0.013	0.013	0.011	0.019	0.014
mass flow (lb/yr)	1,391	394		771	478	103	9,832	729	1,251	261
Cu max.	0.039	0.015	0.010	0.010	0.011	0.018	0.018	0.010	0.024	0.010
min.	<0.010	<0.010	<0.010	<0.010	<0.010	<0.010	<0.010	<0.010	<0.010	<0.010
mean	0.012	0.010	0.010	0.010	0.010	0.010	0.010	0.010	0.011	0.010
Mass flow (lb/yr)	263	226		114	172	84	7,835	666	712	181
Cd max.	0.010	0.010	0.010	0.012	0.010	0.013	0.010	0.010	0.010	0.015
min.	0.000	0.000	0.000	0.000	<0.001	0.000	0.000	0.000	0.000	0.000
mean	0.007	0.006	0.006	0.006	0.006	0.007	0.006	0.006	0.006	0.007
Mass flow (lb/yr)										
Mn max.	0.550	0.054	0.011	1.82	0.680	0.012	0.010	0.018	0.840	0.020
min.	<0.010	<0.010	<0.010	<0.010	<0.010	<0.010	<0.010	<0.010	<0.010	<0.010
mean	0.062	0.014	0.010	0.128	0.111	0.010	0.010	0.010	0.052	0.011
Mass flow (lb/yr)	1,363	312		1,459	1,895	82	7,534	692	3,455	202

Source: From Hardie, M. G. and J. C. Jennett. Water resources problems and solutions associated with the new lead belt of S.E. Missouri, in *Water Resources Problems Related to Mining* (Baltimore, MD: American Water Resources Association, 1974), pp. 109–122. With permission.

TABLE 6.15

Metal Concentration of Streams at Sampling Sites in the New Lead Belt Study Area—Total Metal Content

Element		Station No.									
		1	2	3	4	5	6	7	8	9	10
Pb	max.	0.200	0.100	0.100	0.100	0.830	0.100	0.100	0.680	0.120	0.920
	min.	<0.005	<0.005	<0.005	<0.005	<0.005	<0.005	<0.005	<0.005	<0.005	<0.005
	mean	0.029	0.011	0.010	0.013	0.081	0.011	0.014	0.035	0.019	0.009
	Mass flow (lb/yr)	643	249		151	1,389	111	10,315	2,316	1,246	158
Zn	max.	0.180	0.153	0.026	0.280	0.175	0.032	0.018	0.026	0.208	0.057
	min.	<0.010	0.008	<0.010	<0.010	<0.010	0.008	0.008	<0.008	<0.010	<0.008
	mean	0.043	0.017	0.011	0.060	0.043	0.011	0.011	0.011	0.018	0.012
	Mass flow (lb/yr)	966	369		686	743	88	8,107	751	1,196	221
Cu	max.	0.025	0.027	0.029	0.036	0.036	0.035	0.028	0.032	0.026	0.030
	min.	<0.010	<0.010	<0.010	<0.010	<0.010	<0.010	<0.010	<0.010	<0.010	<0.010
	mean	0.011	0.011	0.011	0.011	0.012	0.011	0.011	0.011	0.011	0.011
	Mass flow (lb/yr)	246	237		128	206	88	8,000	716	702	201
Cd	max.	0.010	0.016	0.010	0.010	0.016	0.010	0.010	0.012	0.010	0.012
	min.	0.000	0.000	0.000	0.000	<0.001	0.000	0.000	0.000	0.000	0.000
	mean	0.008	0.008	0.007	0.007	0.008	0.007	0.007	0.008	0.007	0.008
	Mass flow (lb/yr)	168	169		83	134	59	5,484	500	489	139
Mn	max.	0.640	0.620	0.042	0.070	0.690	0.028	0.200	0.064	0.960	0.034
	min.	0.010	0.010	0.010	0.010	0.019	0.010	0.010	0.010	0.010	0.010
	mean	0.062	0.037	0.012	0.032	0.133	0.011	0.017	0.012	0.049	0.012
	Mass flow (lb/yr)	1,391	821		367	2,279	88	13,125	807	3,238	225

Source: From Hardie, M. G. and J. C. Jennett. Water resources problems and solutions associated with the new lead belt of S.E. Missouri, in *Water Resources Problems Related to Mining* (Baltimore, MD: American Water Resources Association, 1974), pp. 109–122. With permission.

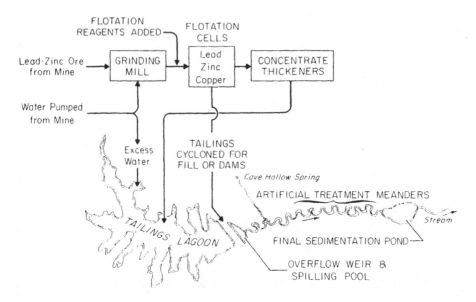

FIGURE 6.5 A mine and mill separation using a lagoon followed by a meander section and final sedimentation. (From Boyles, M. J. Impact of Argo tunnel acid mine drainage, Clear Creek County, Colorado, in *Water Resources Problems Related to Mining* (Baltimore, MD: American Water Resources Association, 1974), pp. 41–53. With permission.)

escaping the system. The meanders have been successful in eliminating the algal problem downstream. Lead concentration in aquatic vegetation has significantly decreased (Figure 6.5).

6.52   ENVIRONMENTAL CONTROL MEASURES AFTER THE CLOSURE OF A LEAD-ZINC MINE IN GREENLAND

The Black Angel Mine is situated at Maarmorilik, a remote area in the Bay of Uummannaq on the west coast of Greenland about 500 km north of the Arctic Circle. The main orebodies occur in an 1,100-m high mountain with very steep slopes along two fjords. The mountain is partially covered by the Greenland ice cap. The orebodies are located about 700 m above sea level.[8]

The concentrator, the mine town, and all the services were located at sea level in Maarmorilik, separated from the mine by the fjord Affarlikassaa. The only access to the main orebodies is via two cableways, with a span of 1.5 km.

The aims of the demolition and cleanup in Maarmorilik are:

- It shall be possible for persons to stay in the area without risk of life and health due to former mining activity.
- The town area shall be left neat and tidy. Equipment and visible waste shall be removed from former areas of prospecting.
- Existing and potential sources of pollution shall be eliminated or limited so that the original environmental condition before the mining activities began can be recreated as completely as possible within a few years.

The main sources of pollution from the mining activities at Maarmorilik are:

- Tailings, in particular, the plume of tailings in the fjord Affarlikassaa
- Dispersed metal-rich minerals in the surroundings

- Remains of ore and concentrates in the industrial area
- Dust from various places where ore and concentrates are handled
- Waste rock dumps; in particular, the one reaching the fjord (North Face Dump)

The first source will cease when the flotation plant stops the discharge of tailings in seawater suspension. Undisturbed tailings on the bottom of the fjord will release insignificant small amounts of lead as shown by laboratory experiments.

The pollution from the second source is believed to be insignificant and relatively quickly terminating.

The pollution from the third source will be minimized by clearance and cleaning.

Pollution from the fourth source, dust, will decline significantly when the mining activities stop and the mine is abandoned.

The fifth source, the waste rock dumps, however, will be unaffected by the closure of the mill and mine, and will be the most serious pollution source in the future. Since one dump, the North Face Dump, is already an important source of pollution, a decision has been made to remove it. Taking practical possibilities into consideration, the safest place to deposit the approximately 400,000 tons of waste rock is thought to be the bottom of the fjord Affarlikassaa, where approximately 8 million tons of tailings have already been deposited. Laboratory experiments have shown that the amounts of lead released over time after depositing the waste rock on the fjord bottom will be

$$kg\, lead = 7.53 \times \sqrt{days} \qquad (6.10)$$

The removal of the North Face Dump from the mountain slope to the bottom of Affarlikassaa is the most expensive single operation of the clean-up after the mining activities. There are other waste rock dumps in the surroundings of Maarmorilik, but none of them reaches the sea. It has been decided to await the results of some years' monitoring before determining if it is necessary to take action against their pollution.

In order to fulfil the third aim, that existing and potential sources of pollution shall be eliminated, all places and equipment contaminated with ore or concentrates must be cleaned as described above, and waste rock dumps with an unacceptable environmental impact must be removed.

The tailings discharge at a depth of 30 m in the fjord Affarlikassaa results in a high degree of pollution of the fjord by zinc, cadmium, and lead. Typical concentrations are seen in Table 6.16. Unpolluted seawater collected and analyzed with the same equipment typically gives 0.5, 0.03, and 0.2 μg/kg for zinc, cadmium, and lead.

Obviously, the tailings release metals to the receiving seawater. In September, when the water in the fjord is stratified, the metals are kept mainly below 30 m (a sill lies between Affarlikassaa and the outer fjord Qaamarujuk at a depth of approximately 25 m); however, in March, when the density stratification is broken down, the metals are brought up to the upper part of the fjord and spread out into the neighbouring fjords. The amounts of metal dissolved in Affarlikassaa and the adjacent fjord Qaamarujuk are seen in Table 6.17.

Blue mussel, *Mytilus edulis*, and seaweed, *Fucus vesiculosus* and *F. disticus*, were collected and analyzed annually; i.e., 1 year before the start of the mining activity.

It soon became obvious that a strong influence was exerted by the mining on the zinc and lead content of these intertidal organisms. The geographical distribution of lead in *Mytilus* and *Fucus* is seen in Figures 6.6 and 6.7. It is important to note that the maximum for lead (and also zinc) contamination in *Mytilus* and *Fucus* is not where the tailings are discharged in Affarlikassaa nor where the concentrations are shipped, but below the North Face Dump. This dump contains approximately 400,000 tons of waste rock from the mining in the above mountain. It is mainly composed of marble and dolomite and contains 0.8% Pb and 2.5% Zn. The dump is situated on a mountain slope from an altitude of 260 m to below sea level. It is believed that it is the action of waves and tidal movements

TABLE 6.16

Dissolved Metals in the Fjord Affarlikassaa (µg/kg) (Average of 6–8 Samples)

	Zn	Cd	Pb
Below 30 m March 1989	44.0	0.370	47.1
Above 30 m March 1989	17.4	0.191	18.8
Below 30 m September 1989	138.0	2.45	242.0
Above 30 m September 1989	4.1	0.043	2.5

Source: From Asmund, G. Rehabilitation and demolition after the closure of the zinc and lead mine Black Angel at Maarmorilik, Greenland, in *POLARTECH, Proceedings of the 3rd International Conference on Development and Commercial Utilization of Technologies in Polar Regions* (Copenhagen, Denmark: Danish Hydraulic Institute, 1990), pp.744–759. With permission.

TABLE 6.17

Tonnes of Dissolved Metals in the Two Fjords Near Maarmorilik, Average of September and March, 1985—1988

	Zn	Cd	Pb	Vol 10^6 m^3
Affarlikassaa	6.6	0.050	4.4	70.3
Qaamarujuk	8.4	0.077	2.8	1,338.0

Source: From Asmund, G. Rehabilitation and demolition after the closure of the zinc and lead mine Black Angel at Maarmorilik, Greenland, in *POLARTECH, Proceedings of the 3rd International Conference on Development and Commercial Utilization of Technologies in Polar Regions* (Copenhagen, Denmark: Danish Hydraulic Institute, 1990), pp. 744–759. With permission.

that results in mobilization of lead and zinc from the dump. A similar effect has been observed on polluted sediments at Sørfjord, Norway.

The lead pollution of fish caught in the immediate vicinity of the mine was at its maximum in the late 1970s. Work done by the mining company to abate the pollution has probably resulted in a decrease in fish pollution; therefore, in 1989 the only lead-polluted fish parts were capelin, liver of shorthorn sculpin, and head and shells of prawn.

The spreading of lead in dust outside the mining town has been estimated by collecting and analyzing the lichen *Cetraria nivalis*. By using the correlation

$$\text{Lead deposition mg/m}^2 \text{ year} = \text{Lead Concentration mg / kg / 2.7} \qquad (6.11)$$

one can correlate that between 1.6 and 1.4 t/year of lead has been spread outside Maarmorilik Town and up to 23 km towards the west.

The guidelines for demolition and cleanup have been written in order to fulfil the aims mentioned above.

Accessible mine entrances and shafts will be closed or fenced in to prevent entrance to the mine by humans and possible falling accidents.

All installations and equipment will be left in the mine. Explosives and other dangerous goods must be removed.

To prevent pollution from seepage, ore plants and silos will be cleaned, and fuel, lubricants, and other toxic wastes removed. Furthermore, a concrete wall must be cast in order to hold back intrusive water in a drift under the ice cap.

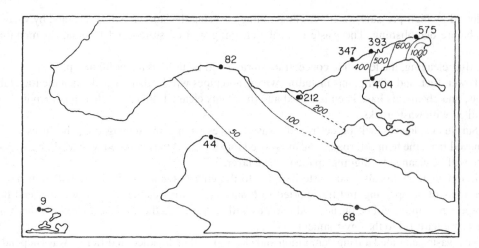

FIGURE 6.6 Geographic distribution of lead in the soft part of the blue mussel (mg/kg dry weight). (From Asmund, G. Rehabilitation and demolition after the closure of the zinc and lead mine Black Angel at Maarmorilik, Greenland, in *POLARTECH, Proceedings of the 3rd International Conference on Development and Commercial Utilization of Technologies in Polar Regions* (Copenhagen, Denmark: Danish Hydraulic Institute, 1990), pp. 744–759. With permission.)

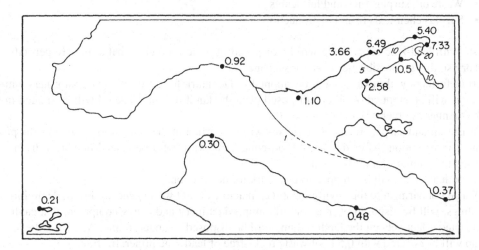

FIGURE 6.7 Geographic distribution of lead in new growth tips of seaweed, mg/kg dry weight. (From Asmund, G. Rehabilitation and demolition after the closure of the zinc and lead mine Black Angel at Maarmorilik, Greenland, in *POLARTECH, Proceedings of the 3rd International Conference on Development and Commercial Utilization of Technologies in Polar Regions* (Copenhagen, Denmark: Danish Hydraulic Institute, 1990), pp. 744–759. With permission.)

Cables, machinery, and terminals in Maarmorilik will be removed. Cables will be removed by blasting and dumped into the fjord.

Ore silos, ore conveyors, concentrator, and concentrate conveyors will be cleaned either by sweeping, vacuuming, or high-pressure washing. The water used in cleaning will be filtrated and recirculated. The lead-zinc-containing sludge will be disposed of underground, as mentioned below.

Pits and other major cavities will be filled with waste from demolitions of buildings or with waste rocks without lead-zinc mineralization.

The storage for lead-zinc concentrates is located in an old marble quarry. After shipment of concentrate, the old marble surface will be cleaned by scraping and brushing with heavy equipment followed by high-pressure washing. The building covering the storage will be cleaned by vacuuming before demolishing. The waste from the cleaning will be stored underground, as mentioned below.

After cleaning, the lead-zinc concentrate storage will be used as a landfill dump.

Power, fuel, and other plants including wiring and pipes will be demolished when the fuel, lubricants, and chemicals have been removed and the plants cleaned. Pits and other major cavities will be filled with waste rocks.

Surplus explosives will be destroyed. Waste, fences, signs, and damaged crash fences will be removed from the heliport, roads, and areas used for storage. Areas polluted with lead-zinc concentrate will be cleaned by digging up the surface layer.

Equipment, chemicals, and waste harmful to the environment will be collected from the former areas of prospecting and transported to Maarmorilik for treatment as waste generated here. Prospecting camps will be burned and remains will be buried in the area together with other waste that is not harmful to the environment.

The waste generated during demolition and cleanup will be handled in different ways depending on its harmfulness to the environment.

Waste harmful to the environment consists of:

- Waste with lead-zinc concentrate or ore
- Transformers with toxic oil
- Waste oil, surplus fuel, and lubricants
- Surplus chemicals from the production

Waste with lead-zinc concentrate will be disposed of underground in drifts that are permafrozen and dry so that no possibilities exist for seepage.

Transformers with toxic oil will be shipped to Denmark for destruction. Transformers without toxic oil will be emptied of oil and disposed of in the landfill dump. The oil will be treated in the same manner as other oil-containing wastes.

Waste oil and surplus fuel and lubricants will be burned in the generators as long as the power plant is in operation. After that period, oil-containing waste will be placed into steel drums and shipped to Denmark for destruction.

Chemical waste will be shipped to Denmark for destruction.

Waste not harmful to the environment, i.e., machinery and waste generated from demolitions of buildings, will be disposed of in a landfill dump. Machinery and heavy equipment that cannot be compacted effectively on the landfill dump will be disposed of underground. After use, the landfill dump will be covered with 0.5 m of waste rocks free of lead-zinc mineralization.

The North Face Dump will be removed and deposited on top of the marine tailings depot on the bottom of the fjord Affarlikassaa.

Environmental investigations during close down will measure the effect of the waste dump removal. Several seawater samples will be collected before the depositing of waste rock begins and several samples will again be collected 1 week afterwards. Two months later a new collection of seawater will be made. Samples of seawater will be taken in order to estimate the release of metals from waste rock to seawater during the deposition in Affarlikassaa.

Fish will be collected and analyzed both during the cleanup operations and 6 months later. Furthermore, liquid effluents from the concentrate storage and the North Face Dump will be sampled during work at the respective locations. The spreading of solids in the fjord system will be investigated by sediment traps. Later, studies will be made of sediment cores. Whenever activity occurs at Maarmorilik, two fixed stations in Affarlikassaa and Qaamarujuk will be sampled every month at approximately ten different depths.

When all activities, including waste rock removal, have ceased, the environment will begin its restoration process. This process will be monitored once a year for 10–15 years. The sampling frequencies will be incredibly longer concurrently with the rejuvenation of the investigated items to their natural states. It is expected that fish will return to the natural lead values in a few years, while blue mussels and seaweed will be polluted for many years to come. It is expected, however, that the zone in which collection and consumption of blue mussels have been restricted during the monitoring period can be gradually reduced from the 37-km distance west of Maarmorilik to a smaller area.

6.53 MINE ENVIRONMENTAL REHABILITATION

Bersbo is a mining town in the southern part of Sweden. Except for a short period in the 1930s, mining activities ceased, after 600 years or more, during the first years of the 20th century. In the mid-1800s, the mine had its heyday and was, for a decade, when the productivity of the Falun Mine was temporarily in decline, the greatest copper producer in Sweden. Compared to present conditions, the copper production was not very impressive: 600–900 t/year. The ore was handsorted on site and smelted in a plant some 8 km away. Therefore, the only waste left in Bersbo is waste rock, as well as a small tailings pond from the experimental activities in the 1930s to concentrate for zinc.[9]

The waste rock was disposed of in several varying-sized heaps spread over a large area (Figure 6.8). The most important are specified, together with the tailings, in Table 6.18, along with data on size and areas. Figures 6.9 to 6.11 show the discharges of copper, zinc, and cadmium from the major waste units. They have been calculated from a variety of investigations carried out in Bersbo.

More than 200,000 m³ of the waste rock masses were dumped in the waterfilled shafts; i.e., submerged disposal was applied. All the small heaps and the shallow deposits were concentrated in two large heaps that were contoured to fit a suitable covering. In this way the total deposit area was reduced from around 19 to about 9 ha.

Some of the shafts (about 1 dozen) were encompassed by the new deposits, but most could be avoided. The former had to be sealed off by a special reinforced-concrete, self-bearing roof in order

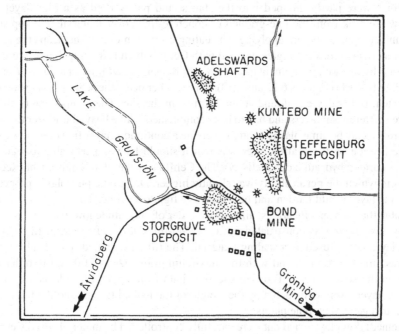

FIGURE 6.8 The siting of the mine waste units before remedial action. (From Lundgren, A. T. Bersbo: The first full scale project in Sweden to abate acid mine drainage from old mining activities, in *Acid Mine Drainage, Designing for Closure, GAC/MAC Annual Meeting* (Vancouver, Canada: BiTech Publishers, 1990), pp. 241–253. With permission.)

TABLE 6.18

The Major Units of Solid Mine Waste Before Remedial Actions Took Place, Together with Size and Volume

Mine Waste Units	Amount (kton)	Cone, (kg/ton)		Metal Source (tonnes)	
		Copper	Zinc	Copper	Zinc
Storgruve deposit	500	2.9	7.0	1,450	3,500
Steffenburg deposit	800	2.9	7.0	2,320	5,600
Kuntebo deposit	10	2.0	6.0	20	60
Fillings (roads, etc.)	20	2.9	7.0	58	140
Grönhög Mine	30	2.5	19.6	75	588
Adelsvärd deposit	10	2.5	19.6	25	196
Tailings pond	15	2.0	8.6	30	129
Tailings delta	5	2.0	8.6	10	43
Small heaps in the forest	10	1.5	5.0	15	50
Total (tonnes)	1,400,000	—	—	4,003	10,306

Source: From Lundgren, A. T. Bersbo: The first full scale project in Sweden to abate acid mine drainage from old mining activities, in *Acid Mine Drainage, Designing for Closure, GAC/MAC Annual Meeting* (Vancouver, Canada: BiTech Publishers, 1990), pp. 241–253. With permission.

not to extend the settlements of the rock masses in the shafts to the covering layers, which almost certainly would have destroyed their sealing properties over time. The latter shafts were sealed by a thick, unreinforced plug of concrete or Cefill (cement-stabilized fly ash).

The tailings were partly dumped into the shafts, and partly used as a filter layer separating the coarse waste rock from the fine-grained or concrete-like cover material. The top surface was smoothed and compacted before applying the sealing cover. On one of the heaps the sealing layer consisted of a crushed rock aggregate grouted with Cefill, which is a pumpable mixture of coal fly ash (about 91% by dry weight), cement (about 8% by dry weight), additives (about 1%) and water (an additional 35% to 40%) (Figures 6.12 and 6.13). On the other heap, the sealing layer was constructed of a compacted, rather dry clay found about 2 km from the site. The compaction of the clay took place in three different sublayers and had to be accomplished with a heavy bulldozer that made 6–15 passes before the clay became plastic enough to create a continuous layer free of hollows and cracks.

The specification on the sealing layer was that it should be less permeable to water than 1×10^{-9} m/sec (saturated hydraulic conductivity). The Cefill layer met that requirement except in some minor areas in which the permeability was slightly higher. Due to the puzzolanic properties it was judged that the layer would harden and become more tight over time. The clay layers became even less permeable than the specifications, almost one order of magnitude lower.

The protective layer was constructed out of a 2-m-thick layer of a glacial till found close to the native clay that was used as a sealing material. The till was applied in two sublayers, the first of which was only 0.5-m thick and with the maximum grain size of 300 mm in order not to create impacts in the sealing layer. The size specification of the upper 1.5-m layer was a maximum 600 mm. The layers were compacted by the weight of the loaded trucks, and the final surface was smoothed and vegetated by pine plants.

All the remedial works carried out were carefully controlled. The materials used were examined continuously or on a daily basis; for example, every square metre of all three sublayers of clay was videographed, and every shipment of fly ash was controlled by laboratory tests. Several hundred permeability tests in the field and in the laboratory were performed. During the field tests many permeability cells were installed permanently in the sealing layers (15 cells in the cover of each heap).

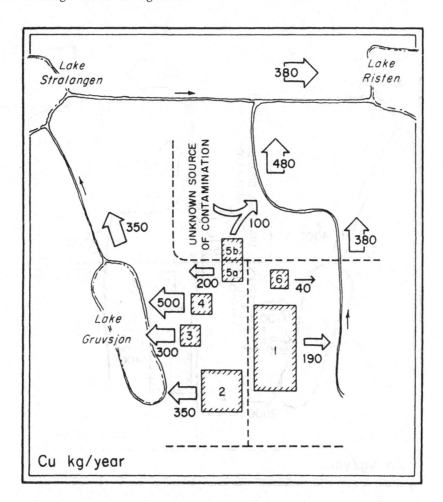

FIGURE 6.9 The discharge of copper from major waste units. (From Lundgren, A. T. Bersbo: The first full scale project in Sweden to abate acid mine drainage from old mining activities, in *Acid Mine Drainage, Designing for Closure, GAC/MAC Annual Meeting* (Vancouver, Canada: BiTech Publishers, 1990), pp. 241–253. With permission.)

Some lysimeters were installed directly under the sealing layer to collect the percolate through this layer. Some 15 lysimeters were installed in direct contact with the sealing layer to act as oxygen diffusion cells. They were carefully sealed and thoroughly flushed with nitrogen before the oxygen concentration was measured. After some months the oxygen measurements were repeated.

Lysimeters have also been installed under the reclaimed areas where all the waste rock was removed. The objective has been to monitor the quality of the percolating water before it reaches the groundwater.

If the sealing layer is as impervious as it should be according to the specifications, i.e., if the full-scale permeability follows the small-scale permeability without interferences of defects, such as cracks and hollows, then the sealing would be fully saturated with moisture and the oxygen diffusion rate would be very low. Hence, monitoring the moisture in the cover would be the perfect control of the efficiency of the sealing layer.

Tests were made with neutron/neutron probes in tubes that were drilled through the layers and 0.5 m into the waste rock. However, the measurements were not found to be reliable. Instead, calculations showed that the lateral transport capacity of the protective layer was too small to drain a normal rain percolation, and if the sealing layer acted as it was supposed to, it should result in a free groundwater table in the cover. Therefore, about 30 ordinary groundwater tubes were installed

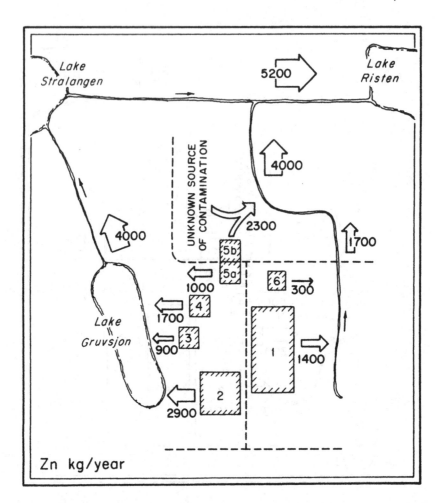

FIGURE 6.10 The discharge of zinc from major waste units. (From Lundgren, A. T. Bersbo: The first full scale project in Sweden to abate acid mine drainage from old mining activities, in *Acid Mine Drainage, Designing for Closure, GAC/MAC Annual Meeting* (Vancouver, Canada: BiTech Publishers, 1990), pp. 241–253. With permission.)

in the protective cover of the two heaps down to their respective sealing layer. The 300-mm bottom parts of tubes were perforated and dressed with fabrics.

The primary objective of covering the waste rock was to significantly reduce the transport of oxygen to the waste. Monitoring the oxygen concentration within the heaps is therefore of greatest importance because this full-scale project would act as a pilot project for the rest of the sulphidic waste deposits. Thus, a total of 11 small tubes was installed at different depths in the two heaps before the covering took place. Cocks were installed on the tubes after they had been brought through the protective cover.

Water quality monitoring has been performed in two separate programmes, covering the surface water and the groundwater separately.

The covering operation was completed in March 1989. Due to the great variations in water quality over time, it is still too early to draw definite conclusions about the surface and groundwater conditions. However, it is already clear that the discharge conditions are different at the two deposits. The discharge of the Storgruve deposit (which was covered by Cefill plus till) is also the drainage of the mine, as well as that of the waste that was dumped in the mine. The water composition of this stream seems to have been changed only slightly. The major change is the increase of iron discharge and the heavy precipitation of ferrous iron in the stream.

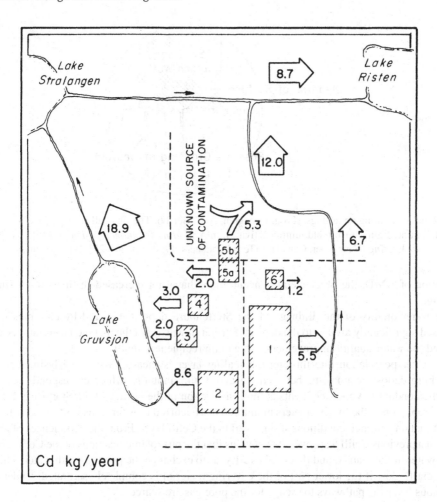

FIGURE 6.11 The estimated discharge of cadmium from the major waste units. (From Lundgren, A. T. Bersbo: The first full scale project in Sweden to abate acid mine drainage from old mining activities, in *Acid Mine Drainage, Designing for Closure, GAC/MAC Annual Meeting* (Vancouver, Canada: BiTech Publishers, 1990), pp. 241–253. With permission.)

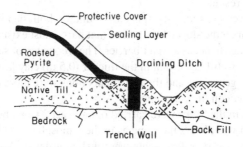

FIGURE 6.12 Sealing on tailings. (From Lundgren, A. T. Bersbo: The first full scale project in Sweden to abate acid mine drainage from old mining activities, in *Acid Mine Drainage, Designing for Closure, GAC/MAC Annual Meeting* (Vancouver, Canada: BiTech Publishers, 1990), pp. 241–253. With permission.)

Obviously, covering the waste has caused a reduction in the ferric weathering products, which results in an increased solubility of iron and a possibility of secondary oxidation of sulphides. The ferrous iron then oxidizes again, resulting in the iron precipitation outside the heap, and thereby causing a secondary lowering of the pH in the effluent water. Despite this secondary and temporary

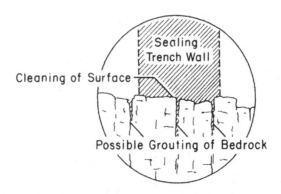

FIGURE 6.13   Sealing on tailings. (From Lundgren, A. T. Bersbo: The first full scale project in Sweden to abate acid mine drainage from old mining activities, in *Acid Mine Drainage, Designing for Closure, GAC/MAC Annual Meeting* (Vancouver, Canada: BiTech Publishers, 1990), pp. 241–253. With permission.)

production of AMD, the other metal concentrations have not increased in this stream since the covering.

The water quality of the drainage of the Steffenburg deposit (covered by clay plus till) has decreased significantly around to the deposit. Farther downstream the metal concentrations are still high and the water quality improvement in the main recipient is small.

The water percolation rate through the sealing layer is measured in three lysimeters on the Steffenburg deposit (clay) to be between 0.5 and 3 mm annually, which corresponds well to the hydraulic conductivity measurements at the construction of the layers. At the Storgruve deposit no water percolation to the five lysimeters monitored has occurred thus far. This is believed to be due to the fact that unsaturated conditions still prevail in the Cefill layer. From the experiences of an excavation in a previous Cefill test cover, it is obvious that the puzzolanic reactions of the Cefill consume all the water in the matrix, and that it takes >1 year to exchange the pore gas with water. This must be especially slow if the lower part of the covering material is completely saturated. Under such a continuous layer the pathways to escape for the pore gas are scarce.

The groundwater levels in the protective layer were measured after the layer was completed. It should be noted that the summer and fall were extremely dry, which is verified by the descending groundwater table. The table did not rise significantly until November of that year. The clay layer seems to be more tight than the Cefill layer, which conforms with the permeability measurements, but not with the lysimeter results.

It is also evident that wet spots and dry spots are found on both deposits that cannot always be explained by their position on the slope. This means that parts must exist in both sealing layers that are less efficient than average to act as a water barrier. The frequency of such parts may be higher in the relatively thin (0.25 m) Cefill layer than in the thicker (0.5 m) clay layer. On the other hand, the slope gradients are higher, generally, and the rim connections were more delicate to get well sealed off on the Cefill deposit than on the clay deposit.

Oxygen measurements in the heaps were carried out at least 2 years before the remedial actions started. The concentration was always the same as in the atmosphere at all depths, which indicated that the heaps of rather coarse rock fragments were well ventilated. Also, during the works proceeding the covering, the oxygen concentration was above 20%, with the exception of some calm days when the filter layer was almost completed and the concentration temporarily dropped a few percent.

A new series of oxygen measurements in the heaps started as soon as the coverage was completed on the Steffenburg deposit (clay). At that time the Cefill cover at the Storgruve deposit was not fully completed, yet a substantial decrease in oxygen concentration took place in the deposit. Some of the small holes through the Cefill cover were not sealed off until several months later, and this probably explains the slower response in this heap. The apparatus used for the measurements has an uncertainty of 0.2%, and the measurements showed values sometimes well below zero (despite calibration with nitrogen). Therefore, all measurements below 0.5% are uncertain. The variation of the values over time is also rather great. However, the measurements have shown that the sealing layers are very tight against oxygen transport.

What does a decrease in oxygen concentration to values below 0.5% mean? According to the laboratory experiments, the metal concentrations should be reduced, as a consequence of the decreased oxidation of sulphides, by a factor of at least 5. Together with the reduction of water percolation, which seems to be on the order of at least 99%, this should lead to a total reduction of the metal transport of at least 99.8%. This is probably too optimistic, as the leaching experiments use a lot more water than that corresponding to the percolation in the covered heaps. Regardless, the result should be very good, good enough to be followed up for several years to come.

6.54 DESIGNING CLOSURE OF AN OPEN PIT MINE IN CANADA

Based on present ore reserves, mine closure at Equity Silver Mines Ltd. is currently set at year-end 1992. As waste materials are predominantly acid generating, mine closure will not be a "walk-away" situation but, rather, will require ongoing maintenance of environmental and treatment facilities for an indefinite period.[10]

In developing the various scenarios for closure and anticipated cost liabilities, Equity Silver submitted a conceptual closure plan to MEMPR (Ministry of Energy, Mines & Petroleum Resources) in October 1988. The plan has been under intense review since submission. The review process initially went through the Equity Silver Surveillance Committee, an independent body consisting of federal and provincial government and interest groups in the area. This process took the better part of a year with recommendations from the review being used to set the conditions of a revised reclamation permit issued by MEMPR. These conditions formed the basis for future studies and refinement to the plan.

One specific condition of the permit is to establish a bond of size sufficient to meet environmental cost in perpetuity. To ensure no erosion of this fund base, annual surplus incurred is based on real interest rates; i.e., interest less inflationary increases. The ultimate bond size will be established on the annual projected maintenance cost to sustain environmental fixtures times the real interest-generating potential of the portfolio in which the fund is held.

Equity Silver actively pursues system improvement and research to optimize environmental systems. Data from research programmes developed onsite are reported annually and are used to update closure concepts.

Until adequate mitigative measures can be implemented, collection and treatment of AMD are required to safeguard the surrounding environment, hence, the bond placement to sustain the operation over the long term.

Equity Silver employs conventional AMD treatment using quicklime as a base, followed by agitation, then clarification within settling ponds. Approximately 800,000 m^3 of AMD is treated annually, producing approximately 80,000 m^3 of sludge consisting of 6% to 7% solids by weight. Sludge is presently pumped from the ponds and mixed with tailings. As long as the mine is in operation, this is considered to be an acceptable method for handling the material. However, with mine closure imminent this disposal will no longer be available and will require an alternate practice. High-density sludge (HDS) treatment is being considered and is at the feasibility stage. By using this process, sludge volumes can be reduced to 15% to 20% of that presently produced. Disposal options being considered include placement on filter beds for further dewatering or subaqueous disposal in one of the flooded open pits. The latter option carries some risk as pH depressions below 6.0 can redissolve and mobilize metals (especially zinc) within the sludge. Testwork does, however, exhibit that sludge has adequate buffering capacity to hold pH in the 7.5 range, with no mobilization of soluble metal.

Quicklime purchased from southern British Columbia represents a major portion of the cost to treat AMD. Equity Silver has staked local claims of limestone that has been kiln tested and confirmed to produce a good grade of quicklime. A study has been awarded to a consultant to investigate the viability of constructing a lime kiln locally to offset the high cost of transporting lime to the mine site.

Overburden is used as a cover and growth medium for sealing acid-generating wastes. This overburden, which consists primarily of glacial till is placed on dump surfaces at an average thickness of 0.75 m. Covers are graded and seeded to complete the reclamation process. Because it is virtually impossible to assess the mitigative effect these covers have on the waste dump acid-generation process, assessments are carried out using scaled-down test plots. These plots are constructed of well-compacted till basins with drained underflows terminating in 45-gal plastic drums to capture runoff. Seven of these plots have been constructed. Four contain oxidized waste material from the Southern Tail Pit, with the remaining three consisting of unoxidized material from the Main Zone Pit.

Each set of tests for the two waste materials is comprised of control piles (uncovered) to gauge water infiltration and metal loading vs that from piles covered with glacial till. The covered piles were not compacted, but have been seeded to establish a vegetation cover.

Samples are taken from the collection vessels after each major rainfall with volumes recorded and samples analyzed for pH, sulphate, acidity, and dissolved and total copper, iron, and zinc. Total loading of metal out of the pits is calculated on the basis of volume times qualitative analysis.

After tabulating the annual loading results for the Southern Tail Pit, we find that water infiltration is reduced by about 50% by the mill covers, metal loading by about 75%, and sulphate and acidity loading around 67%. Little change was observed in the Main Zone Pit material as acid-generation processes have not fully developed.

Although strong evidence exists that clay covers reduce oxidation and metal loadings out of waste piles, concern remains as to the longevity of the process. In order to eliminate metals leaching from wastes, infiltration of precipitation must be completely eliminated. A test programme is being designed to evaluate the practicality of covering a section of waste dump with a high-density PE membrane. Parallel tests will be run to assess the practicality of dump resloping, followed by till placement and compaction on steep slopes. The outcome of the test results may dictate an alternate cover technique to that presently used.

The Southern Tail Pit, which was the first zone to be mined, has been backfilled with acid-generating waste from the Main Zone Pit. Disposal of waste into this pit not only provides a readily available dump site, but also provides an area in which a portion of the waste can be disposed below water.

Upon completion of mining, water quality deteriorated in the Southern Tail Pit with pH dropping to 3.0 and copper and zinc values increasing to 13 and 11 mg/L, respectively. This deterioration was a product of oxidation of the pyrite in wall material and fractured waste in two slide areas. Prior to backfilling the pit, tests were carried out to assess the impact of waste rock disposal on final water quality. Results indicated that pH would recover to 7.5–8.0 and metal values would significantly decline.

Backfilling of the pit proceeded in three stages. The first step was to backfill to a horizon 1 m below the projected floodplain, followed by placement of a 2-m layer of inert non-acid-producing waste. This layer serves as a buffer zone to decrease AMD production at the water interface. The third stage of backfilling involved placement of waste above this barrier and water table. These wastes are also acid generating and are reclaimed to reduce oxidation rates. Water has been discharged to the environment from mid-1987 to the present and has not shown any deterioration in downstream water quality. An increase in zinc values has been noted over the last 2 years which is believed to be a product of AMD inflow from slide material. The pH of the receiving pool of water is not sufficient to fully precipitate zinc. Dump resloping and sealing tests were projected to be carried out during the summer of 1990. The objective of this testwork was to eliminate seepage through the wastefills, at least to a degree to which AMD input can be buffered by alkalinity within the system.

When mining of the Main Zone was completed, the waste from the adjacent Waterline Pit was placed within the Main Zone. The pit was flooded to cover wastes to eliminate acid generation.

As a portion of the wall within the Main Zone Pit is acid generating, a dam was constructed at the entrance to the pit to raise the water level. Postclosure water quality in the Main Zone Pit was modelled under a programme. The study concluded that pit water would turn slightly acidic during flooding, but would return to a neutral condition shortly thereafter due to the influence of alkaline groundwater. The model did not include the option of backfilling the pit with mine waste, which will further enhance alkalinity of the ponded water.

Tailings, like the greater distribution of waste rock, have the potential to produce acid given proper conditions. Fortunately, Equity Silver elected to build water-containing dams that allow tailings to be flooded through the operating stage and after the closure of the mine.

Two tests were set up to kinetically evaluate the acid-generating properties of tailings submerged below water. In the first test, tailings were placed in an aquarium, then flooded with freshwater. The water was allowed to fluctuate through evaporation, with tailings exposed to the air for months at a time. After being subjected to numerous wet and dry cycles, it was found that zinc and, to a lesser extent, copper appeared in solution. The supernatant remained neutral throughout the test.

With the second test, tailings were covered with water throughout the test. To simulate wave action and oxygen entertainment, air was continuously introduced to the water cover using a bubbler system. Zinc values reached 8 mg/L in a short time frame and remained at this level throughout the remainder of the test. The pH remained neutral.

Results of this testwork indicate that acid-generating tailings must be covered with water at all times but do not necessarily preclude oxidation and mobilization of soluble metals (zinc).

Although no formal testwork has been adopted to fully evaluate passive treatment options, strong evidence has come to light that swamps in the immediate area of the mine site have the capacity to remove metal from solution. In one particular swamp, to which treated water is discharged, removal of dissolved zinc is at 70%. Reliance on this swamp to remove trace values of metal, particularly zinc, may lead to the development of cultured swamps downstream of the mine site.

6.55 METAL CONTENTS AND TREATMENT OF MINE WATER

The Berkeley Pit is a large abandoned open pit mine in Butte, MT. Since 1982 it has been filling with water; the level is currently over 213 m (700 ft) deep. At the present rate of inflow, the pit water level will rise to the low point in the rim by the year 2011. Before that time, around 1997, water rise will reach exposed alluvium in the pit walls, creating a threat to groundwater quality in the valley south of the pit. In terms of contained volume of water and quantity of metal pollutants, the Berkeley Pit is unmatched by any acid-producing mine in the United States and possibly the world.

The Berkeley Pit intersects a multitude of openings to the old underground workings to the north and west, and below the pit.

The Butte area is drained by Silver Bow Creek, which is the headwaters of the Clark Fork River, a major tributary of the Columbia. The pit is at an altitude of 1,676 m (5,500 ft). Because of the altitude and latitude, the climate is cool, with harsh winters.

The deposits at Butte were huge. Prior to 1983, Butte had produced over 20 billion lb of copper, 4.9 billion lb of zinc, 3.7 billion lb of manganese, 850 million lb of lead, 750 oz of silver, and 2.9 million oz of gold.

6.56 WATER TYPES AND CONTENTS

Essentially two types of mine water are found in Butte: underground and pit. The two have the same elements in solution, but most of them are present in much higher concentrations in pit water than in underground water. The underground solutions are more reducing than the pit water, and the ionic species present in the two waters are somewhat different.

Table 6.19 shows the composition of some underground water samples and pit water samples, as well as federal monthly discharge standards. The Kelley Mine is near the Berkeley Pit and the

TABLE 6.19

Compositions of Various Mine Waters of the Butte Area

(mg/L)	Pit Water[a]		Kelley[b] 120 m	Travona[b] Surface	Federal Std.[c] Monthly
	1 m	100 m			
Al	103	193	3.26	<0.040	1
Total As	0.031	1.15	1.04	0.106	0.5
As (III)	0.001	0.087			
As (V)	0.005	0.768			
Cd	1.08	1.87	0.11	<0.005	0.05
Ca	433	482	407	179	
Cu	133	203	0.14	<0.004	0.15
Total Fe	202	1,020	285	4.64	1
Fe (II)	60	958			
Fe (III)	142	140			
Mg	153	280	156	51.1	
Mn	73.7	162	52.4	12.84	2
K	10.4	18.7	37.4	5.4	
Na	73.1	70.8	49.8	66	
Pb	0.112	0.522	N/A	<0.04	0.3
Zn	212	497	114	0.11	0.75
Cl	9.82	22.1	20	37.7	
SO$_4$	4,850	6,760	N/A	495	
SiO$_2$	85.6	111	N/A	17.9	
pH	2.700	3.150		6.88	6–9
Ion storing	0.100	0.160			

[a] Davis, 1988. SiO$_2$ at 100 m was analyzed and provided by Dr. J. Sonderegger at the Montana Bureau of Mines.
[b] Duaime, 1990.
[c] *Federal Register*, 1988.
Source: From Huang, Hsin-Hsiung.[11]

two are intimately connected by underground workings. Water flows from the underground mines to the pit. The Travona Mine is 2,440 m (8,000 ft) west of the pit and is not directly connected to the Kelley/Berkeley system, as most of the underground connections between the two areas were bulkheaded before mining stopped.

Pit water is more oxidized than underground water and is highly acidic. Water at the surface has an Eh of 820 mV and a pH of 2.8, while the water at 130 m below the surface has an Eh of 468 mV and a pH of 3.14. In general, metal concentrations in near-surface waters (0–10 m) are lower than in deeper waters, representing the bulk of the solution in the pit.

About 40% of the water inflow to the Berkeley Pit is from underground mines. The other 60% is from surface sources, both natural flow from the adjoining perched alluvium, and leakage from the active tailings pond and leach pads through the alluvium. At present, no way has been found to precisely calculate the relative amounts from the two alluvial sources, but the leakage from the tailings pond is believed to contribute much more water than the natural alluvial flow.

Table 6.20 is an estimated average composition of the surface flows to the pit. It was calculated by subtracting the underground contribution of the various elements from the measured quantities in the pit.

Table 6.21 shows the estimated amounts and percentages of the various elements contributed by the underground flow and the surface sources to the pit. Clearly, surface sources supply most of the metal ions.

TABLE 6.20

Estimated Composition of Surface Water Flows to the Berkeley Pit

Substance	Ca	Mg	Na	K	As	Al
mg/L	495	314	81.7	4.2	1.08	295
Substance	Cd	Cu	Fe	Mn	Zn	
mg/L	2.74	318	1,238	205	662	

Source: From Huang, Hsin-Hsiung. Characteristics and treatment problems of surface and underground waters in abandoned mines at Butte, Montana, in *Mining and Mineral Processing Wastes* (Littleton, CO: Society of Mining, Metallurgy, and Exploration, 1990), pp. 261–270. With permission.

TABLE 6.21

Amount and Sources of Berkeley Pit Constituents

Substance	Pit Content Tons	%	Bedrock Contribution Tons	%	Surface Contribution Tons	%
Al	6,400	100	4.79	0.01	6,395	99.9
As	25	100	1.53	6.12	23.5	93.9
Cd	61	100	1.62	2.66	59.4	97.3
Ca	16,700	100	5,970	35.7	10,730	64.3
Cu	6,900	100	2.06	0.03	6,898	99.97
Fe	31,000	100	4,180	13.5	26,820	86.5
Mg	9,100	100	2,290	25.2	6,810	74.8
Mn	5,200	100	769	14.8	4,431	85.2
K	640	100	549	85.8	91	14.2
Na	2,500	100	731	29.2	1,769	70.8
Zn	16,000	100	1,670	10.4	14,330	89.6

Source: From Huang, Hsin-Hsiung. Characteristics and treatment problems of surface and underground waters in abandoned mines at Butte, Montana, in *Mining and Mineral Processing Wastes* (Littleton, CO: Society of Mining, Metallurgy, and Exploration, 1990), pp. 261–270. With permission.

Water from the underground mines, even though it is in contact with large amounts of sulphides, remains relatively low in metals because no source of air or other oxidants are found in the flooded mines. The sulphides are not being oxidized and therefore do not contribute ions to the solution. The percentages of the various metals present in the tailings are much lower, but the collective tonnage is vast, the particles in the ponds are very small in size (as they have all been through crushing and grinding circuits), and well-aerated solution is steadily supplied to the ponds. Under these conditions, sulphides are continuously oxidizing in the tailings pond and the waste and leach dumps, contributing metals to solutions that eventually deposit in the Berkeley Pit.

The same reactions that contribute metals to solution provide sulphate from oxidized sulphide minerals. The resulting high concentrations of sulphate (4.2 g/L at the pit water surface and 12 g/L at 130 m) promote metal dissolution and hinder precipitation. Most of the metal ions present in pit water complex strongly with sulphate. For instance, 90% of the zinc dissolved at depth is present as $ZnSO_4$ (aqueous). This greatly lessens the activity of Zn^{2+} and prevents precipitation of the zinc compounds. Although stable metal-sulphate solids are seen, their stability is not high enough to offset the increased solubility of the metals.

Some minerals are near saturation in pit water and may now be precipitating from solution. The most common measure of saturation is the Saturation Index or SI, defined as the difference between the logarithm of the solubility constant of the mineral of interest and the logarithm of the product of the activities of the constituent ions in solution. This expression, however, does not correctly predict the degree of saturation. A better measure is the Individual Degree of Saturation or IDS, which is the actual amount of mineral that must be added to or removed from the solution to achieve saturation.

Table 6.22 shows both the SI and the IDS for a number of minerals suspected of being near saturation in pit water. Both surface and deep pit waters are near saturation with the ferric minerals iron hydroxide and jarosite, the aluminium minerals alunite and kaolinite, the calcium mineral gypsum, and amorphous silica. Thus, if ferric, aluminium, calcium, or silica ions are added to pit water without a corresponding addition of water, the result should be mineral precipitation rather than increased concentration in solution.

The concentrations of the metals in the surface water are much lower than in deep water (as shown in Table 6.22), even though the surface water is more acidic. This is due, at least in part, to the oxidative behaviour of iron. Near the water surface, atmospheric oxygen, aided by sunlight, causes iron to be oxidized from the ferrous or 2+ state to the ferric or 3+, which promptly forms a solid hydroxide, such as $FeOOH$ or $Fe(OH)_3$. As the solid particles form, they tend to adsorb or incorporate other metal ions into the ferric hydroxide structure. Eventually, the particles sink, carrying any incorporated ions down with them. This would account for the surface iron concentration of 202 mg/L, as compared to 1,020 mg/L at depth. Adsorption can explain the lowering of the deep arsenic concentration of 1.21 mg/L to 0.031 at the surface more completely than the formation of the most likely arsenic mineral, scorodite ($FeAsO_4$), which would require an arsenic concentration of around 100 mg/L. Manganese may behave in a manner similar to that of iron, being oxidized near the water surface from the 2+ state to 3+ or 4+, forming a solid oxide, and coprecipitating with the iron.

In the first few years of pit flooding, winter freezing may have played a role in lowering surface metal concentrations. When part of a solution freezes, the ice tends to have a lower concentration of soluble ions than the original solution. The resulting unfrozen portion of the solution is enriched in soluble ions, is denser than the original solution, and sinks, while the relatively clean ice, of course, floats. This process has been invoked to explain certain natural brines derived from seawater. It is certainly not a factor at present in the Berkeley Pit, as no ice cover formed in the winters of 1988–1989 and 1989–1990. The pit did freeze in previous years.

A practical process is needed for treating pit water to produce water meeting applicable drinking water/discharge standards and to recover the valuable metals. Because the volume to be treated is large and a plant must run for many years, simple processes are preferred over those more advanced requiring careful control and highly skilled operators.

A major concern is the disposal of unstable waste sludges produced by pit water treatment. The ideal disposal site for sludge is the area in which the sludge constituents originated: abandoned mine workings. The inactive underground mine workings in Butte would be very difficult to maintain access to and do not have sufficient volume to be a viable long-term disposal site, but the pit itself could be a very satisfactory permanent sludge pond. Any sludges placed there should be well understood in terms of composition and chemical interactions with pit water. Unstable ions in the sludges will certainly concentrate in the pit water, and ions introduced during treatment may interact with the native solution.

Metal concentrations in the pit solution can be lowered significantly by adding lime to raise the pH and sparging the solution with air to oxidize the metals in solution. First, the lime dissolves

$$CaO + H_2O = Ca^{2+} + 2OH^- \qquad (6.12)$$

and raises the pH. With the rise in pH, aluminium precipitates as the hydroxide

$$Al^{3+} + 3OH^- = Al(OH)_3 \qquad (6.13)$$

TABLE 6.22

Individual Degrees of Saturation and Saturation Indices of Pit Waters

	Water at 1-m Depth: Initial pH 2.7					
	Ferrihydrite	Geothite	Jarosite	Na-Jarosite	Hematite	Alunite
Initial metal						
Conc (mM)	3.51	3.51	3.51	3.51	3.51	3.74
Final pH	2.551	2.311	2.586	2.682	2.312	2.910
IDS (mM)	0.718	2.492	0.262	0.037	1.240	−0.343
SI	0.580	2.843	2.991	0.165	5.486	−1.923
	Jurbanite	Gypsum	Quartz	SiO$_2$ (am)	Kaolinite	Scorodite
Initial metal						
Conc (mM)	3.74	10.70	1.42	1.42	3.74 (Al)	0.005 (As)
Final pH	2.537	2.705	2.700	2.700	3.865	3.007
IDS (mM)	2.328	−1.778	1.213	−0.770	−0.748	−1.375
SI	0.571	−0.070	0.829	−0.188	−16.891	−5.361
	Water at 100-m depth: initial pH 3.15					
	Ferrihydrite	Geothite	Jarosite	Na-Jarosite	Hematite	Alunite
Initial metal						
Conc (mM)	0.251	0.251	0.251	0.251	0.251	7.26
Final pH	3.040	3.006	3.067	3.168	3.000	2.909
IDS (mM)	0.193	0.263	0.072	−0.014	0.131	0.226
SI	0.893	3.156	2.801	−0.294	6.111	1.710
	Jurbanite	Gypsum	Quartz	SiO$_2$ (am)	Kaolinite	Scorodite
Initial metal						
Cone (mM)	7.26	12.06	1.85	1.85	7.26 (Al)	0.001 (As)
Final pH	2.547	3.153	3.150	3.150	3.933	3.496
IDS (mM)	5.536	−1.563	1.638	−0.346	−0.175	−0.720
SI	1.248	−0.056	0.942	−0.074	−10.091	−2.915
	CuSO$_4$:5H2O	FeSO$_4$:7H2O		MgSO$_4$:7H2O	ZnSO$_4$:7H2O	MnSO$_4$
Initial metal						
Conc (m/W)	3.20	17.15		1.52	7.60	2.95
Final pH	3.518	3.587		3.801	3.787	
IDS (mM)	−546	−763		−2.431 (M)	−3.025 (M)	<−5.0 (M)
SI	−2.513	−1.962		−2.688	−2.892	−7.967

Source: From Huang, Hsin-Hsiung. Characteristics and treatment problems of surface and underground waters in abandoned mines at Butte, Montana, in *Mining and Mineral Processing Wastes* (Littleton, CO: Society of Mining, Metallurgy, and Exploration, 1990), pp. 261–270. With permission.

and the increase in calcium concentration precipitates gypsum

$$Ca^{2+} + SO_4^{2-} + CaSO_4 \qquad (6.14)$$

Other metal-sulphate complexes dissociate as free sulphate is removed from the solution

$$MSO_4 = M^{2+} + SO_4^{2-} \qquad (6.15)$$

so that still more sulphate is available for gypsum formation.

Exposure to air and light oxidizes iron in solution

$$2Fe^{2+} + 1/2O_2 + H_2O = 2Fe^{3+} + 2OH^- \qquad (6.16)$$

but at low pH the rate is slow. The rate is strongly dependent on the pH and a major concern is the pH level needed to have a reasonable rate of iron oxidation.

Oxidation is slower when sulphate, as Na_2SO_4, is increased. However, even though the sulphate concentration in pit water is high, iron oxidation is faster even than in the simple solution. This is due to the presence of other metal ions in the solution. The catalytic effects of at least some of them are enough to more than offset the extra sulphate.

Raising the pH does not significantly change the iron oxidation rate. Tests at a pH of 8.5 have shown a rate similar to that observed at 6.25. Other influences on iron oxidation might include photooxidation and iron bacteria.

As gypsum and ferric hydroxide coprecipitate, they tend to adsorb other ions from solution. Zinc, for instance, should theoretically remain in solution until the pH rises well above 5, but about 10% of the zinc is removed from solution at a pH of 5. Other metals behave similarly. In particular, arsenic is removed from solution by ferric hydroxide precipitation.

As the pH rises above 7, metals other than iron start to precipitate as hydroxides:

$$M^{2+} + 2OH^- = M(OH)_2 \qquad (6.17)$$

This process should be effective at removing most of the objectionable metals from solution at a reasonable pH. An exception is manganese. The stable oxide of manganese is MnO_2, which should form readily if the manganese is present in solution in the 4+ state. Because manganese in the pit solution is primarily present in the 2+ stage, an oxidation step for manganese is needed. Manganese can be oxidized by atmospheric oxygen, but the rate is slow below pH 9. That pH turns out to be the upper limit of water discharge standards, so it would be desirable to remove manganese from the solution at a lower pH. Other possible approaches include the formation of $MnCO_3$, which would not require manganese oxidation, solvent extraction, or use of an alternative oxidant.

The most practical approach to Berkeley Pit water treatment appears to be the use of a multistage process. The first step would raise the pH of raw pit water, while simultaneously aerating the solution. The objective would be to remove as much aluminium and iron as possible from solution as hydroxides. The large quantity of iron oxide produced would adsorb almost all of the arsenic present, while leaving most of the zinc and manganese in solution to be dealt with in later steps. The first-stage sludge would carry a large portion of the initial sulphate, as well as arsenic, iron, and aluminium and would be suitable for disposal.

The second stage of treatment would have the objective of precipitating zinc and possibly manganese, along with the remaining sulphate. The sludge from the second stage could then be treated to recover these metals, with residual sludge being recycled to stage 1. The primary means of treatment would be to raise the pH.

The reagents being considered to raise the pH in the first stage are lime (CaO), calcium carbonate ($CaCO_3$), and soda ash (Na_2CO_3). These are all commonly available and reasonably safe and cheap. Both lime and calcium carbonate are effective at raising the pH to over 5 and with both iron is air oxidized and removed as ferric hydroxide. The major difference is the behaviour of zinc. The use of lime resulted in larger zinc losses from the solution than with the use of calcium carbonate. When powdered lime was used to raise the pH to 5.38, 67% of the iron in solution precipitated, while 50% of the zinc was lost to the sludge. When calcium carbonate was used to raise the pH to 5.34, 84% of the iron precipitated and only 7% of the zinc.

This may occur because although calcium carbonate raises solution pH, it does so in a gentler manner than lime. As calcium carbonate reacts in acid solution

$$2CaCO_3 + 2MSO_4 = 2CaSO_4 + 2M^{2+} + 2H_2CO_3 \qquad (6.18)$$

it does not produce local pockets of high pH. Lime, added as the solid or slurried, dissolves faster than gypsum can form and creates local areas with much higher pH than the average solution pH. Reactions can occur in these regions that are not in general equilibrium with the system, such as the formation of zinc hydroxide. Thus, zinc and other metals coprecipitate to a lesser degree in a calcium carbonate system than in a lime or soda ash system. Calcium carbonate has other advantages over lime in that it is cheaper and has been reported to produce better settling and less voluminous sludges than lime.

The primary means of treatment would be to raise the pH. Calcium carbonate cannot raise the pH to a level that is high enough to be useful in the second stage. Unless the calcium carbonate particles are scrubbed, surface films inactivate them at a pH of about 6.5. Both lime and soda ash can raise the pH high enough to work. Lime would be a calcium source to precipitate the remaining sulphate, but soda ash would be a carbonate source for manganese and possibly zinc precipitation. It may be that a mixture of the two will turn out to be most effective.

It would be desirable to keep the final pH below 9, as discharge would have to be acidified if it is necessary to go higher for thorough metals precipitation. Methods that might be used to remove residual metals after second-stage treatment include zeolite and ion exchange and solvent extraction.

Other metals of concern in Berkeley Pit water include copper, lead, and cadmium. High levels of copper are present in pit water, and the recovery of copper is of economic interest. Cementation with iron

$$CuSO_4(aq) + Fe(s) = Cu(s) + FeSO_4 \qquad (6.19)$$

as a pretreatment before stage 1 is an obvious method to be tried, and preliminary tests indicate that copper in the pit solution will respond well to this approach.

Lead and cadmium are present in pit water at rather low levels, although their concentrations exceed discharge standards. Lead and possibly cadmium will probably precipitate in the stage 1 and 2 sludges.

As mentioned previously, when water freezes, ions in solution tend to stay in the unfrozen solution, resulting in a relatively clean ice. This principle works, to a degree, on Berkeley Pit water. Concentrations of metals in solution were lowered by about 50% in melted ice produced by freezing 10% of the solution. Freezing thus shows some promise as a treatment step. Freezing could also play a role in sludge disposal. It has been used to lower the water content and volume of sewage sludges and might be useful in lowering the volume of treatment sludges.

The climate in Butte is semiarid and the air is quite dry much of the time. Evaporation has been suggested as a means of releasing clean water to the atmosphere rather than to a stream. It turns out that an evaporation pond would be prohibitively large. Studies of evaporation rates in the vicinity indicate that 414 acres of pond area would be required to evaporate an average of 1 million gal/day, taking into account that a pond in Butte would freeze over in the winter and be successful only in the warm months. Treatment of 7.5 million gal/day would require a pond, presumably lined, of over 7,200 acres.

BIBLIOGRAPHY

1. British Columbia Acid Mine Drainage Task Force Report. *Draft Acid Rock Drainage Technical Guide*, Vol. 1 (Vancouver, Canada: BiTech Publishers, 1989).
2. Errington, J. C. and K. D. Ferguson. Acid mine drainage in British Columbia, in *Proceedings of the Acid Mine Drainage Seminar/Workshop, Halifax, Nova Scotia* (Ottawa: Environment Canada, 1987), pp. 67–87.
3. Morin, K. A. and J. A. Cherry. Migration of acidic groundwater seepage from uranium tailings impoundments, *J. Contam. Hydrol.* 2:323–342 (1988).
4. Lapakko, K. Prediction of AMD from Duluth complex mining waste in North Eastern Minnesota, in *Proceedings of the Acid Mine Drainage Seminar I Workshop, Halifax, Nova Scotia* (Ottawa: Environment Canada, 1987), pp. 187–221.

5. Asmund, G., M. Hansen, and P. Johansen. Environmental impact of marine lake tailings disposal at the lead-zinc mine at Maarmorilik, West Greenland, paper presented at the International Conference of Environmental Problems from Metal Mines, Roros, Norway (1988).

6. Boyles, M. J. Impact of Argo tunnel acid mine drainage, Clear Creek County, Colorado, in *Water Resources Problems Related to Mining* (Baltimore, MD: American Water Resources Association, 1974), pp. 41–53.

7. Hardie, M. G. and J. C. Jennett. Water resources problems and solutions associated with the new lead belt of S.E. Missouri, in *Water Resources Problems Related to Mining* (Baltimore, MD: American Water Resources Association, 1974), pp. 109–122.

8. Asmund, G. Rehabilitation and demolition after the closure of the zinc and lead mine Black Angel at Maarmorilik, Greenland, in *POLARTECH, Proceedings of the 3rd International Conference on Development and Commercial Utilization of Technologies in Polar Regions* (Copenhagen, Denmark: Danish Hydraulic Institute, 1990), pp. 744–759.

9. Lundgren, A. T. Bersbo: The first full scale project in Sweden to abate acid mine drainage from old mining activities, in *Acid Mine Drainage, Designing for Closure, GAC/MAC Annual Meeting* (Vancouver, Canada: BiTech Publishers, 1990), pp. 241–253.

10. Patterson, R. J. Mine Closure Planning at Equity Silver Mines, in *Acid Mine Drainage, Designing for Closure, GAC/MAC Annual Meeting* (Vancouver, Canada: BiTech Publishers, 1990), pp. 441–445.

11. Huang, Hsin-Hsiung. Characteristics and treatment problems of surface and underground waters in abandoned mines at Butte, Montana, in *Mining and Mineral Processing Wastes* (Littleton, CO: Society of Mining, Metallurgy, and Exploration, 1990), pp. 261–270.

7 Hydrologic Impact

7.1 INTRODUCTION

Surface mining affects surface streams through increased runoff and subsequent channel erosion as a result of reduced infiltration rates. The streams may also be affected by decreased surface runoff, which may be good or bad, depending on the specific regional, climatic, and geological setting. Decreased runoff results from diversions and increased infiltration rates where more permeable rock strata become exposed by surface mining.

Modifications of local or regional recharge zones by surface mining involve the alteration of infiltration rates by the removal of vegetative covers, the alteration of soil profiles, and compaction. Reduced infiltration rates decrease groundwater storage and reduce water availability.

Disruptions of groundwater systems and flow patterns by surface coal mining are of particular concern in the semiarid western region of the United States due to the importance of groundwater to livestock and irrigating agriculture in the region. Shallow and coal seam aquifers can be drained by mining activity, causing the temporary or permanent loss of existing wells near mined areas. The impact may not be restricted to the mined area and could extend several miles away from the site. Direct distribution of aquifers may be a very definite hazard where coal extraction is planned on a regional basis.

In south-eastern Montana the Decker Mine, located in an area of local discharge, was opened in the summer of 1972.[1] The Dl coal aquifer is about 15 m thick and provides ample water to stock and domestic wells. A study 1 year later reported that water levels in wells declined up to 8 m. The greatest water level decline occurred close to the mine in wells that were cut off from recharge areas. The rates of decline did not begin to recede after the first year, reflecting the constantly increasing stress on the groundwater flow system (Figure 7.1).

With underground mining, subsidence and fracturing of the overlying strata may cause surface runoff to be diverted underground and may disrupt aquifers, causing local water level declines and change in the direction of groundwater flow near the mine. An aquifer currently discharging in a single spring used by livestock and wildlife may, as a result of subsidence, discharge over a larger area by diffuse seepage.

In hydrologic systems in which multiple level aquifers exist, breach of the confining layer separating the aquifers will cause an increase in vertical flow and hydraulic connection between the affected aquifers. The effects will vary according to the location of the breach relative to regional and local flow patterns.

Contamination of shallow aquifers may also take place through downward seepage of poor quality mine water. The removal of considerable amounts of overburden could have an effect on the artesian pressure in the underlying aquifers. Reduced overburden pressure could cause upward movement of water and increase discharge into the mine pit.

The disruption of groundwater aquifers during mining results in a spoil that is generally less permeable than the original overburden. Reduced permeability inhibits lateral movement of groundwater through a previously undisturbed aquifer. Any water passing through such spoil will contain increased mineralization.

7.2 HYDROLOGIC IMPACT OF PHOSPHATE MINING

Each year, Florida mines approximately 40 million tons of phosphorite, most of which is ultimately converted to various commercial fertilizers. The phosphorite is mined from within the bottom

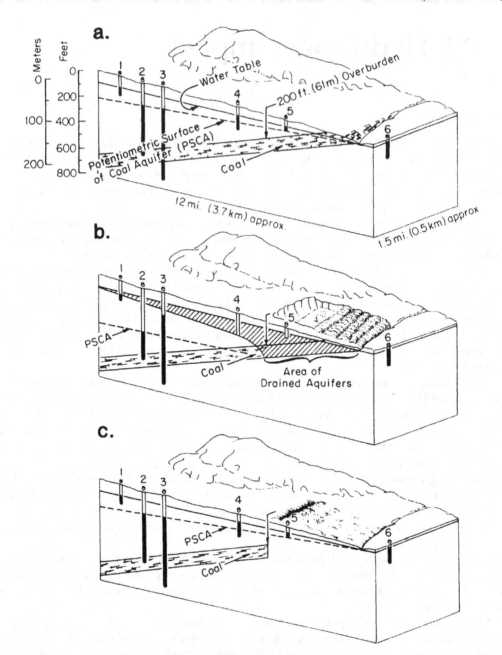

FIGURE 7.1   Impact of surface coal mining on the water table. (a.) Initial water level before mining, (b.) Lowering of water level due to mining, and (c.) Final water level after completion of mining (From El-Ashry, T. M. *Impacts of Coal Mining on Water Resources in the U.S.* (Denver, CO: Environmental Defense Fund, 1976).)

portion of the 40- to 60-ft thick surficial aquifer, using open-pit mining techniques. The ore zone or "matrix" is generally overlain by relatively permeable fine sand to silty and slightly clayey fine sand. The ore itself is highly variable in nature, but, in central Florida, it typically consists of less permeable clayey phosphatic sands and sandy clays with stringers of permeable sand. The ore is much sandier in the north Florida mining area.[2]

Prior to and during mining, the surficial aquifer is dewatered to at least the top of the ore zone to improve mine-cut stability and to improve ore recovery. The overburden sands are excavated and cast into the previous mine cut, after which the ore is excavated, slurried, and pumped to a beneficiation plant where the clay, sand, and phosphate "concentrate" are separated. Until the mined areas are reclaimed, generally 2–3 years postmining, the surrounding unmined areas continue to dewater.

Mined land reclamation in the Florida phosphate industry generally takes one of the following three forms: waste clay area, sand tailings fill area, and land and lakes area. In the first reclamation option, the mine pit is filled with waste clay, which is pumped directly from the beneficiation plant at approximately 3% solids. Embankments, up to 40 ft high, are constructed around the mine pit to contain the clay slurry. The clay is stored above ground to allow for long-term consolidation of the clay and to make certain that the surface of the clay does not ultimately settle below the surrounding topography. After the clay has consolidated and the upper 2–3 ft of clay becomes desiccated, the surrounding embankments are pushed down, sometimes onto the surface of the clay and sometimes onto the surrounding land, and the area is vegetated to complete reclamation.

The second reclamation option consists of pumping sand tailings from the beneficiation plant into the mine pit up to the level of the surrounding ground. Because the sand tailings dewater rapidly as they are deposited, no surrounding embankments are necessary. The surface of the sand tailings is ultimately vegetated.

The third reclamation option consists of flattening the spoil piles and allowing the mine pit to fill up with rain- and groundwater, creating a land and lakes area. Wetland-type vegetation is usually planted around the shore of the lake before the water reaches its ultimate level.

One of the primary objectives of reclamation is to return the mined areas, to the greatest practical extent, to their natural premining condition, particularly with respect to surface water and groundwater resources.

To determine the impact of mining and subsequent restoration of the surficial aquifer of the ground and surface water resources of the affected watershed, it is first necessary to model the premining condition. Figure 7.2 illustrates a typical phosphate mining area with a surface stream on one side and an unmined area on the other. The average hydraulic conductivity and thickness of the surficial aquifer in the study area were selected as 0.9 m/day and 20.7 m as determined from several aquifer pump tests and numerous test borings. A deep recharge rate of 25 mm/year under a hydraulic head difference of about 25 m (from published data) was judged representative for the study area. The surficial soils were considered to have a fair grass cover, typical of natural pastureland, and a minimum infiltration rate of about 10 mm/hr. A Soil Conservation Service (SCS) curve number of 66 was selected based on the minimum infiltration rate and the vegetative cover.

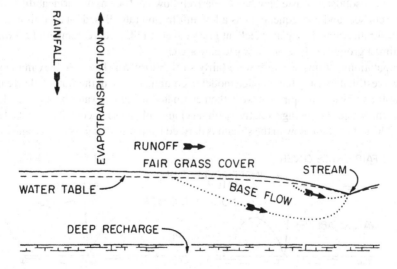

FIGURE 7.2 Natural conditions. (From Garlanger, J. E. Ground water restoration in mined areas, in SME Annual Meeting Preprint 90-76 (Denver, CO: Society of Mining Engineers, 1990).)

The first step in modelling the premining condition was to estimate the net recharge to the surficial aquifer, i.e., the base flow to surface streams. This was done both analytically, using a simple seepage model, and empirically, using measured stream flows. Once the base flow component was estimated, daily water balance calculations were performed using the computer model to determine the daily, monthly, and average annual runoff and evapotranspiration. These calculations were performed for each of the three sections of the affected watershed, i.e., the unmined area, the future mining area, and the stream setback area.

Note that as the distance to the stream decreases, the average annual surface runoff and evapotranspiration values decrease while the base flow increases. The base flow component is negligible in the unmined area of the watershed because as shown by the flow lines, most of the base flow is derived from the land closest to the stream. The weighted average annual surface runoff and evapotranspiration values agree quite well with published data for the modelled region.

Figure 7.3 illustrates a typical mined area prior to reclamation. The overburden soils are redistributed within the mined area as rows of spoil piles. The surrounding areas remain dewatered (as shown) until the mined area is reclaimed, generally 2–3 years after mining. Seepage and other water collected in the mine area are consumed within the beneficiation process.

In this scenario, the net recharge to the surficial aquifer, i.e., the amount of annual rainfall that reports as seepage into the mine cut and as deep recharge, was estimated using the simple analytical seepage model. The seepage model was also used to calculate the amount of base flow that is diverted from the stream to the mine cut. The water balance model was then used to determine the other hydrologic parameters, i.e., surface runoff and evapotranspiration.

As expected, the mined area experiences the greatest impact with respect to all of the hydrologic quantities. Surface runoff is virtually zero because of containment in the mine pit. Evapotranspiration decreases substantially because of the lack of vegetation. The base flow becomes negative as groundwater flows towards the mine both from the unmined area and the stream setback area. Furthermore, because the water table is drawn down to the pit bottom, the deep recharge is reduced significantly. As can be seen, the impact of mining on the adjacent unmined area and the stream setback area is also significant.

The waste clay generated during the beneficiation process is pumped to a mine pit, which is surrounded by earthen embankments constructed of the overburden soils. A typical waste clay pond is schematically shown in Figure 7.4. The waste clay usually contains a high level of nutrients essential for good vegetative cover. An excellent grass cover was considered in the modelling, as per normal practice. The consolidated waste clay has a relatively low coefficient of permeability, on the order of 10^{-0}–10^{-7} cm/sec, and consequently, has a low infiltration rate. Based on a minimum infiltration rate of 1 mm/hr and considering an excellent grass cover, an SCS curve number of 80 was assigned to the reclaimed ground surface of the waste clay area.

The computations for this scenario were fairly straightforward. The base flow in the stream setback area was estimated using the seepage model in combination with the water balance model. The surface runoff and evapotranspiration were then estimated using the water balance model.

Note that the waste clay, being a relatively impervious soil, does not contribute to the base flow of the stream. Most of the base flow to the stream is derived from rain that falls on the embankment and

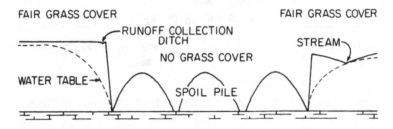

FIGURE 7.3 Mined scenario, 2 years after mining. (From Garlanger, J. E. Ground water restoration in mined areas, in SME Annual Meeting Preprint 90-76 (Denver, CO: Society of Mining Engineers, 1990).)

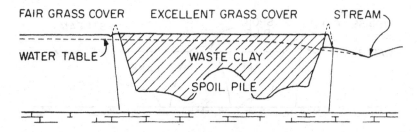

FIGURE 7.4 Final reclaimed scenario, waste clay area. (From Garlanger, J. E. Ground water restoration in mined areas, in SME Annual Meeting Preprint 90-76 (Denver, CO: Society of Mining Engineers, 1990).)

within the stream setback area. Because groundwater inflow from the clay pond area is decreased, the water table is lowered, the net recharge is increased, and evapotranspiration is reduced significantly in the stream setback area. In the reclaimed clay pond and unmined area, the evapotranspiration and surface runoff are close to premining values. Interestingly, the weighted average hydrologic parameter values for this reclaimed scenario are also close to the premining values.

Sand tailings generated during the beneficiation process are also deposited in some of the mined areas, generally up to the surrounding ground level. A typical sand tailings area is illustrated in Figure 7.5. The sand tailings generated in Florida are uniformly graded quartz particles with a fine content (per cent passing the U.S. No. 200 sieve size) usually <8%. Consequently, the sand tailings have a high coefficient of permeability, on the order of 10 cm^2/sec, and hence, a high infiltration rate. A minimum infiltration rate of 11.4 mm/hr was selected for this study. The sand tailings also contain some nutrients essential for vegetative growth. However, because of a low water retention capacity (i.e., available water for vegetative demand), sand tailings cannot sustain good vegetative growth. Therefore, poor grass cover was selected for the modelling of the sand tailings areas. An SCS curve number of 73, based on the high minimum infiltration rate and the poor grass cover, was assigned to the reclaimed ground surface in the sand tailings area.

The computation procedures for this scenario are similar to those used for the waste clay scenario. The seepage model was first used to estimate base flows. The water balance model was then used to determine surface runoff and evapotranspiration.

The results of the modelling indicate that evapotranspiration from the reclaimed tailings pond is reduced significantly from the premining condition. Even though the sand tailings are permeable enough for most rainwater to infiltrate, the available storage in the sand tailings is used up early in the rainy season, subsequently causing relatively high surface runoff. Because of the high transmissivity, base flow to the stream under this scenario is also relatively high. Note that the recharge from the tailings pond area keeps the water table in the stream setback area at a higher level. Because the water table has not changed significantly from the premining level, deep recharge remains at the premining rate. On average, base flow is higher, but evapotranspiration is lower than for the premining condition.

After reclaiming most of the mined areas as either waste clay ponds or sand tailings areas, some of the mined areas are reclaimed as land and lakes. In land and lakes reclamation, the perimeter of

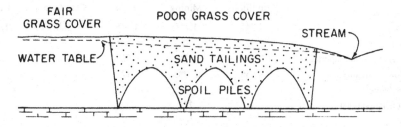

FIGURE 7.5 Final reclaimed scenario, tailings area. (From Garlanger, J. E. Ground water restoration in mined areas, in SME Annual Meeting Preprint 90-76 (Denver, CO: Society of Mining Engineers, 1990).)

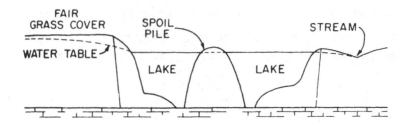

FIGURE 7.6 Final reclaimed scenario, land and lakes. (From Garlanger, J. E. Ground water restoration in mined areas, in SME Annual Meeting Preprint 90-76 (Denver, CO: Society of Mining Engineers, 1990).)

the mine pit is graded to a slope of 4 horizontal to 1 vertical or flatter and is vegetated. The mine pit is then allowed to fill up with rain and groundwater. A few tall spoil piles form islands within the lake. In some cases, artificial wetlands may also be formed along the perimeter of the land and lakes. A typical land and lakes reclamation scheme is illustrated in Figure 7.6.

As in the other scenarios, seepage into the lake and base flow to the stream were estimated using the analytical seepage model and runoff and evapotranspiration were estimated using the water balance model. Lake evaporation in Florida is typically on the order of 1,270 mm/year. Evapotranspiration from the land portion of the land and lakes area was considered to be an insignificant relative to the lake evaporation. The surface runoff was calculated by simply subtracting evaporation, base flow, and deep recharge from the rainfall.

As shown, the surface runoff from the lake (i.e., the overflow to the stream) is small compared with the surface runoff from the unmined areas. The evapotranspiration and surface runoff from the unmined areas are smaller than, but fairly close to, the natural background levels. On average, however, the evapotranspiration and surface runoff from the affected watershed are significantly different than the premining levels. The base flow component is also relatively high. The deep recharge rate remains at the premining level.

Table 7.1 summarizes the average annual hydrological quantities for the affected watershed under premining conditions, during mining, and for the various reclaimed conditions. After reclamation, however, the waste clay scenario provides hydrological values closest to premining conditions, indicating virtually no impact on the ground and surface water resources of the watershed. The reclaimed sand tailings scenario results in higher average base flow and lower evapotranspiration than premining levels. The reclamation scenario with the largest difference between premining and postreclamation scenarios is the land and lakes scenario. This scenario has a higher base flow to the stream and higher evapotranspiration/evaporation than the premining watershed. Consequently, surface runoff is relatively small as compared with the premining level.

TABLE 7.1
Weighted Average Hydrological Quantities (mm) for the Affected Watershed

			Postreclamation		
	Natural Conditions	During Mining	Tailings Area	Waste Clay Area	Land and Lakes
Rainfall	1,445	1,445	1,445	1,445	1,445
Surface runoff	371	51	391	363	160
Evapotranspiration/evaporation	993	726	902	1,011	1,151
Base flow to stream	56	<173>	127	46	109
Deep recharge	25	10	25	25	25
To beneficiation plant	—	775	—	—	—

Source: From Garlanger, J. E., Ground water restoration in mined areas, in SME Annual Meeting Preprint 90-76 (Denver, CO: Society of Mining Engineers, 1990).

Because all three reclamation options are generally used within a particular watershed, the overall impact of reclamation on surface and groundwater resources is intermediate between the range of impacts presented earlier for each option.

Simple analytical seepage models in conjunction with a computerized water balance model and some evaluation have made it possible to rationally measure the impact of mining and subsequent restoration of the surficial aquifer on the ground and surface water resources of the watershed affected by mining. The water balance program used for the analyzes presented in this chapter was developed for evaluation cover requirements for sanitary landfills. Considerable experience and judgment were required in applying it to the mining and reclamation scenarios in this chapter. A more comprehensive model capable of integrating groundwater seepage calculations with water balance calculations would be very helpful for studies of the type presented herein. It is to be emphasized, however, that a computer model can never replace actual field observations in evaluating the impacts of mining and reclamation on the surface and groundwater resources of an area.

7.3 HYDROLOGIC IMPACT OF PHOSPHATE GYPSUM DISPOSAL AREAS IN CENTRAL FLORIDA

Large quantities of phosphate rock are mined and processed in central Florida. The phosphate rock from the mines is further processed in "chemical plants" to produce phosphoric acid. A by-product of the processing of the phosphate rock to produce fertilizer chemicals is an impure form of gypsum referred to as phosphogypsum. For each ton of phosphate rock processed, approximately 1.5 tons of phosphogypsum is produced. The typical method to dispose of this by-product gypsum is to stack it in large piles, locally referred to as gypsum stacks or gypsum fields.[3]

The gypsum stacks have been the focus of many studies in recent years in an attempt to identify their potential for creating groundwater and/or air pollution.

As expected, most gypsum disposal stacks are located as close as possible or practical to the chemical plant in order to keep pumping costs to a minimum. They are often located next to the mining area. A typical gypsum stack is 400–600 acres and has an associated cooling water pond of approximately 250 acres. The gypsum slurry is transported from the chemical plant to the top of the stack using acidic process water. The gypsum slurry is deposited on the top of the stack, the gypsum settles out, and the process water is reused. The process water used to transport the gypsum to the top of the stack is recirculated to the plant generally via the cooling water pond. The process/cooling water is acidic, containing sulphuric and phosphoric acid from the digestion of phosphate rock with sulphuric acid.

In most cases runoff from the side slopes of the stacks is collected in ditches surrounding the perimeter of the stacks. The process water is returned to the chemical plant for reuse in unlined ditches or pipelines.

Typically, once the stack reaches a height of 100–150 ft, another stack is started in a new location and/or the existing stack is expanded. However, due to difficulties in obtaining permits, recently some stacks have been proposed for heights of up to 200 ft.

In the past these plant facilities were generally located in areas away from population centres. However, in recent years, Florida has experienced unprecedented growth and areas which were once remote and removed from population centres are now being surrounded as suburbs grow.

The upper surficial aquifer and the Floridian aquifer are the principal groundwater sources in central Florida. In most instances, these two aquifers are separated by a confining bed which may have an intermediate aquifer system. Underlying the lower Floridian aquifer is another confining bed. The upper surficial or water table aquifer is principally composed of sand, clayey sands, and, in some areas, shell and gravel beds.

The Floridian aquifer consists principally of porous limestones. The confining units are generally sandy or silty clays, clays and marls, and/or dense limestones and dolomites or dolosilts.

The surficial aquifer is unconfined and rises or falls in response to rainfall and discharges to streams and underlying aquifers. The water level of the surficial aquifer lies below the land surface generally from about 4 to 10 ft in the area of most of the gypsum stacks.

The water in the Floridian aquifer is generally confined. Recharge to the Floridian aquifer is principally by lateral flow, leakage through confining beds, and recharge in Karst regions of Florida. Fortunately, most gypsum stacks are located in an area of low recharge to the Floridian aquifer. The general natural flow of groundwater in the central Florida phosphate district is south-westward towards the Gulf of Mexico. Since about 1975 the U.S. Geological Survey (USGS) has monitored and mapped the wet and dry season potentiometric level of the Floridian aquifer. During the winter months agricultural pumpage in south-central Florida can reverse the discharge flow.

The sandy surficial sediments which comprise the water table (surficial) aquifer are typically 5–50 ft thick. These surficial sediments are underlain by 20–80 ft of inner bedded phosphatic, sandy, shelly, clayey, marley sediments that comprise the Pliocene Bone Valley formation. The Miocene Age Hawthorn Formation underlies the Bone Valley Formation. The Hawthorn is an impure marine dolomitic limestone which contains varying concentrations of phosphate and quartz sands, clay, marl, and dolomite and reaches thicknesses of 100 ft. In many areas, the lower portion is an inter-mediate aquifer-producing zone. Underlying the Hawthorn is the Miocene Tampa Formation. The Tampa is similar to the Hawthorn but contains less dolomite and has more clay beds. The Tampa ranges from a few feet thick to 100 ft thick. Portions of the upper Tampa are a thick sequence of Oligocene to Eocene aged limestones. These limestones are hundreds of feet thick and comprise the principal Floridian aquifer. Granular evaporites generally underlie the Floridian aquifer.

The processed water from the chemical plants that is used to slurry the gypsum to the disposal area is highly acidic (pH of 1.4–1.8) and has a high dissolved-solids concentration at about 28,000 ppm. The predominant contaminants are sodium, phosphate, fluosilicates, hydrogen, and sulphate.

Native groundwater has a dissolved-solids concentration of approximately 500 ppm with a pH of generally <7.0.

Migration of radionuclides, fluosilicates, phosphates, and trace metals is easily precipitated as the acid is neutralized by the carbonate in aquifer fabrics.

Recent monitoring data for some operating plants indicate the chemical front is slowly creeping out from the field as the "carrying or absorptive properties" of the aquifer fabric are reached. As a result of the increasing chemical fronts, regulatory agency personnel are putting increasing pressure on the operators to contain/prevent the leaking of process water from the gypsum stacks.

In the past, the gypsum disposal fields were constructed either on natural unmined land or in many cases directly in the mined lands associated with the phosphate mining process. This meant that the gypsum was deposited directly upon the existing land surface or on the top of the Hawthorn Formation (Figure 7.7). However, during the past 10–15 years, several approaches have been taken to locate the stacks in areas which would alleviate the potential for groundwater contamination. Initially, to protect the surface water, ditches were dug around the stacks to collect the runoff and

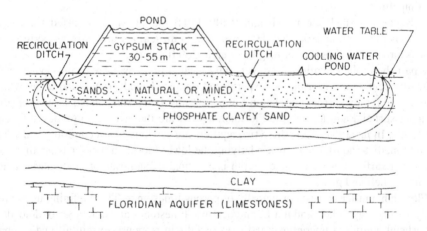

FIGURE 7.7 Gypsum stack over existing surface. (From Garlanger, J. E. Ground water restoration in mined areas, in SME Annual Meeting Preprint 90-76 (Denver, CO: Society of Mining Engineers, 1990).)

seepage from the slopes of the stacks. This worked to collect the surface water runoff from the gypsum disposal areas.

In the early 1980s attempts were made to site stacks in areas in which naturally occurring thick clays could be used as a natural liner. In some areas of the central Florida district, the Hawthorn Formation is very impermeable and is quite thick. In the early 1980s, USS Agri-Chemicals used a modification of this approach in an area in which the Hawthorn was very impermeable and waste clays existed. In addition, a ditch was dug around the stack to prevent lateral migration of leachate. However, the water level in the ditch had to be carefully controlled to prevent migration of contaminated groundwater from the stack into the surrounding surficial aquifer.

In the mid-1980s Gardinier Chemical proposed an extensive system consisting of a compacted clay liner and underdrains overlying a thick sequence (1,520 ft) of naturally occurring Hawthorn clays in their permit application for a new gypsum stack. This was a very elaborate system of underdrains, liners, slurry walls, etc. Due to the Gardinier Chemical Plant's location on Tampa Bay and proximity to nearby population centres, these measures were required to ensure that the stack would be permitted, and that the groundwater would be protected. More recently, IMC Fertilizer (IMCF) has proposed the construction of a new gypsum disposal stack. Initially, IMCF proposed more conventional stack construction techniques in which the stack would be built directly upon the Hawthorn Formation in a mined-out area. Recently, due to increasing pressure from the regulatory agencies, it has revised its plans and proposed the installation of a synthetic liner beneath the stack and a slurry cutoff wall along portions of the cooling water pond (Figure 7.8). The State of Florida is presently considering a measure requiring all new gypsum stacks constructed in Florida to be built with a liner to protect the groundwater.

In summary, recent studies have indicated that some potential groundwater impacts are associated with phosphogypsum disposal areas in Florida. Most of these studies have indicated that the lateral groundwater impact to the surficial aquifer system extends beyond the existing non-lined gypsum disposal stacks for approximately 1,500 ft. In some cases, contamination has been reported in the intermediate aquifer system. The various regulatory agencies including the Environmental Protection Agency (EPA), FDER, and other state and local governments have continued to increase the pressure for permit applicants to design gypsum stacks which will protect the groundwaters of the state. In the past 10 years, gypsum stacks have been designed and sited in order to use the natural confining layers and buffering sediments that occur in Florida. Artificial compacted clay liners and slurry walls, and, more recently, synthetic membranes overlying the natural confining carbonate sediments to mitigate and control the leachate from gypsum disposal systems have been designed. The results of groundwater monitoring for these newly proposed stacks, once constructed, will be used to determine the next generation of controls and constraints to be applied to gypsum stack permit conditions.

7.4 HYDROLOGIC EFFECT OF SUBSURFACE COAL MINING IN THE APPALACHIAN REGION

The bedrock in the region consists of interbedded sandstone, siltstone, shale, limestone, and coal. Sandstones are usually <6 m thick and limestones are usually <1 m thick. Coal beds typically are <3 m thick.[4]

The principal economic seams that have been developed include Pittsburgh (PA), Sewickley (PA), and Waynesburg (WV). The eastern part of the Pittsburgh coal reserve has been removed largely by subsurface mining.

Fractures and joints occur in most of the rock types. Surface joints that are often exhibited tend to follow a regional structure.

Precipitation varies from 1 to 1.3 m annually. A part of the precipitation evaporates and transpires, another part flows overland into streams, and the remaining part seeps downward through soils and rock to the zone of saturation. Within the zone of saturation, the movement continues downward and laterally towards discharge locations.

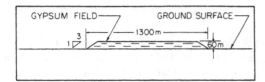

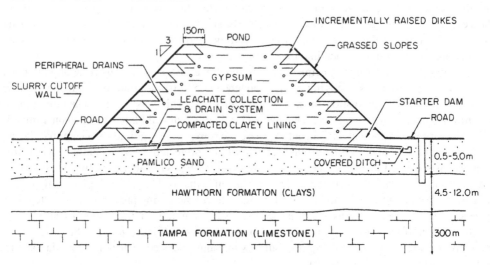

FIGURE 7.8 Modified gypsum stack to protect groundwater. (From Garlanger, J. E. Ground water restoration in mined areas, in SME Annual Meeting Preprint 90-76 (Denver, CO: Society of Mining Engineers, 1990).)

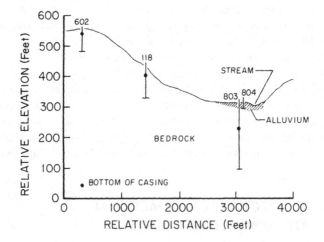

FIGURE 7.9 Water-level hydrographs. (From Stoner, J. D. Probable hydrologic effects of subsurface mining, *Ground Water Monit Rev.* Winter: 128–136 (1983).)

Water-level hydrographs in the four wells shown in Figure 7.9 showed the greatest fluctuations in the hilltop wells and the least change in the valley wells. The differences were primarily attributed to variations in potential infiltration aquifer storage capacity, permeability, and location with respect to the groundwater flow system.

A conceptual model of groundwater flow based on the data analysis from the region is shown in Figure 7.10. Most groundwater circulation occurs within 45 m of the land surface where the more fractured shallow bedrock aquifers occur. Downward movement beneath hills is retarded by low permeability confining layers. At shallow depths below the water table, such confining layers can

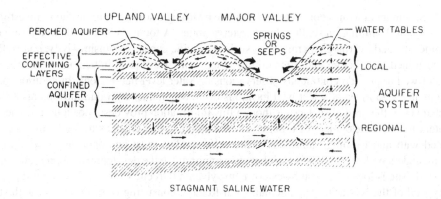

FIGURE 7.10 Conceptual model of groundwater flow. (From Stoner, J. D. Probable hydrologic effects of subsurface mining, *Ground Water Monit. Rev.* Winter 128–136 (1983). With permission.)

cause perching of groundwater and lateral flow to hillsides where the water discharges at springs or seeps. At greater depths, the hydraulic head is commonly large enough to force vertical leakage through confining layers into the underlying confined aquifer unit.

Water loss due to the mining of the Pittsburgh coal seam has been reported in a number of cases. Groundwater levels have been recorded in wells located over three active underground mines. The type of mining and the distance between the well bottom and the mined coal seam differs at each site. Generally, the magnitude of water level decline is expected to be inversely proportional to the thickness of bedrock between the mine and the well bottom. In addition, retreat (pillar removed) and longwall methods of mining may induce fracturing above the coal seam and hydraulically connect the shallow aquifers to the deep mine. Such problems may explain why no decline is seen in well GR-118, which is 73 m (240 ft) above a room and pillar type mining operation, and a 3-m (10 ft) decline is seen in well GR-543, which is 185 (608 ft) above a longwall face.

A simulation of probable hydrologic effects of underground mining was conducted using a two-dimensional cross-section and a digital finite difference programme. The simulation results indicated the following:

- Water levels could decline as much as 4.6 m (15 ft) in 46-m (150 ft) deep wells located along undermined villages. The maximum effects of water level decline would occur within 1 year of mining.
- Springs and shallow wells above drainage probably will not be affected.
- Streamflow may be reduced by 6.6 L/sec/km^2 (0.6 ft^3/sec/mi^2) in 1 year after completion of mining. Larger reductions could occur with higher permeability vertical fracture zones.
- The presence of vertical fracture zones could magnify and accelerate the drawdown effects and mine inflow.

7.5   EFFECTS OF LONGWALL MINING ON HYDROLOGY

Longwall mining relies upon overburden caving to restabilize the large rock mass that is disturbed during coal extraction. Caving characteristics are known to play a very important role in the surface subsidence and the overall pressure redistribution underground. Since the advent of high-extraction mining techniques, concern for local, domestic, and agricultural water supplies has increased. Detailed assessments of the impact of high-extraction mining techniques to local domestic water supplies are very limited. Prior investigations have addressed changes in the well yield and water quality, but few have included a complete hydrological study addressing results both before and after undermining occurs.[5]

The U.S. Bureau of Mines conducted a complete hydrologic study at a mine site to investigate the effects of longwall mining on shallow local water supplies. A total of eight water wells were drilled at the mine site, and hydrologic parameters, such as specific capacity, transmissivity, and well yield, were determined both before and after mining. Electronic water level indicators were installed on each water well to provide continuous observations of water level fluctuations. In addition, periodic measurements were made at select well locations of mining-induced ground movements to characterize changes in the ground surface. Although this information provides a portion of the necessary parameters for the development of a better understanding of the changes to the hydrologic regime associated with undermining, numerous site-specific studies must be conducted before all of the effects are understood. Such studies are inherently important to the development of predictive models.

The study site is located in southwestern Pennsylvania (Greene County). The topography of the site is typical of the Northern Appalachian Coal Region, consisting of hilly terrain, with steep to moderate slopes. The land use in the vicinity of the study site is primarily agricultural. Other areas are relatively inaccessible due to dense vegetation.

The study area overlies two adjacent longwall panels. Panel 1 is approximately 720 ft wide and 7,600 ft long, and panel 2 is approximately 750 ft wide and 7,700 ft long. The overburden or depth of cover over the study site varies from approximately 975–585 ft.

According to the USGS, the Greene County groundwater flow system is complex and strongly controlled by secondary permeability in the form of fracture zones and bedding plane fractures. The annual precipitation for Greene County ranges from 38 to 41 in., where 22–25% of the mean annual precipitation circulates through the aquifer systems via these fracture zones and bedding plane fractures. Stoner reports that fracture zones or openings tend to diminish as the overburden depth increases and aquifers beneath valleys exhibit larger average hydraulic conductivity or permeability than aquifers beneath hilltops. Also, shallow groundwater systems have been documented to be dependent on seasonal climatological data in relation to flow quantity.

A total of eight (6-in. diameter) wells were drilled and placed strategically above the two longwall panels. The wells were not completed in a unique, individual aquifer; rather they were drilled in a manner similar to that of the local area and were thus completed in the relatively shallow saturate zone. To ensure an open wellbore for the life of the study, Schedule 80 (4-in. diameter) PVC casing was installed to total depth. The wells were positioned roughly perpendicular to the trend of the longwall panels. This alignment permitted observations of mining effects along a profile line extending across both longwall panels. Well Nos. 2, 4, 6, and 8 were located at quarter-panel width. Well Nos. 1 and 5 were placed at the centre of each respective panel. Well Nos. 3 and 7 were located above the gate roads between the two panels. The drilling depths of all wells are shown in Table 7.2.

Data were collected from all wells to monitor the hydrologic parameters associated with shallow aquifers before, during, and after mining. In addition, various hydrologic parameters were determined and included well yield, transmissivity, specific capacity, and water level fluctuations. Initial data collection began 8 months prior to undermining of Well No. 1 to determine baseline data on water levels and hydrologic parameters.

Drawdown and recovery tests were performed before and after undermining on all wells to measure significant hydrologic parameters.

Electronic recorders that monitor continuous water level fluctuations were installed on all wells. The wells were configured to permit continuous recording of water level fluctuations. The set-up consisted of a pulse generator, encoder, and a 12-V battery. Fluctuations of water levels are monitored via the pulse generator and are transmitted to the encoder. The encoder is a battery-operated instrument for the automatic real-time monitoring of water level data and is capable of computing a data point every 10 min up to once every 24 hr. For this study, the encoder was programmed to monitor water levels every 4 hr. Remote communication of the encoder is provided via a modem. The instrument can be accessed by telephone. On-site communication with the encoder was performed utilizing a portable computer.

Table 7.2 displays the initial or static water levels, well yields, specific capacities, and transmissivity values for the respective wells. Well yield is simply the volume of water per unit of time

TABLE 7.2

Premining Data

				Well No.			
1	2	3	4	5	6	7	8
Well depth (ft)							
268.0	227.0	167.0	167.0	167.0	167.0	167.0	167.0
Initial water level (ft)							
140.50	202.25	103.00	111.00	71.75	68.00	28.25	14.00
Well yield3 (gpm)							
4.85	NA	5.53	12.07	6.22	4.94	7.34	9.46
Specific capacity[a] (gpm/ft drawdown)							
0.071	NA	0.128	1.069	0.115	0.085	1.471	0.076
Transmissivity (gpd/ft)							
10.67	NA	17.57	127.79	39.80	29.64	712.23	32.43

Source: From Owili-Eger, A. S. C., Geohydrologic and hydrogeochemical impacts of longwall coal mining on local aquifers, in SME Annual Meeting Preprint 83-876 (Denver, CO: Society of Mining Engineers, 1983).

[a] All values were obtained utilizing a pump test.

NA = not applicable due to limitations of equipment used for testing.

discharged from a well, either by pumping or by free flow. It is measured commonly as the pumping rate in gallons per minute (gpm). For this study, all well yields prior to undermining were determined by tests. Well yield was calculated as the volume of water that was released or pumped from storage in a given length of time. The specific capacity of a well is its yield per unit of drawdown, usually expressed as gallons per foot of drawdown. Dividing the yield by the drawdown, each measured at the same time, gives the value of the specific capacity. Specific capacities for all wells prior to mining are shown in Table 6.2.

Transmissivity is the rate at which water will flow through a vertical strip of water-bearing formation 1 ft wide and extending through the full saturated thickness, under a hydraulic gradient of 1.00 or 100%. The transmissivity of a formation is especially important because it indicates how much water will move through the formation. It was originally decided to collect transmissivity information via standard pumping tests and an observation well. However, the low yield rates observed precluded the use of the standard pumping test. As an alternative, a procedure was used for testing. The time-recovery data from the pumped well were used for determining the transmissivity of the formation. The time-recovery data for the pumped well are more accurate than the time-drawdown curve because during the recovery period water-level measurements can be made without interference from pump vibration and without the effects of momentary variations in the pumping rate. The premining transmissivity values are shown in Table 7.2.

Postmining well yield data for Well Nos. 5, 6, 7, and 8 were determined through pump tests. Well Nos. 1, 2, 3, and 4 experienced effects or casing restrictions due to the mining of panel 1. Well yield values for these wells could not be determined using a pump test due to these conditions. Table 7.3 displays postmining well yield data for the four remaining wells.

The specific capacity of Well Nos. 1, 2, 3, and 4 could not be determined (well yield information was needed and such postmining information was unavailable). The postmining specific capacities of Well Nos. 5, 6, 7, and 8 are shown in Table 6.3.

Postmining transmissivity parameters for all wells were determined utilizing the pump test procedure and a "slug" test procedure. This "slug" test method of determining transmissivity values was used on Well Nos. 1, 2, 3, and 4. In this method a known volume or "slug" of water is suddenly injected into or removed from a well and the decline or recovery of water level is measured at closely

TABLE 7.3

Postmining Data

				Well No.			
1	**2**	**3**	**4**	**5**	**6**	**7**	**8**
Final water level (ft)							
248.07	NA	109.41	107.00	67.50	69.00	22.50	26.00
Well yield (gpm)							
NA	NA	NA	NA	4.34	5.10	6.91	5.57
Specific capacity (gpm/ft drawdown)							
NA	NA	NA	NA	0.094	0.097	1.256	0.056
Transmissivity[a] (gpd/ft)							
17.08	NA	47.92	207.16	60.82	96.55	1,462.00	45.16
Transmissivity[b] (gpd/ft)							
NA	NA	NA	NA	20.64	56.24	776.35	10.95

Source: From Owili-Eger, A. S. C., Geohydrologic and hydrogeochemical impacts of longwall coal mining on local aquifers, in SME Annual Meeting Preprint 83-876 (Denver, CO: Society of Mining Engineers, 1983).

[a] Values obtained utilizing a "slug" test procedure.

[b] Values obtained by utilizing a pump test.

NA = not applicable due to offset in well casing limiting installation of downhole instruments.

repeated intervals. Table 6.3 displays postmining transmissivity data for all wells utilizing both the pump test and "slug" test procedures.

Water level fluctuations were observed to varying degrees during the life of the study. A portion of the variation in fluid level can be attributed to the regional drought that occurred during the study. Some of the effects, however, were caused by mining. It was believed that the effects of mining on the remainder of the well array would be minimal due to the proximity of the wells to the mine workings. Furthermore, it was impossible to isolate the effects of mining in the remainder of the wells due to the masking effects of the regional drought.

As mentioned earlier, postmining values for well yield on Well Nos. 1, 2, 3, and 4 were not performed due to downhole restrictions. Well yield tests on Well Nos. 5, 6, 7, and 8 were performed and showed no considerable change that can be attributed to the undermining of panel 1. A minor change in well yield occurred in Well No. 8. This observation, however, was considered to be insignificant.

A comparison of the pre- and postmining specific capacity measurements for Well Nos. 1, 2, 3, and 4 is limited due to downhole restrictions. Measurements for Well Nos. 5, 6, 7, and 8 were obtained and showed no major increases or decreases due to undermining of panel 1.

Tables 7.2 and 7.3 provide transmissivity values for all wells pre- and postmining, respectively. As mentioned earlier, postmining transmissivity values for Well Nos. 5, 6, 7, and 8 were performed utilizing both a pumping test and a "slug" test procedure. Therefore, a direct comparison utilizing one method of determining transmissivity was not obtainable for these wells. Given that two different testing methods were utilized, Well No. 1, located at the centre of panel 1, experienced minimal change in transmissivity. Perhaps subsurface fracturing, which may have occurred as a result of mining, had little effect on Well No. 1. Well Nos. 3 and 4 experienced small increases in transmissivity. Well Nos. 5, 6, and 7 showed slight increases in transmissivity after the undermining of panel 1, whereas a small decrease in transmissivity was observed at Well No. 8 (corresponding to the decrease in well yield and specific capacity previously mentioned).

A dramatic fluctuation in water level occurred in Well No. 1, located over the centreline of the panel, when the longwall was in the vicinity of the well array. In fact, the fluid level began to change when the longwall was about 100 ft from the well. The fluid level began to rise, then fell abruptly when the longwall passed directly below the well. This change is exactly the same as that observed at another test site. Subsequent measurements showed the fluid level beginning to rebound. However, a downhole problem (perhaps an offset or constriction of the casing) occurred when the longwall face was 425 ft beyond the well. This condition limited further data collection. A measurement made 659 days after undermining showed the fluid level to have recovered 20 ft.

Well No. 2, positioned approximately one half the distance between the centreline and the gate road area, experienced a decline through a large portion of the study. The decline was perhaps related to the regional drought. In fact, the rate of fluid level decline did not substantially change as the longwall passed below. Approximately 27 days after undermining occurred (or when the long-wall face was 535 ft beyond the well), the water level dramatically declined below the bottom of the well. Shortly thereafter, the casing offset and further measurements were impossible. It is unknown whether the fluid level recovered because the casing offset occurred at a position some 120 ft above the highest recorded water level for this well.

Well No. 3, located over the gate road area between longwall panels 1 and 2, showed a decline in fluid level; however, the effects of mining and the regional drought could not be separated.

The following conclusions were drawn:

- Well yield and specific capacity for wells located beyond the current panel (i.e., the panel being mined) appear to be unaffected by mining.
- Transmissivity did not change for the well located in the centre of the current panel. Changes in transmissivity values for the remaining wells in the array did not appear to correlate with mining.
- The largest water level fluctuations occur in wells that are directly undermined. Changes in water level position appear to be minimal for wells located in adjacent longwall panels.
- Water level fluctuations for wells located in the current panel begin to occur when the longwall face is less than one overburden thickness from the well.
- Water level fluctuations and ultimate water loss occur before the process of subsidence is completed. Fluid level recovery also appears to begin prior to completion of the subsidence process.
- A loss of water may only be a temporary condition; fluid level recovery was observed in all wells affected by mining (with the exception of the well, where measurements were impossible). Furthermore, a loss of water does not imply that all of the overburden rock mass has been lost to the gob zone. Water may indeed be at some level below the bottom of the well (Table 7.4).

TABLE 7.4
Initial and Final Water Levels

Well No.	Initial Water Level (ft)	Final Water Level (ft)[a]	Relative Change (ft)
1	140.50	248.07	−107.57
2	202.25	NA	NA
3	103.00	109.41	−6.41
4	111.00	107.00	+4.00
5	71.75	67.50	+4.25
6	68.00	69.00	−1.00
7	28.75	22.50	+6.25
8	14.00	26.00	−12.00

Source: From Owili-Eger, A. S. C., Geohydrologic and hydrogeochemical impacts of longwall coal mining on local aquifers, in SME Annual Meeting Preprint 83-876 (Denver, CO: Society of Mining Engineers, 1983).

[a] Final values are those water levels observed (where possible) on day 400.

NA = Not applicable due to offset in casing restricting measurement of water level.

BIBLIOGRAPHY

1. El-Ashry, T. M. *Impacts of Coal Mining on Water Resources in the U.S.* (Denver, CO: Environmental Defense Fund, 1976).
2. Garlanger, J. E. Ground water restoration in mined areas, in SME Annual Meeting Preprint 90-76 (Denver, CO: Society of Mining Engineers, 1990).
3. Gurr, T. M. Hydrologic impacts of phosphate gypsum disposal areas in central Florida, in SME Annual Meeting Preprint 90–193 (Denver, CO: Society of Mining Engineers, 1990).
4. Stoner, J. D. Probable hydrologic effects of subsurface mining, *Ground Water Monit. Rev.* 3(1): 128–136 (1983).
5. Owili-Eger, A. S. C. Geohydrologic and hydrogeochemical impacts of longwall coal mining on local aquifers, in SME Annual Meeting Preprint 83–376 (Denver, CO: Society of Mining Engineers, 1983).

8 Erosion Sediment Control

8.1 EROSION SEDIMENT CONTROL

Sediment is one of America's greatest pollutants. More than 1 billion tons of sediment reach the major streams of the United States annually.[1] Damages are reflected in the reduced carrying capacity of streams, clogged reservoirs, destroyed habitat for fish and other aquatic life, filled navigation channels, increased flood crests, degraded facilities for water-based recreation, increased industrial and domestic water treatment costs, and premature ageing of lakes by enrichment of the water with silt-carried fertilizer that promotes algae growth, destroys crops, and reduces the productivity of floodplain soils.

Erosion and sedimentation are natural processes that are usually gentle actions releasing a controlled amount of silt from watersheds to receiving streams. Surface mining activities accelerate these natural processes and short-duration, high-intensity storms can become a violent force moving thousands of tons of soil in a brief period of time. Cover is a very important factor. With the removal of ground cover, water moves across the denuded area on its own terms, picking up soil particles as it flows and leaving gullies behind. The susceptibility of strip-mined land to erosion depends on:

- Physical characteristics of the overburden
- Degree of slope
- Length of slope
- Climate
- Amount and rate of rainfall
- Type and percentage of vegetative ground cover

Many of the detrimental effects of strip mining can be prevented, via the development of erosion and sedimentation control plans before the disturbance of the area.

8.2 PRELIMINARY SITE EVALUATION

The preliminary site evaluation includes a thorough surface reconnaissance of the potential development site and familiarization with local geology, hydrogeology, and soil characteristics. Wherever extensive grading is anticipated, a preliminary subsurface investigation is desirable to provide information on the geologic, soil, and groundwater conditions.[2]

For a good preliminary site evaluation, topographic, geologic, soils, and zoning maps and aerial photography should be utilized. These maps provide information on physical features relating to erosion and sediment control and are often used as base maps on which locations of critical physical features and preliminary layouts of the potential development can be indicated.

In addition to maps, available publications on soil and geologic conditions should be utilized. Engineering characteristics generally considered in sediment and erosion control include:

- Depth to bedrock
- Depth to water table
- Soil classifications
- Grain size gradations
- Permeability
- Available water capacity

- Reactions (pH)
- Shrink-swell potential
- Moisture-density relationship

In the soil survey reports, the following aspects are generally covered:

- Suitability as the source for topsoil
- Pipelines (construction and maintenance)
- Road and highway locations
- Pond and reservoir sites
- Dikes, levees, and urban embankments
- Drainage systems
- Irrigation
- Terraces and diversions

Assessment of prime physical features critical to erosion and sediment control is required in the preliminary site evaluation. These features should be studied and delineated on a site map.

8.3   LAND TYPE

The proposed mine development site should be categorized into three basic land types: barren area, agricultural area, and woodland area.

Barren areas are nearly or totally void of any vegetation. These areas often require considerable grading and elaborate vegetative and structural measures to control erosion and sedimentation.

Agricultural areas are open areas under cultivation or are potentially cultivable. Unless these areas are planted in grain crops, agricultural areas generally support a stand of grasses, legumes, or herbaceous plants.

Woodlands are described as mature stands, pole stands, and mixed stands. Mature stands generally contain trees with trunk diameters of 10 in. (25 cm) or greater. Pole stands are thick stands of tall, small, round trees with trunk diameters between 6 and 10 in. (15–25 cm). Mixed stands contain both mature- and pole-sized trees. Mixed stands are generally found in woodlands that have been selectively lumbered, whereas pole stands generally occur in tree plantations or in woodland areas that have been previously nearly cut or burned over.

Mature trees with full crowns have the highest aesthetic value. However, they are less likely to recover from injury than smaller and younger trees. Mature trees, due to their crown size, bark structure, trunk strength, and broad root structure are more capable of resisting changes in wind and sun exposure resulting from extensive clearing. Crowded pole-sized trees often have constricted crowns and root systems in proportion to their height. This condition reduces their ability to withstand exposure to intense sun and wind.

8.4   SOIL AND ROCK

The presence of highly erodible soils is a very important physical feature, especially if these soils occur on moderate to steep slopes. It is not always possible to easily recognize a highly erodible soil horizon, as a stand of vegetative cover can mask the soil layer or the erodible soil can occur beneath a surface soil of different character.

Highly erodible soils are usually characterized by a deficiency of soil particles that have cohesive strength. This cohesive strength is usually a function of the clay-sized (colloidal) fraction of the soil horizon. The presence of highly erodible soils should be confirmed at an early stage of the site survey.

The occurrence of rock outcrops in the proposed development site should be noted. When a rock is excavated, the rock can be stockpiled for use in erosion and sediment control.

8.5   STREAMS

Streams are very often the recipient of sediment from the development site as well as the means of transporting it to private and public properties downstream from the mine development site. Streams themselves can contribute to the sediment load through channel degradation and bank erosion.

Three major factors contribute to increased stream erosion:

- Restriction of the stream channel due to sedimentation
- Increased runoff due to decreased infiltration in the runoff area
- Destruction of the natural vegetation along streambanks

For small streams flowing through major development areas, the erosional effect of increased runoff is certainly a major consideration in sediment and erosion control.

The stream gradient will, to a large extent, affect its sediment transport capabilities. Wide floodplains, meandering courses, and sediment build-up in the channel indicate shallow gradient. Increased sediment load will cause additional sedimentation of the channel.

Streams are dynamic entities in nature. Room for normal channel migration and adjustment to newly imposed runoff stress must be maintained.

Where the streambanks are high and steep, additional runoff from the watershed causes serious streambank erosion problems. Every attempt should be made to preserve or enhance the vegetative cover of streambanks, especially grasses, sedges, and woody shrubs with dense fibrous root systems. In poorly vegetated areas, it may be necessary to flatten the slopes and establish a good vegetative stand in order to control streambank erosion.

8.6   FLOODPLAINS

During periods of intense runoff, floodplains become inundated by flood flow and act as an extension of the stream itself. The integrity of the floodplain must be preserved.

8.7   IMPOUNDMENTS

Impoundment structures can be utilized for stormwater retention and sediment collection. They should have sufficient area and capacity. Natural impoundments such as lakes are aesthetically valuable physical features and should be protected against sediment damage. During the preliminary site evaluation, the existing conditions of the lake and shoreline should be evaluated in order to determine possible effects of sedimentation on the ecological and physical features of the impoundment.

In designing man-made ponds, the pond capacity should be sufficient to handle the anticipated water flow. When the impoundment capacity is impaired by sediment build-up, the sediment should be removed and disposed in a manner that will not reintroduce it into the system.

8.8   GROUNDWATER CONDITIONS

Groundwater conditions affect erosion and sedimentation. Groundwater seepage prevents the development of a vegetative cover and causes soil to slough into the ditches where it is directly introduced into the drainage system. The presence of springs and mottling in the soil indicates a high-water table.

8.9   VEGETATIVE COVER

A dense vegetative cover of grass, weeds, shrubs, vines, or trees is very effective in preventing erosion on steep slopes, swales, and along drainage ways and impoundment waters. Vegetation should be evaluated in terms of its benefit to erosion and sediment control.

Vegetative cover along waterways and impoundments must be protected, as it is both a soil stabilizer and a filter for sediment-laden water flowing into watercourses.

8.10   PLANNING

In the planning stage, a course of development is formulated. The planning stage is divided into four progressive steps:

- Preliminary site investigation
- Preliminary design
- Subsurface investigations
- Final design

8.10.1   Preliminary Site Investigation

As discussed earlier, during preliminary site investigations, the critical physical features must be evaluated in terms of their relationship to erosion and sediment control. These features should be delineated on a base map.

The complete drainage system should be shown on the topographic map. Important woodland tracts must be delineated.

A complete detailed topographic map should be utilized to record the critical physical features.

8.10.2   Preliminary Design

During the preliminary design phase, the mine development plan should be made in a manner that will minimize damage to physical features critical to erosion and sediment control. Grading damage should be minimized as avoiding steep slopes, which result in high cuts and fills, and by following natural ground controls as closely as possible.

In locating drainage ways, care must be taken to ensure that the resulting channel gradient and related discharge velocity will not cause erosion of the vegetative drainage way liner.

8.10.3   Subsurface Investigations

The subsurface investigation on geological features and soil characteristics should be utilized to determine the erodibility of the soils and their capability of sustaining a long-term vegetative cover.

As a general rule of thumb, a 50% (2:1) slope is assumed to be the maximum slope upon which vegetation can be satisfactorily established and maintained. However, maximum vegetative stability cannot be attained on slopes steeper than 33% (3:1). Optimum vegetative stability requires slopes of 25% (4:1) or less. The maximum slope should be applied only to ideal soil conditions in which the soil is not highly erodible and has an adequate moisture-holding capacity.

For droughty soils (those that exhibit a poor moisture-holding capacity due to excessively high permeability and a low percentage of fines) and for highly erodible soils, the maximum permissible slope should be considerably <50%.

Droughty soils generally have <30% fines. When these soils are encountered in cut areas and where reconditioning by the addition of fines or suitable topsoil is not planned, the cut slopes should not exceed 35% (3:1) so that a suitable vegetative cover is established. Furthermore, these soils must be planted with warm-season, deep-rooted, drought-resistant grasses and legumes suited to that particular region. For more drought-resistant soils with >30% fines, conventional cool-season grasses and legumes of the region can be utilized.

Where fills are to be constructed with droughty soils and where some finer-grained, drought-resistant soils are available, a portion of the drought-resistant soil should be segregated for later use in topdressing the fill.

Soils containing excessive amounts of fines, especially clay-sized particles such as clays and clayey silts, can also be difficult to stabilize with vegetation. The dense structure of many of these soils inhibits root development and moisture penetration. Cut slopes in these types of soils should be kept as flat as possible in order to enhance infiltration. On flatter slopes, where sloughing will not occur, the slopes should be dressed in topsoil or other suitable soil, or the existing soil can be upgraded by the addition of organic material and fertilizer. On steep cut slopes, the existing soil should be reworked as the cut progresses and while the slope is accessible to scarification, spreading, and compaction equipment.

The vegetation growth on any soil slope can be enhanced by roughening the soil surface. This practice helps germination by reducing sheet erosion and increasing water infiltration. The soil surface should be roughened along the contour in order to reduce the chance of drilling.

Since a long-lived vegetative erosion control cover on critical slopes can be achieved via the establishment of locally adaptable ground covers and shrubs, it is desirable to include seeds of these plants along with the conventional hydroseeding of grasses and legumes. Overplanting grassed slopes with ground covers or shrubs before vegetative deterioration occurs results in slope erosion. Planting directly to ground covers any shrubs using erosion control mattings or marshes to prevent erosion during the period is also a good practice.

Soil erodibility depends upon several physical features. The relative proportion of sand, silt, and clay in the soil, the organic content of the soil, the soil structure, and the permeability of the soil are major factors. After delineating these parameters, the grass erosion measure of soil erodibility is expressed as tons of removed sediments per acre of land surface area.

Well-graded soils generally exhibit relatively high resistance to erosion because they have both cohesive and intergranular strength. Loose, granular, fine-grained soils are highly erodible when exposed on steep slopes. Some types of clay soils are less erodible because they have greater cohesive strength. However, many of the indurated clays and silty clays that contain expansive clay minerals are susceptible to excessive erosion by slaking and alternate wet and dry cycles.

The soil testing programme should include the determination of pit and nutrient levels of soils brought to the surface by mining activities, as these soils will support vegetation. In these areas in which toxic soil compounds are commonly encountered, testing must be performed to determine these compounds so that corrective measures can be taken.

Problems involving pH values in the soil are common and must always be investigated. Excessively acidic soils will require regular applications of crushed or pulverized limestone so as to support a vegetative cover. In some instances, the use of vegetation with acid-tolerant characteristics is possible. The nutrient level of the soil is influenced by its nitrogen, potassium, and phosphorous content; soils deficient in these nutrients will require regular applications of proper fertilizers selected on the basis of soil tests.

Groundwater seepage is caused by the exposure of the groundwater table and can cause serious erosion and sediment control problems. Where subsurface investigations reveal severe high-water table conditions, every effort must be made to minimize disturbance of these potential problem areas. It may be better to avoid disturbance where these conditions exist.

This is especially true with fluid clay formations. On steep cut slopes, seepage can cause sloughing in erodible soils. Excessive seepage also prevents the maintenance of a satisfactory long-term vegetative cover.

When seepage is encountered in cuts, costly structural measures may be required to control erosional problems. Where the seepage is confined to a small localized area, the water generally can be trapped below the surface by using perforated drainpipe, graded stone, and sand filters; it should lead to a disposal area. Where the seepage extends over a considerable distance along a slope, a longitudinal pipe and stone under drain might be necessary.

For use as topsoil, it must contain at least 30% fines (–#200 mesh) and should meet the state standards for organic content, seed content, and noxious weed content. If the topsoil quality does not meet the standard, additional nutrients and chemicals to the soil surface will be necessary. Nutrient and chemical additive quantities must be established on the basis of soil tests.

8.10.4 FINAL DESIGN

Stabilization of major waterways is an important task in the final design process. Major waterways include all natural or constructed waterways that can be described as either permanent or intermittent streams.

Increased runoff, channel constriction caused by siltation or construction, and destruction of natural vegetation can accelerate waterway erosion.

In large development work, the corrosive effect of increased runoff can be controlled through the construction of storage ponds that will collect and store runoff water during periods of heavy precipitation. The ponds should be constructed to allow the gradual release of the stored runoff during low flow periods. Storage ponds also affect infiltration and evaporation, both of which reduce the total runoff. They also collect sediment which would otherwise be deposited in the waterways.

Filling of floodplains must be avoided except at roadway crossings. Properly sized conduits should be placed at crossings. All natural vegetation, whether it be grass, brush, or tree, adjacent to natural waterways must be protected from excavation activities and preserved in its natural conditions.

Vegetation along waterways helps erosion and sediment control in three important ways: the dense root mat helps hold the soil in place; the foliage (grasses, legumes, and other low-growing plants) and dead litter such as leaves filter out the sediment from the overland flow; and vegetation dissipates the erosive energy from falling raindrops.

In the event the natural conditions cannot induce satisfactory erosion control, induced vegetative and structural practices will be needed.

Vegetative measures are adopted when one or more of the following conditions exist:

- Poor quality vegetative cover
- Relatively flat terrain
- Low stream bed
- Tangential flow
- Low flow velocity
- Fertile soil

Prior to planting the vegetation, the banks should be graded to a fairly flat slope, preferably 25–33% grade or flatter. Such excavation will destroy any existing vegetation in the bank, but the bank excavation will increase the chemical capacity. The grading should be performed during periods of low precipitation, and the soil exposure time should be minimized.

Vegetation for planting in the streambank stabilization process should be selected after considering the following factors:

- Erosive forces
- Soil moisture
- Sedimentation
- Soil conditions

In most cases strip planting techniques should be used. The technique involves planting a strip of wet soil-tolerant, highly erosion-resistant vegetation in the critical area of the waterline, and conventional robust-rooted grasses and legumes above the critical zone. For added protection in selected locations, wet soil-tolerant bushes and trees can be planted near the waterline. For the protection of graded and planted areas until a strong stand of vegetation is established, an erosion control melting, or blanket can be utilized in addition to normal mulching practices. The local soil and water conservation district, the state forest service, and fish and wildlife agencies can help in selecting local types of vegetation for use in waterway stabilization.

Obstructions such as logs and boulders should not be removed indiscriminately, as they are required by fish and other aquatic life. These obstacles act as natural energy dissipaters. Their removal may increase the velocity of flow and thereby can intensify erosion at critical areas. Straightening may be undesirable because the stream gradient can be steepened. Steepened gradient can increase the rate of downcutting in the channel, or it may rejuvenate the downcutting cycle in a stable channel. Headwater gully erosion can increase. The removal of recently deposited sediments from the stream channel is beneficial as it returns the stream to a more stable and more material alignment and channel configuration.

Structural measures for the protection of natural waterways against erosion are divided into two types: grade control structures and bank protection structures. Grade control structures are used to control the gradient of the waterway channel so that the velocity of flow is reduced, thereby reducing both channel and bank erosion.

The most common type of grade control structure is the check dam. Check dams are short dams constructed of a variety of materials such as logs, treated lumber, stone, concrete, and synthetic materials. The check dam flattens the slope of the stream and dissipates energy. Stone or concrete is placed in the high energy area at the downstream of the check dam so as to prevent undercutting of the structure. In streams susceptible to flooding, check dams should be used with caution as they reduce flow rates and increase the chance of flooding.

Two types of bank protection structures are used: revetments and deflectors. Revetments comprise a wide variety of both rigid and flexible structures which are used as erosion-resistant facing on streambanks and lakeshores. The flexible type of structure is preferred and is generally more economical for streambank protection. Flexible revetments, such as riprap, fabiform mats, and gabions, have an advantage over rigid revetments such as asphalt paving or monolithic concrete because they are capable of adjusting to minor changes in foundation conditions without losing their integrity. The most common type of flexible revetment used for streambank protection is randomly placed stone riprap, composed of loose stone placed on sand/gravel filter and/or filter cloth. Other types of flexible revetments, although not nearly as flexible as stone riprap, are gabions, fabiform mats, interlocking concrete blocks, and steel or concrete tetrahedrons. Selection of the revetment type for a particular bank condition will depend upon the strength requirements, length of required service, and aesthetic factors. In areas in which the extreme durability of randomly placed stone riprap is unnecessary or in areas in which rock is not readily available, other types of revetments may be necessary, considering also fish and wildlife habitat or aesthetic reasons.

Common types of rigid to flexible revetments include concrete or asphalt paving, grouted stone riprap, and sacked concrete. To be effective, a rigid-type revetment requires a firm, stable foundation, and careful construction. Where fills are being protected a high degree of compaction is required beneath the revetment to prevent excessive settlement. To prevent undercutting at the toe, all revetments should be placed below the existing ground surface. In addition, adequately sized loose stones should be placed at the toe for additional safety.

All flexible and most rigid revetments are placed on a grade of approximately 50% (2:1). Some types of rigid revetments, including some varieties of gabions, wood sheet piling, and metal sheet piling, are built with a vertical face. These types of revetments are commonly used where water access, e.g., boat traffic, is essential or are used as retaining walls involving the filling of floodplains.

The other type of bank protection structure, the deflection structure, usually consists of a stone, concrete, or wooden groin that angles outward from the shore in a downstream direction and deflects the current away from a critical area of the streambanks. This type of structure should be used in wide streams in which the deflected current will not damage the opposite streambanks.

Many material types covering a wide range of costs are available for the construction of bank and shore protection structures. The following factors are considered for the selection of materials:

- Ability of the material to withstand the stress conditions at the site
- Initial cost and availability of construction material
- Maintenance and replacement costs
- Aesthetic considerations

Stabilization of minor waterways is an important requirement in sediment and erosion control. Minor waterways include all natural and constructed waterways, such as roadway draining ditches, drainage swales, or diversion ditches, which are not classified as either permanent or intermittent streams.

The location and design of minor waterways are of considerable importance. The waterways that collect and transport the surface runoff to the streams in the watershed can be major sources of sediment pollution if they have been poorly constructed or inadequately maintained.

Whenever possible, the natural drainage system should be preserved. When natural waterways are utilized, the natural vegetation system should be preserved. Traffic should not be allowed in the waterways. The natural vegetation may not, by itself, be able to resist the additional erosive force from increased runoff caused by mine excavation and development activities. Additional reinforcement with structural measures and additional planting may be necessary.

The vegetated waterways facilitate the loss of surface runoff through infiltration. Whenever vegetation cover is not adequate, a series of short check dams or a stone lining can be used. The use of vegetative waterways can be limited by factors, such as high soil erodibility, steep slope, high-water table, excessively droughty soils, and high soil toxicity. Such problems can be alleviated by using a series of check dams to flatten the gradient of the waterway and dissipate the flow energy.

Wet soil conditions due to a high-water table can often be resolved by using pipe underdrains. In droughty or toxic soil conditions, undercutting and backfilling or topdressing with nontoxic drought-resistant soil can be utilized. Care should be taken to establish a good bond between the native soil and the placed soil and to compact the placed soil.

Robust-rooted grasses that germinate rapidly and grow fast are most suitable for stabilizing waterways. With adequate maintenance they form a dense root mat and a dense uniform surface cover which does not restrict the movement of water and benefits both surface water infiltration and transpiration loss of near-surface groundwater.

For protection against erosion in grassed waterways, the following factors should be considered:

- The erodibility of the soil for the proposed slope
- The flow velocity limitation for the vegetation
- The flow resistance of the selected vegetation
- The method of vegetation establishment required to accommodate the volume and velocity of the design flow

In general, seeding is practised in waterways in which the designed water velocity is < 1.3 m/sec. Sodding is used in waterways when the flow velocity is expected to be between 1.3 and 2.6 m/sec. Seasonal considerations may sometimes rule out seeding. If the soil is erodible at the above-mentioned velocities, additional structural measures will be required.

Temporary stabilization measures are required in seeded as well as sodded waterways to protect against erosion until the vegetation is firmly established.

Jute netting is commonly used for temporary stabilization of waterways. The jute netting is placed directly over the prepared seedbed, and with proper anchoring, it minimizes soil erosion. Due to its thick fibrous composition, the jute also functions as a mulch.

Plastic, paper, and fibreglass nettings are also available. They have a longer service life than that of jute netting. However, due to the dense structure of the individual material strands used in forming the nettings, they do not function as a mulch. Therefore, the plastic and fibreglass nettings are used over a long fibre mulch such as straw or hay.

Subsurface drainage is a common problem in the establishment of vegetation in waterways when using nettings and mulch. Permeable, granular soils often cause piping and subsequent loss of soil from beneath the mulch and netting. Periodic erosion checks must be established across the waterways and beneath the netting.

Soil slopes include natural soil slopes and all denuded cut and fill. Man-made cut and fill slopes are usually constructed at a grade of 0–50%. Under some exceptionally good soil and hydrological conditions, the grade can be steepened to 67%. A slope of 33% is considered to be the maximum for the safe operation of excavation equipment. Factors usually considered in designing a slope include slope stability, soil erodibility, and the ability of the soil to support vegetation.

Slope stability is determined by soil mechanics. The stability of slopes is analyzed to predict the possibility of landslides.

Soil erodibility is a function of:

- The quality of vegetative and cover
- The soil gradation and permeability
- The degree of soil compaction
- Clay mineralogy
- The slope grades
- The slope length
- The quantity of water collected by the slope

As the length of the slope increases the erodibility of the slope and the water quantity collected increase. The effect on erosion can be controlled by different types of diversions such as terraces, benches, top of cut ditches, temporary diversion dikes, and interrupter dikes. Benches and terraces break the length of cut and fill slopes and collect runoff water and direct it to a disposal point.

Diversion dikes are temporary berms of the soil placed along the top of cuts and fills or at intervals along graded natural slopes for diverting runoff water away from the denuded slope. The runoff is diverted to a stabilized disposal point, such as a level spreader, temporary flexible down drain, temporary sectional down drain, or flumes. Diversion dikes are utilized during the development phase, and in the case of fills, they should be maintained until an adequate vegetative cover is established. In cuts, diversion ditches are established at the top of the cut.

Compaction of fills is a critical factor in erosion control. Poor compaction can cause serious problems. As a minimum criterion for erosion and sediment control, the upper 1 ft of all fills should be compacted to at least 85% of optimum. The surface of the cut and fill should be roughened at a perpendicular direction to the flow to reduce the flow velocity and enhance water infiltration. Discing and light scarification will accomplish this effect. Where the slope is too steep to allow parallel movement of vehicular traffic, cleated dozers traversing the slope can achieve a satisfactory texture on newly placed soil.

The gradation of the soil on the surface of the slope affects both the erodibility of the soil and its ability to support vegetation. For example, many types of well-drained silty sands are highly erodible and may be droughty. When the soil exhibits either or both of these conditions, the configuration of the slope should be adjusted to accommodate this condition, or the slope should be topdressed with an erosion and drought-resistant soil. As an alternative to topdressing, suitable soil can be mixed with the existing soil. When topdressing is applied, the dressing soil must be firmly bonded to the existing soil surface in order to prevent slippage downslope. This bonding can be increased by scarification of the slope.

The quality of the vegetative soil is determined both by the physical characteristics of the soil (density, permeability, moisture-holding capacity) and by its chemical composition (the pH of the soil, the presence of nutrients in the soil, and toxic elements). All those factors affect the quality of the vegetative cover.

When the pH and nutrient levels of the soil are known, it may be possible to adjust these values to the desired level by the addition of lime, fertilizers, etc. When the vegetative condition of the soil cannot be modified, topsoil should be placed.

Seasonal factors also influence vegetation quality. Cool, moist periods of the year are favourable for seed germination and plant growth. Different types of vegetation exhibit a widely differing ability to tolerate certain climatic conditions. In a given region, the most suitable vegetation can be determined with the guidance of the local soil and water experts.

In some cases fast-growing annuals are planted as a temporary ground cover until climatic conditions are favourable for the germination and growth of more desirable perennial grasses and legumes.

Soil stabilization measures include both short- and long-term vegetative measures. They are utilized to control water and wind erosion of soil during and after grading operations.

Interim stabilization measures are used to retard erosion over the short term, i.e., hot summer and winter months, until conditions are more favourable for long-term vegetative stabilization. Such measures include mulching and use of nettings, blankets, etc., along with the seeding of annual grasses. Fibre mulches such as straw, hay, and woodchips, as well as chemical soil binders, are commonly used to stabilize graded areas prior to seeding for permanent vegetation. The chemical soil binders can penetrate and bind the near-surface soil or bind the surface of the soil. Chemical soil binders are used primarily to protect denuded soil from wind and water erosion during delays in grading operations or until permanent seeding is possible.

Some chemical soil binders also function as mulches (e.g., hay) to aid in germination and the growth of seeded vegetation as they conserve moisture in the soil and provide temporary protection against erosion.

The use of woodchips for short-term soil stabilization and mulching has become common. Woodchips are long lasting, due to their weight and shape, and they require little or no tacking to keep them in place.

Fireglass mulch is also available. All organic and inorganic fibre mulches require some form of attachment in order to prevent them from being blown or washed away. Three methods are commonly used to secure fibre mulches. The first method, called crimping, is used on straw and hay mulches and is carried out by a crimping machine which partially punches the mulches into the soil. The second method of securing mulches, known as tacking, applies an asphalt or chemical binder to the mulch. The third method includes the use of various types of nettings made of jute, plastic, paper, and fibreglass.

Sediment retention structures are designed to remove sediment from the runoff water. The basic function of these structures is to hold the runoff water for sufficient lengths of time so that the sediment settles out of suspension.

The most common type of retention structure is the sediment pond which is excavated along a natural or man-made waterway. Sediment basins are usually considered short-term structures. The capacity of the sediment basin is determined upon the area of its watershed, the topography of the watershed, the infiltration rate of the soils in the watershed, and the regional hydrological factors.

Most states have regulations covering the design and construction of all sizes of sediment basins. These regulations must be reviewed before designing the impoundment structures.

8.10.5 Formulation of an Erosion and Sediment Control Plan

A detailed erosion and sediment control plan should be developed along with mine plans. The plan must be presented for review, approval, and certification to all cognizant agencies empowered by sediment and erosion control legislation.

During mine development the amount of land to be exposed at any time should be minimized. The site should be developed in stages. All clearing, grading, and stabilization operations in an area should be completed before moving into another area. The sequence and scheduling in which development will occur must be clearly established in maps.

The same map that delineates the sequence of development must show the location of all sediment retention structures. The plan documents must indicate a sequence of the construction of these structures. All sediment control structures for natural waterways should be installed before any clearing or excavation work is initiated.

Traffic control is important in woodland developments where uncontrolled traffic can cause severe tree damage. Vegetative filter strips along waterways and all undisturbed open spaces should be delineated on a site map and designated as "off limits" areas for all vehicular traffic. For woodland areas, all vehicular traffic will stay within the roadways, access corridors, or utility rights-of-way. All traffic should be restricted from crossing streams or stabilized drainageways except at approved stabilized crossing locations.

All critical areas along streams must be marked on the site maps, and the recommended method of stabilization must be indicated. Stream stabilization work should be scheduled for periods of low precipitation during the growing season and should be performed prior to the initiation of clearing and grading operations in the watershed.

The erosion and sediment control plan must clearly define vegetative practices, both temporary and permanent. The plan must state and show where and when sod, temporary seeding, and permanent seeding are to be used. Specifications should be provided regarding ground preparation, sod quality, seed type and quantity, fertilization, and mulching.

The construction specifications must clearly define the maximum length of time that a graded area can be left uncovered after the completion of grading and also after grading shutdowns as is common practice during the winter months. Short-term stabilization practices during any lengthy grading delays should also be specified.

Figures 8.1 through 8.4 illustrate sedimentation control practices.

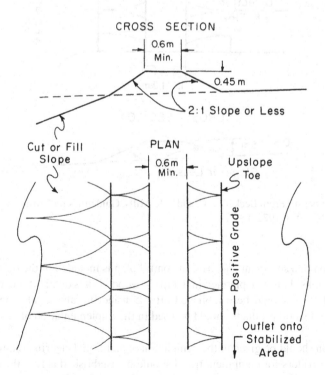

FIGURE 8.1   Diversion dike. (From Becker, B. C. and T. R. Mills. Guidelines for Erosion and Sediment Control, U.S. EPA Report R2-72-015 (1972).)

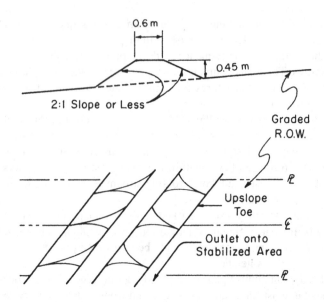

FIGURE 8.2 Interceptor dike. (From Becker, B. C. and T. R. Mills. Guidelines for Erosion and Sediment Control, U.S. EPA Report R2-72-015 (1972).)

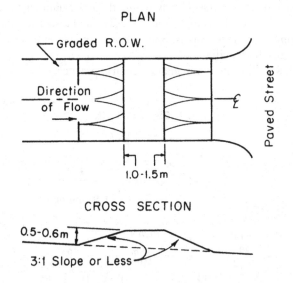

FIGURE 8.3 Filter berm. (From Becker, B. C. and T. R. Mills. Guidelines for Erosion and Sediment Control, U.S. EPA Report R2-72-015 (1972).)

8.10.6 Operation

The implementation of a sediment and erosion control plan is included in the operation procedures.

Clearing operations should be planned in detail. In wooded sites detailed procedures for removal and disposition of trees should be established after considering any legal aspect of the clearing operations. Salvaged wood products should be used in the implementation of sediment and erosion control practices.

Special attention should be given to the completion of additional clearing required for equipment travel corridors. Corridors for equipment travel should be established across the areas that will not be immediately denuded, especially in denuded areas. The natural filter strip areas should be given careful attention. Disturbances to vegetated floodplains should be avoided.

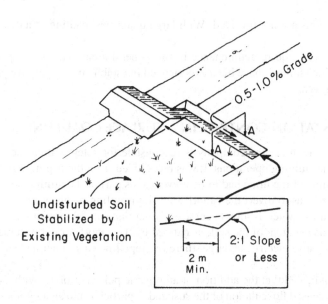

FIGURE 8.4 Level spreader. (From Becker, B. C. and T. R. Mills. Guidelines for Erosion and Sediment Control, U.S. EPA Report R2-72-015 (1972).)

Stockpile areas should be selected and designated on a grading plan. These areas should be allocated after considering sediment and erosion control requirements such as direct delivery of sediment to waterways, damage to vegetation, and unnecessary destruction of trees. Temporary or interim stabilization of soil stockpiles should be instituted. Critical slopes on stockpiles should be avoided. Stockpiling immediately adjacent to watercourses should not be allowed because the stockpiled material will provide a direct and a high-volume source of sediment to storm runoff.

Temporary vegetative measures planned for the implementation on major stockpile areas should be established immediately after the stockpile operation is completed. Proper mulching and soil stabilization in conjunction with these seeding operations should be carried out.

As the rough grading operations near completion, the installation of structure and vegetative practices must be promptly accomplished. Diversion dikes, interceptor dikes, filter berms, etc. should be constructed according to plan. The use of woodchips on a cut and fill plan should begin as soon as slopes are available to receive them. Once chemical soil stabilization has been planned, it must be applied as soon as the slopes are completed. The establishment of temporary vegetative cover should be initiated as soon as slopes are available.

Timely implementation of sediment and erosion control practices is essential. Each day a potential sediment source remains unstabilized, it exists as a source of pollution.

Drainageway protection must be an integral part of the grading operations. Traffic should not be allowed to cross waterways except at specified locations.

As soon as the sediment and erosion control measures begin to function, their maintenance process must be initiated. Sediment removal from structures designed to trap and filter must begin. Prompt replacement of items, such as straw bales, woodchips, and seedbeds is vital.

8.11 MAINTENANCE

For timely clean-out and stable disposition of trapped sediment from sediment retention basins, specific schedules should be established. A rule of thumb is to clean out a basin when it has lost 50% of its storage capacity due to sediment deposition. The removed sediment should not be indiscriminately piled or dumped because the sediment can move back into the storm drainage system by successive storms. Disposition behind a protective berm or filter strip can often suffice if large quantities of sediments are not involved. With large quantities, hauling to a disposal area may be required.

Vegetative practices require maintenance in two general areas. The first is periodic refertilization. In areas in which failures have been experienced in establishing vegetative protection, prompt attention must be given.

8.12   SEDIMENTATION CONTROL IN A SURFACE COAL MINE

As the operator of the mine, the company is responsible for safeguarding the environment from the impact of surface mining operations. It was recognized from the beginning that siltation of the streams in the vicinity of the mine had to be prevented as part of the mining operations.

The plan adopted was to divert the runoff water into settling ponds where natural gravitational settling would clear the water. It soon became apparent that natural gravitation alone would not do the job. The high clay content of the soil, as carried by the runoff, would become colloidal. These particles carry a negative electrical charge, therefore repelling each other and causing turbid water conditions.

Bench testing indicated that the addition of an organic polyelectrolyte, with a positive electrical charge, effectively caused flocculation of the suspended particles into an agglomeration of particles. Gravitational settling out then occurred. The next phase of the programme was to devise a system to accomplish this under field conditions.

The questions that needed answers were:

1. What was the rainfall pattern?
2. What were the watershed capacity and probable runoff?
3. What was the correct design level for dams and outlet structures, 10-year, 25-year, and 100-year flood frequency?
4. What was the anticipated silt loading?
5. What was the range of effectiveness of the polyelectrolyte?
6. What hardware was available to continually meter the correct volume of chemical to cause flocculation and water clarification?

A unique system was developed in a coal-mining district in Washington State that effectively answers these questions on an operational basis. The system continually monitors the water flow and automatically meters the correct parts per million of the chemical into the turbid water. Rapid mixing follows, then flocculation. Settling out and clarification take place in the quiet water of the second pond. Clear water was decanted into the receiving water.

The coal mine is within the Centralia Coal District from which coal has been mined since the late 1800s, the heaviest period of activity being the days of the coal-burning railroad locomotives. Earlier mining involved different methods. The actual coal field dedicated to this project is contained within a 21,000-acre property. During the projected 35-year life of the operation, approximately 7,000 acres were projected to be disturbed.

The land is typical southwestern Washington State rural low-elevation woodland. The topography is a rolling landscape with rounded hills that rise somewhat steeply from flat, poorly drained valley bottoms. Second-growth conifer and hardwood stand cover the hills. Douglas fir is the primary commercial tree species. Red alder is the most plentiful hardwood species, found either mixed with the conifer or in nearly pure stands. Alder is of minor commercial importance. From a forest classification standpoint, the area is classed as a Douglas fir area, with the usual plant species commonly associated in this type.

The valley bottoms consist of small farm units that pasture beef cattle and raise hay. Farming is marginal in these valleys, with the farmer usually "working out" and working the land on a part-time basis. Many of the farmers who have sold their land to the mine remain on their farms as lessees.

As stated, the mine is in southwestern Washington State, west of the Cascade Mountain range. Mild temperatures and a lengthy winter rainy season is the normal climate of the region. The average annual rainfall is considered to be 45–48 in./year, with autumn rains beginning in mid-October and continuing almost uninterrupted until mid-April. Snowfall occurs rather infrequently; some winters pass without any snowfall at the lower elevations. Snow depth seldom exceeds 4–8 in. and remains on the ground for no more than a few days.

The rainfall pattern follows a reasonably predictable pattern, i.e., long periods of steady rainfall between early October and late April. During these months, a high probability exists for some daily precipitation. This means that the system must be designed to accommodate extended periods of high runoff as well as sudden storms having peak periods of precipitation.

The main watercourse in the mine locale is Hanaford Creek, which rises to the east of the mine, approximately 15 mi away, in the low-elevation timbered mountains. It flows due west, draining approximately 56 mi^2 before emptying into the Skookumchuck River at Centralia. During periods of high precipitation and runoff, when the Skookumchuck's banks are full, the Hanaford overflows its channel and floods the adjacent pasture land. Winter flows may exceed 500 cfs, while summer flows can be as low as 2 cfs.

Modest as it may appear, Hanaford Creek is important to the area and a significant migratory fishery is located there. Near its headwaters and in its mountain locale it becomes a sparkling stream with spawning areas for salmon and steelhead. It is for this reason, and also because it has a very limited assimilative capacity, that such an effort to guard its quality is essential.

The mining operations do not take place in Hanaford Creek proper. However, the mine shop-office, coal preparation plant, power plant, and coal inventory area are located close together on its south bank. Coal stripping takes place along the ridges to the south and eventually to the north.

The minor streams draining these adjacent valleys, with their small watersheds, are tributaries to Hanaford Creek, and these are directly affected by mining activity. Because these tributaries receive the runoff from the mine area, they have been incorporated into a system to prevent siltation of the mainstream, Hanaford Creek.

The mining process at the Centralia mine exposes the parent material (overburden) to the weather in a "bottoms up" operation. This parent material is called the Skookumchuck Formation and is made up of marine and nonmarine deposits. It is a fine-grained, readily erodible material.

From the beginning it was recognized that sedimentation is the primary water quality problem in the mine. Because of the low sulphur content of the coal (0.4–0.7%), the problem of acid mine water did not present itself. Before mining progressed, drainage settling ponds and ditches to intercept any silt-carrying runoff were proposed to be built. As proposed, these settling ponds would cause natural gravitational settling out of the silt, and only clean water would be discharged into Hanaford Creek. In addition, a programme of as early as possible grass seeding was needed to hold the soil in place and also to prevent siltation. The proposal included the possibility that the water would have to be treated "chemically" to correct any condition that may cause degradation of the receiving water's ecosystem. At this juncture the state's newly organized Department of Ecology (reorganized/renamed Water Pollution Control Commission) insisted that the runoff passing through the mine area may become contaminated and thus constitute an industrial waste. A Waste Discharge Permit was negotiated along the lines of intended course of action and policy. The Waste Discharge Permit listed the standards with which the mine was to comply.

Fortunately, the management of both the mine and the power plant had the foresight to initiate a water quality study of Hanaford Creek and its tributaries around the mine before mining began. For nearly 3 years, consistently in each season, monthly water samples were taken from many "stations" in the creek area. The information obtained from this sampling programme was extremely beneficial as the programme progressed from water monitoring to actual water quality control. Cooperating and participating in this pre-mining water monitoring programme were the Department of Ecology and the State Department of Fisheries.

Approximately 600 acres were cleared of all vegetative cover, exposing the mineral soil to oncoming rains. Three settling ponds were constructed to receive the runoff from three small watersheds, the largest of these was 1,100 acres. The decant from these ponds was to come through a half-round culvert riser with removable boards on the flat side; control of the discharge level was regulated by installing or removing these 4-in. boards. Thus, the decant would always be at the surface, allowing for maximum resident time and maximum gravitational settling of the waterborne silt.

A bench test was conducted to determine the physical and chemical properties of the soil when subjected to runoff conditions. The test strongly indicated that gravitational settling alone would be insufficient to clarify the muddy water and showed that a significant fraction of the soil (clay) that would be subjected to runoff would become near colloidal and remain suspended in the water for more than 1 week. The actual empirical time for complete natural settling to occur was not determined because the time required extended far beyond the resident time in any of the settling ponds. Residence time in the ponds was a matter of 8–23 hr, depending upon the obvious factors of rainfall intensity and storm duration.

Because of the bench test results, it would be necessary to go beyond mere diversion of the runoff into settling ponds. To prevent the mining activity from adversely affecting the watercourses, a method of removing the silt suspended in the runoff and thus decanting only clean water was essential. The problem was to find the "something" that would clarify the muddy water quickly and to do so effectively under winter weather field conditions. Bench tests indicated that the soil-water suspensions were quite stable; the particles possessed negative charges and would not settle out unless this negative polarity was altered.

To effect clarification of the runoff water, a chemical, Nalco 634, was introduced by Nalco Chemical Co. Nalco 634 is a high molecular weight organic polyelectrolyte with a high charge density, which is effective as a primary coagulant, providing the positive charge that neutralizes negative colloidal particles and also "bridges" these particles to form a floc. It is designed to flocculate silt or other finely divided matter in aqueous systems, it neutralizes the charge on siliceous material and clays, producing a rapid settling floc, it adheres to the substrate and remains in the settled sludge, and it does not add to the biochemical oxygen demand (BOD) of the wastewater because it adheres to the settled solids. The company was able to demonstrate via bench tests that when introduced into turbid water at the rate of 10–20 ppm, Nalco 634 neutralized the negative charges of the colloidal clay particles, causing them to flocculate and settle out, thereby rapidly clarifying the water.

The chemical was bioassayed by a recognized fisheries biologist and found not to be harmful to the aquatic life found in the streams at the dilution rates anticipated. The Department of Fisheries had authorized the use of this compound.

The first step in clarification of the runoff water was to determine the anticipated silt loading at the point at which the chemical was to be introduced into the runoff water. Settling ponds were arranged in sequence, each sequentially downstream. The heavy silt-laden runoff entered the upstream end of the stream pond. As soon as streamflow velocity diminished, the heavy silt load settled out. Turbidity at this point was over 1,000 JTU (Jackson turbidity units), 10,000–15,000 mg/L, and 1.5–2% solid. By the time the runoff passed through the stream pond, the heavier material had settled out. At the discharge weir, turbidity was measured to be 85–120 JTU, 120–130 mg/L, 0.4–0.7% solids by volume, and 0.012–0.013% oven dry solids.

The Waste Discharge Permit required that the discharge be no more than 5 JTU above normal background. Records indicated that the range in turbidity of Hanaford Creek during the rainy season was between 20 and 55 JTU.

Bench tests indicated that Nalco 634 would effect clarification to the 25–50 JTU range if the dosage levels were as follows:

Turbidity (JTU)	Dosage (ppm)
40–80	5
80–120	10
120+	20

These were to be initial guidelines. Experience and continual bench testing were to be the watchwords.

It was next necessary to estimate the expected flow rate of runoff at the discharge weir. This was necessary so that the chemical dosage to the gallons per minute flow, parts per million were converted to cubic centimetres per minute. The weir resembled, for the most part, a sharp-crested weir. Streamflow in inches over the weir equal to gallons per minute was calculated:

Estimated Flow Across Weir		Dosage Nalco 634 (cm³/min)
cfs	gpm	10 ppm
2.2	1,000	40
8.9	4,000	160
15.6	7,000	270
26.8	12,000	450
40.0	18,000	680

The average flow rate was estimated to be in the 6,000–10,000 gpm range and chemical dosage to be at 10 ppm or 230–340 cm³/min range.

The next step in going from bench test to field operation was to devise a method of introducing the chemical at the discharge weir of the upstream pond. Recommendations were to (1) dilute the chemical with fresh clean water, (2) cause rapid mixing to provide for maximum contact between the chemical and the suspended particles, (3) follow with a period of slow, gentle mixing to allow for the floc building and particulate stripping, and (4) create a period of relatively quiet water to allow for the floc to settle out. These recommendations pertain to an industrial plant complex in which mixing facilities, freshwater, and electrical power are readily available and taken for granted. In the initial stages, no mixing facilities existed except the velocity of the runoff passing through the culvert. The freshwater source was the domestic water supply of a farm tenant. No electrical power supply existed.

The first "chemical treatment station" was set up at a point at which the farmer's water supply could be utilized. A simple open-sided structure resembling a school bus stop was constructed at the site above the stream channel to shelter the feed pump from the weather. In order to effect adequate mixing of the chemical and the turbid runoff, a 50-ft half-round culvert with baffles every 5 ft was placed in the stream channel. All the water from this drainage had to pass through this mixing area.

At the "chemical treatment station" four barrels of Nalco 634 were set in place. The waterline was tapped, and a freshwater supply of approximately 15 gpm was made available. A precision feed pump model 9101–21 was chosen to pump the chemical from the drums. A dilution-tee was installed in the waterline, and the mixture was carried to the mixing culvert in a ½-in. plastic line. The flow of the mixture was forced more by gravity than by pump pressure. A portable gasoline generator was set up to supply the power to the feed pump.

Facilities were soon proved inadequate. The concept of settling ponds in the sequence was a viable concept. The chemical did cause flocculation and clarification but at a higher dosage than expected. The method of introducing the chemical into the water was under-designed, to say the least.

The chemical worked in the field. The ponds affected settling. What was needed was the hardware to meter the chemical to match the flow on a continual basis for 7–8 months each year. The parameters of the problem were:

- As the runoff increased or decreased due to precipitation, the volume across the discharge weir increased and decreased accordingly. This change would occur continually. Part of the system had to sense this change and automatically initiate action.

- The testing of water three times daily showed that the turbidity of the water would remain within the range of 85–120 JTU. Because these turbidities were well within the effective range of the chemical, only changes in flow (inches of water) over the weir and not changes in turbidity be registered. However, because the flow is not proportional to level, a device was necessary to proportion the polyelectrolyte to the flow.
- The feed pump would have to respond automatically and accurately to the continually changing demand.
- The system should be maintenance free and rugged enough to withstand field conditions.
- 220–110 V electrical power was an absolute necessity.

This system consisted of four components:

- Liquid level capacitive probe
- Liquid level monitoring and control instrument with flow proportional output circuit
- Automatic pump control
- Chemical metering pump

Necessary support equipment included a chemical holding tank, various plastic plumbing lines, tees, PVC pipe, etc.

Located at the upstream pond spillway, the liquid level capacitive probe sensed the changes in the level of the outflow. It had no external electronic moving parts, but continually emitted a signal that indicated the exact water level in the pond. The liquid level monitoring and control instrument accepted the signal from the capacitive probe and converted the signal to a 10–50 mA DC output for a 0–100% pond height variation. This variation is equal to 0–24 in. H_2O over the weir. This, in turn, is equal to 0–25,000 gpm. The 10–50 mA analogue output is flow proportional. The automatic pump control accepts the 10–50 mA DC output from the control instrument and amplifies it to produce a signal which is used to control the output of the chemical feed pump. This component is an interrupted stroke metering pump with an output range of from 1 to 10 gal/hr.

As stated before, daily testing demonstrated that the turbidity level would remain in the range of 85–120 JTU and would not be appreciably affected by an increase in flow. The bench testing proved that the polyelectrolyte was quite effective through and beyond this range at 5 ppm. It is then simply a matter of calibrating the system to meter in this amount. After calibration, the correct amount is automatically metered according to the flow. The exact amount of the chemical is pumped from the holding tank through plastic lines into a spreader tube hung across the spillway. The chemical drips into the water that passes below.

Strong, rapid mixing follows immediately; diluting the chemical with non-turbid water is unnecessary. This mixing is the key to good chemical dispersion and must be provided to produce an effective flocculation result. This mixing phase has been accomplished by designing iron obstructions that cause turbulence and mixing into the spillway fins and angle.

The remaining step in clarification is the period of quiet water provided by the second pond. Here, settling out and clarification take place. Clean water is decanted from this pond.

While field testing of the above system was going on, bench testing of other flocculating formulations was being carried out. Bench tests indicated that American Cyanamid's Super Floc 330 would be effective over a wider range of turbidity at half the dosage rate of Nalco 634. The Cyanamid chemical was placed in actual field operation. At the same time two drainages with identical systems were operating, and the Super Floc 330 was used in one of the systems. After some initial problems with viscosity, it was found that Super Floc 330 was a more effective, efficient chemical.

The proof of the system is that it works. Under the regulations of the Waste Discharge Permit, the water above and below the mine is analyzed three times a day. On an operational basis, during periods of high rainfall and resulting runoff, the two-pond chemical flocculating system prevents the siltation of Hanaford Creek. Pertinent numbers for comparison are as follows:

Entering silt load	1.5–2.0% solids by vol	1,000+ JTU
Silt load of chemical station	0.4–0.7% solids by vol	85–120 JTU
Decant from the second pond	Clear water	4–15 JTU

The flocculated silt accumulates quite quickly in the downstream ponds. It is necessary to periodically, depending upon the size of the pond, remove this silt or in some way allow for its accumulation.

8.13 SURFACE MINE SEDIMENTATION CONTROL

Surface mining and reclamation have the potential to at least temporarily increase surface runoff and the resultant erosion and gully development. This increase in erosion is due in part to the compaction of the soil surface with heavy equipment, the creation of large relatively unvegetated watersheds, and the elimination of natural dendritic ephemeral drainage patterns. Methods for stabilizing gully channels which may develop prior to the establishment of sufficient vegetation to control erosion should be adopted. Utilization of various sizes of riprap check dams, reinforcement of riprap check dams, and the relationship of the type of control structure to the area of watershed and resultant flow velocities will be necessary.

Drainage patterns on reclaimed areas depend on the remedial landscaping that is applied to an area and the natural erosional processes that subsequently modify the landscape. Remedial landscaping includes determining the topographic layout of an area, the size of its watershed, the type of vegetation to be used, the need for terraces and/or other drainage features, and the shape and layout of the total drainage system needed. In this area, three basic types of drainage patterns were found:

- Predesigned
- Naturally formed on the reclaimed surface
- Predesigned and subsequently modified by natural erosional processes

Predesigned drainage patterns were characterized by single, linear terraces, and/or waterways. Natural drainage patterns have not been established on a regional scale, but on a local scale, dendritic, parallel, braided, and yazoo drainage patterns exist. Where modification of predesigned drainage features had taken place, it generally occurred by means of yazoo drainage.

A quantitative measure of a basin's drainage characteristics can be determined from the peak rates of runoff that occur in a particular drainage system. The riprap size capable of maintaining a stable waterway is a function of a waterway's corresponding peak rate of runoff.

The peak rates of runoff analysis for the gullies investigated are shown in Table 8.1. The peak rate of runoff ranged between 21.7 and 386.2 cts. The average peak rate of runoff was 109.6 cfs, but only six gullies were determined to have peak rates of runoff >100 cfs.

Most of the material used to construct the riprap structures consists of crystalline limestone and associated chert, most likely from the Mississippian-age Burlington Formation. The chert appeared to contain a type of silica which is susceptible to water absorption, expansion, and subsequent fracturing.

The characteristics of the riprap structures used for gully control essentially consist of three components:

- Shape and layout
- Particle size distribution
- Usage of check dams

In layout, most of the gullies had a linear channel, while some had single bends and others were sinuous.

TABLE 8.1

Time Concentration Calculations for Gullies Investigated at the Prairie Hill Mine, Missouri, Central Missouri Method

Gully No.	Peak Rate Runoff (cfs)	Watershed Location Factor (L)	Soil Infiltration Factor (I)	Topographic Factor (T) (ft)	Watershed Shape Factor (S)	Vegetative Cover Factor (V)	Contour Farming Factor (C)	Surface Storage Factor (P)	Runoff Frequency Factor (F)	Peak Rate Rate Runoff (Q) Q=qLITSSVCPF (cfs)
1	29.8	1.01	1.2	0.79	1.05	2	1	1	1	59.9
2	16.3	1.01	1.2	0.79	1.03	2	1	1	1	32.2
3	166.5	1.01	1.2	0.79	0.99	2	1	1	1	315.7
4	17.1	1.01	1.2	0.86	1.03	2	1	1	1	36.7
5	51.8	1.01	1.2	0.86	1.07	2	1	1	1	115.5
6	22.1	1.01	1.2	0.86	1	2	1	1	1	46.1
7	201.7	1.01	1.2	0.79	1	2	1	1	1	386.2
8	80.2	1.01	1.2	0.79	1.1	2	1	1	1	168.9
9	131.5	1.01	1.2	0.79	0.98	2	1	1	1	246.8
10	15.9	1.01	1.2	0.79	1.05	2	1	1	1	32
12	48.9	1.01	1.2	0.86	0.97	2	1	1	1	98.9
13	47.5	1.01	1.2	0.79	0.96	2	1	1	1	87.3
14	28.2	1.01	1.2	0.79	0.86	2	1	1	1	46.4
15	33.6	1.01	1.2	0.79	0.95	2	1	1	1	61.1
16	38	1.01	1.2	0.79	0.96	2	1	1	1	69.9
17	50.9	1.01	1.2	0.79	0.94	2	1	1	1	91.6
18	51.2	1.01	1.2	0.79	0.85	2	1	1	1	83.3
19	155.7	1.01	1.2	0.79	1.05	2	1	1	1	313.1
20	39.5	1.01	1.2	0.79	1.01	2	1	1	1	76.4
21	11.2	1.01	1.2	0.79	1.05	2	1	1	1	22.5
22	10.8	1.01	1.2	0.79	1.05	2	1	1	1	21.7

Source: From Vories, K. C. and C. D. Elifrits. Controlling large gullies on a Midwest surface mine, in SME-AIME Preprint 86-348 (Denver, CO: Society of Mining Engineers, 1986).

The lengths of the waterways varied, but the width of the channels was generally found to be about 14 ft. Most likely this width is a function of the dozer blade used to grade the waterway. A plan view of the waterways showed channel sides were generally parallel. In some cases, the width of the channel was irregular because the placement of the riprap in the channel was highly variable. As a result, some of the waterways had jagged, elliptical, or cone-shaped channel sides in outline. Ripraplined channels generally were between 1 and 2 ft in depth. In cases in which a series of check dams were used, the channel depth was between 2 and 4 ft.

In general, waterways constructed of uniformly distributed riprap (riprap generally with a range of 2–10 in.) placed with a pan scraper had streamlined channels, whereas gullies dump-filled with well-graded or large riprap had irregular channels. Figure 8.5 illustrates the various components of a channel in cross-section. In some channels, lack of uniformity of size in riprap utilized and differential weathering between the channel riprap and surrounding soil accounted for the formation of ridges along the sides of the channel and the subsequent development of yazoo gullies. The use of a gravel filter bed below the appropriate size channel riprap and a uniformly sized riprap should minimize this problem. Channel riprap, where placed using a pan scraper, consisted of a thin veneer of channel riprap designed to secure the gravel filter bed beneath. In places in which the size distribution was uneven or riprap was not graded using a pan scraper, the thickness of the channel cover was variable, increasing the probability of channel erosion.

Wings on check dams generally were flared upstream. Check dams were usually placed where terraces fed into channels or where special conditions dictated (Figure 8.6).

The particle size distribution of the riprap depended on:

1. Time of emplacement—In general, riprap structures earlier built had wide particle distributions. Also, later structures were placed with a tractor-pulled pan scraper.
2. Method of emplacement—Riprap placed using a pan scraper tended to have uniform particle distributions, while the riprap carried to gullies by dump trucks tended to have non-uniform particle distributions.
3. Structure under consideration—In general the particle size distribution of the channel beds tended to be more uniform than that of check dams.
4. Amount of weathering—Fracturing of the chert increases the percentage of the smaller sized fraction at the expense of the larger sized fraction.

Table 8.2 includes the range and nominal diameter of riprap used in the gullies at the site but does not include sizing characteristics of the filter bed material. Any riprap with a maximum size of <10 in., although uniform in relation to other riprap distributions found, was in itself generally well graded. This type of material was generally found in the newer, pan scraper waterways.

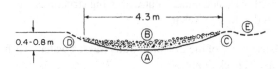

A) Gravel filter bed (5 cm)
B) Protective rip-rap channel cover (7.5-20 cm)
C) Ridge formed by construction or differential erosion
D) Ideal gradation between channel sides and channel
E) Yazoo gully formed from C

FIGURE 8.5 Waterway cross-section. (From Vories, K. C. and C. D. Elifrits. Controlling large gullies on a Midwest surface mine, in SME-AIME Preprint 86-348 (Denver, CO: Society of Mining Engineers, 1986).)

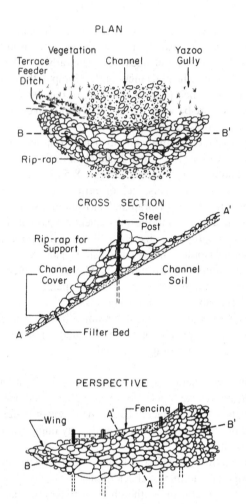

FIGURE 8.6 Components of a check dam. (From Vories, K. C. and C. D. Elifrits. Controlling large gullies on a midwest surface mine, in SME-AIME Preprint 86-348 (Denver, CO: Society of Mining Engineers, 1986).)

Three basic types of check dams were found in the gullies investigated:

1. Rock check dams reinforced with fence panels and wing structures
2. Rock check dams without fence panels, but with wing structures
3. Straw bales supported with steel rebar

Fenced check dams were used on 7 of the 21 riprap waterways; 4 waterways had no check dams; and 10 waterways had nonfenced check dams.

Straw bales were found to be ineffective in controlling erosion in large gullies because they were easily washed out or bypassed. Intact straw bales were found in their original orientation, yet the gully water had either flowed around or undermined the bales. All bales showed significant signs of physical deterioration resulting from the effects of long-term weathering.

In general, the watershed size, the gully length, the gradient, the need for terrace ditches, the extent of yazoo gully formation, and the peak rate of runoff are proportional to the number of check dams needed for a particular gully.

A plot of frequency vs. the number of dams per gully indicates that two dams were used per gully most often at the Prairie Hill Mine. The number of dams used per gully at the mine is a function of

TABLE 8.2
Waterway Analysis for Gullies Investigated at the Prairie Hill Mine, Missouri

Gully No.	Watershed Area (Acres)	Slope Gradient (%)	Gully Length (ft)[a]	No. of Check Dams	Particle Range (in.)	Nominal Diameter (in.)	Peak Runoff Rate (cfs)	Triangular Sectional Channel Area (ft²)	Channel Water Velocity (ft/sec)	Recommended NCSA Riprap Range/Av. (in.)	Waterway Field Status[b]
1	7.4	9	400	2F[c]	3–9	6	59.9	7	8.6	3–12/6	S
2	3.5	9	400	3F[d]	2 to 16+	5	32.2	7	4.6	1–3/1.5	S
3	62.7	2	520	7[e]	2–20+[f]	8	315.7	17.5	18	15–48/24	E
		8	160	0							
		12	120	0							
4	3.7	10	280	2F	3–8	5	36.7	6	6.1	12–6/3	S
		14	160	0							
		10	440	1							
5	14.7	5	560	6e	3–18[f]	8	115.5	10.5	11	5–18/9	S
6	5.1	19	280	2F	2–9	5	46.1	5	9.2	3–12/6	E
7	79.6	4	360	3F	2–24+	12	386.2	12	32	15–48/24	S
8	25.3	25	160	2F	2–10	6	168.9	7	24.1	15–48/24	E
9	46.8	17	120	2F	2–21+	4	246.8	11.3	22	15–48/24	E
10	3.4	5	60	2	3–8	5	32	5	6.4	2–6/3	S
		13	140	0							
12	13.7	8	140	0	2–8	5	98.9	8.5	11.6	5–18/9	S
		19	120	2							
13	13.2	2	120	1	12–24+[f]	20+	87.3	8	10.9	15–18/9	S
		4	100	3							
14	6.9	6	120	1	2–9	5	46.4	5	9.3	3–12/6	S
		8	50	1							
15	8.6	10	80	1	2–24+	9+	61.1	10	6.1	12–6/3	S
		12	40	1							
16	10	3	320	0	2–30+	12+	69.9	7	10	3–12/6	S
		9	100	0							
		12	40	0							

(Continued)

TABLE 8.2 (CONTINUED)

Waterway Analysis for Gullies Investigated at the Prairie Hill Mine, Missouri

Gully No.	Watershed Area (Acres)	Slope Gradient (%)	Gully Length (ft)[a]	No. of Check Dams	Particle Range (in.)	Nominal Diameter (in.)	Peak Runoff Rate (cfs)	Triangular Sectional Channel Area (ft²)	Channel Water Velocity (ft/sec)	Recommended NCSA Riprap Range/Av. (in.)	Waterway Field Status[b]
17	14.4	11	140	0	2–20+	8	91.6	10	9.2	3–12/9	S
18	14.5	11	120	0	2–16	9	83.3	7	11.9	5–18/9	S
19	57.7	7	160	0							
		3	1600	12[e]	3–20+	11+	313.1	14	22.4	15–48/24	E
		9	200	0							
		4	200	1							
		6	280	1							
20	10.5	20	120	3	2–7	5	76.4	10.5	7.3	2–6/3	S
		7	200	0							
		3	240	0							
21	2.2	4	300	2	2–9	4	22.5	6	3.8	No. 8–1.5/.75	S
22	2.1	34	120	0	2–9+	5	21.7	7	3.1	No. 8–1.5/.75	S

a Length of riprap channel and/or check dam series.

b S = stable; E = erosion.

c F = Check dams fenced (not fenced if blank).

d F-=2 check dams fenced, 1 not fenced.

e Series of check dams without riprap channel.

f Distribution applies to riprap in check dams; the majority of the channel was stable, but portions were scoured.

Source: From Vories, K. C. and C. D. Elifrits. Controlling large gullies on a Midwest surface mine, in SME-AIME Preprint 86-348 (Denver, CO: Society of Mining Engineers, 1986).

gully length and gradient. Although a few exceptions exist, points plotted below the line indicate conditions under which one dam or less was used; points plotted above the line indicate conditions under which two dams or more were used. The gully length refers to the length of the channel in the plan view and is limited to the length between gradient changes. For example, a particular gully may be 400 ft in length, but 100 ft may correspond to a gradient of 4%. This plot is useful for determining the number of dams needed on a section of the channel once an estimate of gradient and the corresponding channel length is known for that section of the channel.

Often the spacing of the dams used on a gully was evenly spaced over the length. For example, if a gully were 100 ft long and two dams were to be used, then a spacing of 33 ft would be used between the dams. Check dams were often used in cases in which terraces intersected the gully; however, the design of terraces was such that they were also evenly spaced along the slope. In some of the longer gullies, a series of check dams was used along portions of the gully, while other portions of the gully contained fewer check dams.

Fencing was generally limited to the central portions of the check dams, but in some cases was extended along inner portions on the wings. Fencing consisted of standard, heavy, steel-welded livestock panels cut lengthwise in halves. The fencing was reinforced with between two and four steel T-posts driven into the channel floor. Only a few of the dams used riprap on the downstream side of the fence to help support the structure.

Mined land that is ready to be reclaimed is first graded with the construction of any terraces, levees, ditches, and drainageways, and then is subsequently covered with topsoil and seeded. Terraces generally are designed to have gradients of 1–3%, while ditches can have gradients up to 5%.

Several types of vegetative cover are utilized at the mine. Often the stands are a mixture of vegetative types. One mix includes a warm-season grass: Blackwell switchgrass, birdsfoot trefoil, Korean lespedeza, and crownvetch. Another mix emphasizes cool-season species for erosion control and wildlife habitat, including timothy, orchard grass, smooth bromegrass, alfalfa, and red clover.

Two basic techniques were used in the construction of riprap-lined gullies at the Prairie Hill Mine: (1) grading of the gully using a dozer blade followed by placement of the riprap in the channel by dump trucks, and (2) the same method used as in (1), and the riprap is then spread in the channel using a pan scraper. Gullies that were constructed using the second method had a better-shaped channel than those constructed using the first method, especially when the riprap had a more uniform particle size distribution.

During the course of the gully investigation, it became apparent that a combination of techniques was needed to successfully control erosion and that failure of one of the components of the riprap structure will most likely lower the degree of erosion protection and may even initiate the formation of new gullies. In general, the use of appropriate, uniform-sized riprap for a particular flow velocity is considered the most important parameter in channel design.

The channel shape shown in Figure 8.5 is recommended. V-shaped channels, in contrast, promote scouring. The use of both filter beds and channel riprap were the most helpful in reducing the effects of scouring. Care should be taken to prevent the development of ridges along the channel, as the effectiveness of the waterway in collecting water and helping to create yazoo gullies is reduced. Banking of the channel in a manner similar to that of banked roads may be needed if the channel on the downslope side could be breached.

An estimate of the peak rate of runoff should be determined for the waterway of interest. If a known size of riprap is to be used, then the corresponding velocity can be determined from Table 8.3. Dividing the peak rate of runoff by the velocity determines the needed channel area. From this the appropriate channel depth and width can be determined.

The layout of the waterway can be beneficial, detrimental, or indecisive. The layout of the waterway should be designed to reduce the gradient of the channel. Bends can be designed into the channel layout, but it appears that linear waterways are sufficient in most cases. Waterway lengths are generally small in comparison to the size of the watershed, and, therefore, the effects of the gradient within the channel itself should not be an important factor in determining the peak rate of the

TABLE 8.3
NCSA Graded Riprap Stone

Flow Velocity (ft/sec)	NCSA No.	Size in. (sq. openings)			Corresponding Size of Filter Stone NCSA No.
Max.	Av.	Min.			
2.5	R-1	1 1/2	3/4	No. 8	FS-1
4.5	R-2	3	1 1/2	1	FS-1
6.5	R-3	6	3	2	FS-1
9.0	R-4	12	6	3	FS-2
11.5	R-5	18	9	5	FS-2
13.0	R-6	24	12	7	FS-3
14.5	R-7	30	15	12	FS-3

Source: From Vories, K. C. and C. D. Elifrits. Controlling large gullies on a Midwest surface mine, in SME-AIME Preprint 86-348 (Denver, CO: Society of Mining Engineers, 1986).

runoff. However, if the waterway length is relatively long, then the effects of the channel should be included in the peak rate of runoff analysis.

The peak rate of runoff is mainly affected by the size of the watershed. The size of the watershed may be reduced during the initial design process or by adding more waterways to a particular drainage system. Other factors also affect the peak rate of runoff, and the contribution of any of these factors will help reduce the effects of erosion in a waterway.

The gullies that appeared to be the most effective in controlling erosion utilized gravel filter beds covered by well-graded channel riprap of the appropriate size. Table 8.3 can be used to compare the actual particle size used and the recommended National Crushed Stone Association (NCSA) particle size. In cases in which the waterway experienced erosion, the problem resulted from either a lack of a properly sized riprap for a particular velocity, a sufficient range but gap-graded particle size, or a combination of both. In most cases in which the channel was stable, the actual range of particle sizes was consistent with the guidelines set by the NCSA (Table 8.3). Scouring in gully no. 6 was not caused by inappropriate riprap but resulted from an unfinished check dam that was supposed to capture a feeder ditch.

Check dams worked best with well-graded distributions in which the dam sufficiently prevented the passage of smaller riprap, provided enough permeability to prevent the check dam from impounding water to the point at which water topped or side-tracked the dam, and did not overstress the fencing to the point of failure.

The uniformity of the channel riprap was observed to be an important factor in the construction and the stabilization of the channel. Channels having wide, well-graded particle size distributions or containing larger diameter materials tended to have irregular channel surfaces that allowed the scouring of the channel and filter beds.

No particular pattern is applied to the placement of check dams at this site, but it is apparent that the use of a check dam can be beneficial in reducing scouring in channels. The use of wings on check dams helped channel waters brought in by ditches and reduced the effects of yazoo gullies along the main channel.

When fenced structures are used, it is important to provide proper support on the downslope portion of the check dam to prevent its failure.

8.14   SURFACE MINE DRAINAGE CONTROL

Storm water runoff from mine facilities such as haul roads, waste disposal sites, and orebodies have become recognized as the primary facilities associated with nonpoint source drainage. Depending on the orebody, facility design, and geographical setting, nonpoint source drainage can pollute and have effects therefrom on the local surface and groundwater resources.

Design considerations for plan development must incorporate parameters, including contributing drainage area size, slope, flow velocity, vegetation characteristics, and soil scour velocities.

The success of nonpoint source drainage and erosion stabilization planning is based on implementing control measures that dissect drainage patterns, reduce flow velocities, and disperse runoff toward sediment control sites. Interim and long-range drainage control measures should begin during exploration activities.

Nonpoint source drainage differs from point sources in that the storm water cannot be traced to a specific, identifiable point of entry into the waterway or aquifer.

Impact on water resources from nonpoint sources is considered to occur more frequently or be more significant than point sources. Nonpoint source pollutants include sedimentation, changes in background pH level, alteration to the biota community, reduction to beneficial uses, and introduction of heavy metals or toxic substances. Sedimentation is frequently considered to be the most common form of nonpoint source pollution in surface water resources.

The erosion-transport-sedimentation process affects surface waters by covering stream bottoms, smothering aquatic insects and spawning gravels, clogging inter-gravel spaces used by young fish, and altering the actual water quality. Additional items having an impact on water quality that are related to sedimentation include potential increases in nutrient levels, such as nitrogen and phosphorous compounds, temperature, pH, and a reduction in dissolved oxygen.

The primary source of erosion-related factors associated with mining operations is haul roads. These are typically controlled through the implementation of "best management practices" (BMPs). The development of BMP programmes for control of nonpoint source pollution should include interim and longitudinal water management planning. An effective surface-drainage water management plan incorporates interim drainage stabilization for erosion control with reclamation activities.

The Thunder Mountain Mine is located in central Idaho adjacent to the Frank Church River of No Return Wilderness Area. The project is situated at an elevation of about 2,438 m (8,000 ft) at the headwaters of two primary tributaries to the Middle Fork Salmon River. The location of the mine increased concern and interest within the regulatory and public sectors. Primary concerns associated with project development and operations are related to the impact of sediment on local anadromous fisheries resources and the use of cyanide in the recovery process, as well as the potential impact on surrounding water resources.

Thunder Mountain is operated as an on-off heap-leach gold mine. Leached ore is treated with alkaline chlorination to destroy residual cyanide in the heaps. Following the cyanide destruction process, spent ore is disposed of by individual off-loading pads and placing the material in approved waste disposal facilities.

Water quality monitoring for selected parameters is conducted on a weekly or monthly basis at monitoring wells, springs, and adjacent streams. The water quality parameters currently of most interest to state and federal agencies include pH, cyanide (weak acid dissociable form), turbidity chloride, and key heavy metals.

To address issues and concerns identified during the permit process, an innovative surface water management plan was developed. The plan was initially designed based on evaluating potential flood events associated with rainfall and snowmelt characteristics representative of the project region.

Following a review of several hydrologic model results, a combination of rain-on-snow was selected for drainage control facility design. This approach provided maximum protection from typical snowmelt runoff events, plus an additional conservative safety factor for the occasional rain-on-snow runoff event.

Sediment control facilities were implemented during the exploration phase of project development. Sediment control facilities were constructed during autumn of the first exploration season. Early storm water runoff management structures were designed to prevent concentrations of water

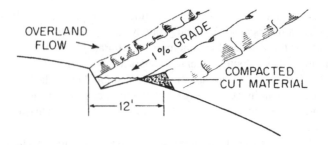

FIGURE 8.7 Dispersion terrace system. (From Mohr, R. Thunder Mountain Mine, *Min. Eng.* February:210–212 (1991). With permission.)

from building up on areas cleared of vegetation for exploration purposes. Dispersion terraces were constructed based on design parameters, including total drainage area, slope, soil scour velocities and runoff potential per acre of drainage area (Figure 8.7).

Location of dispersion terraces was initially conducted via the use of site contour maps generated from the aerial mapping of the project area. Watershed drainage areas contributing runoff to exploration sites were divided into subdrainage areas to control the amount of runoff contributing to any one dispersion terrace.

After the preliminary design of drainage control facilities was completed, dispersion terraces were located in the field by surveying their alignments as laid out in site maps. All terraces were surveyed at a constant slope of 1%. Establishment of this design criteria prevents runoff collected in the dispersion terraces from exceeding a flow velocity of 0.6–0.8 m/sec (2–2.5 ft).

Grade control of terraces was accomplished by placing a stake every 15 m (50 ft) along the length of the facility during the surveying process.

Construction of dispersion terraces was accomplished via a bulldozer. The bulldozer developed an inward-sloping cut, placing soil material from the cut on the outslope side of the terrace, compacting the material as it follows the grade stakes along the length of the facility.

At the end of each dispersion terrace, a level spreader was constructed to enhance the dispersion of runoff onto undisturbed vegetation cover. Through this technique of combining dispersion terraces and level spreaders, the runoff was prevented from concentrating on sites cleared for exploration, minimizing the erosion process. Additionally, dispersion terraces act to collect settleable solids which are the result of the 1% grade. By aligning terraces at this gradient, scour of the terrace surface is prevented and deposition of settleable solids is enhanced.

Dispersion terraces and level spreaders were used in combination throughout the exploration phase of Thunder Mountain Mine.

Project construction began by clearing and grubbing all vegetation from project facility siting areas. As a result of creating large, cleared areas, the soil was exposed to the erosion-transport-sediment deposition process over about 61 ha^2 (150 acres). This exposure of soil to the erosion process on steep slopes created the need for an intensive drainage stabilization and erosion control plan to protect local water resources from sediment impacts.

The basis of the plan incorporated dividing the disturbed areas into small subdrainages with a maximum size of about 2 ha^2 (5 acres). Additionally, all runoff from upland watershed areas was diverted around project facility sites to minimize the contributing area of drainage to project facility sites under construction.

Aerial mapping of the entire project area identified the small subdrainage basins. At the base of each subdrainage area, a dispersion terrace was developed at a 1% slope. Dispersion terraces collected all surface storm water runoff and routed it off and away from the project's area of disturbance.

Actual alignment of dispersion terraces was modified in the field to incorporate existing exploration road cuts to the extent that they were practical. The use of existing exploration roads minimized the amount of additional disturbances required.

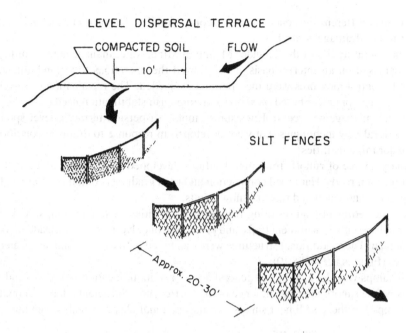

FIGURE 8.8   Level spreader/silt fence sediment trap. (From Mohr, R. Thunder Mountain Mine, *Min. Eng.* February:210–212 (1991).)

Dispersion terraces were developed such that the upper terrace extended farther into the forested area adjacent to the cleared zone than the next downslope terrace. By aligning terraces in this manner, runoff from the uppermost terrace would disperse over forest ground cover vegetation while concentrating runoff to the next downslope terrace. This technique was applied to all terraces as they were constructed down the slope of a cleared area. This enhanced the dispersion of runoff by preventing it from re-concentrating into gully flow conditions.

Due to the large areas of disturbance associated with project construction, additional sediment trap facilities were incorporated at the end of all dispersion terraces and level spreaders (Figure 8.8). Two similar sediment trap techniques were applied based on the anticipated flows expected from each terrace.

The primary sediment trap technique used consisted of placing silt fence structures at the outfall point of dispersion terraces and level spreaders. Silt fences were incorporated due to the minimal costs associated with the material and the life expectancy of the product, thought to be longer than the life of the project.

Placement of the silt fence structures was based on forming the fence material in an arch pattern such that runoff entering the fence would not spill around the uphill edge. Forming an arch out of the silt fence structure in this manner forced the runoff to the pond before spilling over the top of the material at the central point of the arch. Rocks were placed on the downslope side of the silt fence arch at the point at which runoff spilt over the fence. The use of rocks in this location prevents additional scour of soil, minimizing sediment impacts.

A combination of silt fence and log barriers was incorporated as the alternative sediment trap technique for the construction phase of the project. The use of log barriers, in conjunction with silt fences, was applied to the drainage stabilization and erosion control plan in order to evaluate the efficiency of the two techniques and to determine if cost savings could be realized by increased applications of log barrier sediment traps.

After evaluating the efficiency of the silt fences vs. the log barriers, a decision was made to increase the use of silt fences. This was because of the long-life span of the silt fence material in field applications at elevations similar to that of Thunder Mountain. Also, the material is flexible enough to reuse at different sites once an area becomes stabilized or project development precludes the need for future drainage control.

During the operating life of the Thunder Mountain Mine, water management planning and reclamation progressed on an interim basis. This allowed the operations to expand annually, while surface runoff control measures were incorporated as needed. The reclamation of disturbed sites was conducted when practical based on short- and long-term stabilization needs.

Interim water management control measures employ dispersion terraces, level spreaders, silt fences, and several new techniques that were developed in response to drainage conditions generated from major mine facilities.

The primary source of runoff from mine facilities developed for project operations was generated from mine haul roads. Haul road compaction and road gradients combined to create large flow rates during major snowmelt and rain-on-snow events.

To control the sediment load resulting from project facilities, sediment sumps, rock filter sediment traps, roadside culvert sediment traps, and protection of culvert outlets were added to the water management plan. These additional structures were combined with existing and new water management facilities (Figures 8.9 and 8.10).

Sediment sumps were located in two general areas. The main location was at the end of dispersion terraces before runoff entering silt fence or log barrier sites. Sufficiently large amounts of sediment were trapped in these sediment sumps to require annual cleaning with a backhoe. The other

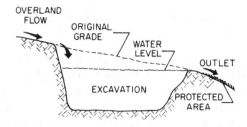

FIGURE 8.9 Excavated sediment sump. (From Mohr, R. Thunder Mountain Mine, *Min. Eng.* February:210–212 (1991).)

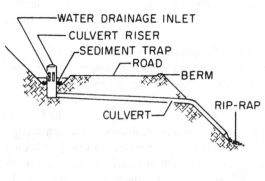

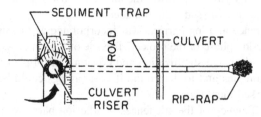

FIGURE 8.10 Roadside culvert sediment trap. (From Mohr, R. Thunder Mountain Mine, *Min. Eng.* February:210–212 (1991).)

location was along the length of dispersion terraces. Sediment sump placement about every 15 m (50 ft) along the terraces acted to extend the functional life of silt fences before maintenance of the structures was required.

Water management facilities, similar to those described, have been constructed throughout the project area. This extensive use of nonpoint source water management techniques has effectively allowed the Thunder Mountain Mine to operate in a wilderness setting without having a major impact on the local water resources. The key aspect of the overall water management plan has been to maintain an interim programme on an annual basis that is designed around the project's yearly progression.

Practical applications to sediment control can be applied to surface mining operations in a cost-effective manner. The key to establishing a functional water management and reclamation programme is to begin with a thorough understanding of the hydrologic regime of the project area.

Once the hydrologic design parameters are set, the next step is to divide the drainage areas within the project area of disturbance into individual subdrainages. The intent here is to minimize the total flow that can be generated from any one area to a level that will not cause erosion of native soils. By dispersing runoff into small components, nonpoint source pollution from surface runoff can be minimized.

Once an effective water management plan has been established, the programme can be modified annually over the life of the mine. The use of cost-effective projects such as silt fences that can be used throughout the life of the project help control costs.

Coordinating water management with reclamation planning helps expedite overall postmining closure requirements. Additionally, interim reclamation speeds the revegetation process such that areas reclaimed early in the project life can reduce the total number of acres under disturbance. This minimizes the total area contributing to the erosion-transport-sediment process.

This approach can also provide an effective programme for meeting anticipated state and federal nonpoint source requirements.

BIBLIOGRAPHY

1. Grim, C. E. and R. D. Hill. Environmental Protection in Surface Coal Mining, National Environmental Research Center, U.S. National Technical Information Service, PB-238 538 (1974), pp. 101–115.
2. Becker, B. C. and T. R. Mills. Guidelines for Erosion and Sediment Control, U.S. EPA Report R2-72-015 (1972).
3. Vories, K. C. and C. D. Elifrits. Controlling large gullies on a midwest surface mine, in SME-AIME Preprint 86-348 (Denver, CO: Society of Mining Engineers, 1986).
4. Mohr, R. Thunder Mountain Mine, *Min. Eng.* 44 February:210–212 (1991).

9 Environmental Impact of Gold Mining

9.1 INTRODUCTION

Dirty gold mining has ravaged landscapes, contaminated water supplies, and contributed to the destruction of vital ecosystems. Cyanide, mercury, and other toxic substances are regularly released into the environment due to dirty mining.

Modern industrial gold mining destroys landscapes and creates huge quantities of toxic wastes. Due to the use of dirty practices such as open-pit mining and cyanide heap leach mining, companies generate about 20 tons of toxic substances for every 0.333-ounce of a gold ring. The waste, usually a grey liquid, is laden with cyanide and toxic heavy metals.

Many gold mines dump their toxic wastes directly into natural water bodies. The Lihir gold mine in Papua New Guinea dumps over 5 million tons of toxic wastes into the Pacific Ocean each year, destroying corals and other ocean life. Companies mining for gold and other metals in total dump at least 180 million tons of toxic wastes into rivers, lakes, and oceans each year, more than 1.5 times the wastes that the U.S. cities send to landfills on a yearly basis.

To limit the environmental damage, mines often construct dams and place the toxic wastes inside. But these dams do not necessarily prevent contamination of the surrounding environment. Toxic wastes can easily seep into soil and groundwater and can be released in catastrophic spills. At the world's estimated 3,500 dams to hold mine wastes, one or two major spills occur every year.

Toxic waste spills have had devastating consequences in Romania, China, Ghana, Russia, Peru, and South Africa. In 2014, a dam collapsed at the Mount Palley gold and copper mine in British Columbia, sending about 25 million cubic metres of cyanide-laden wastes into nearby rivers and lakes, enough to fill about 9,800 Olympic-sized swimming pools. The spill poisoned water supplies, killed fish, and damaged local tourism.

One of the largest rope pit mines is located near Salt Lake City—the Bingham Canyon mine. One of the deepest mines in the world, it is about 4,000 verticals from its rim. To the bottom, the Bingham canyon is known as a copper mine, but the site also yields gold. More than 600 tons of gold have been produced out of this mine since its opening in 1906, and every year, $1.8 billion of metals are produced here.

The Grasberg mine in Indonesia is one of the largest gold mines in the world and is owned by an American company Freeport McMoran. The Grasberg mine is located in the middle of Lorentz National park, creating such a huge scar on the earth that can be seen from space. The mine dumps about 80 million tons of waste debris into the Ajkwa river system every year. Another U.S. company, Newmont, owns the Batu Hijau mine, also in Indonesia. This operation dumps its waste into the ocean near the island of Sumbawa.

9.2 SMALL-SCALE GOLD MINING IN GHANA

Another interesting example of the environmental impact of small-scale gold mining is Ghana. Environmental pollution and many socioeconomic problems have been major issues with the communities where small-scale gold mining is practised in Ghana.

Small-scale mining in Ghana poses serious environmental impact on the surrounding communities. Virgin forest and fertile farmlands located in the middle belt of the country, which includes a part of the Brong Ashante region, western region, and some parts of the eastern regions are disappearing as gold mining has become the dominant activity in these areas. The majority of the challenges posed by small-scale mining in Ghana are experienced in environmental degradation and destruction and its effect on several ecological systems. Land degradation, mercury pollution, and pollution of water bodies are some of the major challenges posed by small-scale surface gold mining. Small-scale surface mining is one of the greatest agents of land degradation, destroying about 13% of the total forested land that surface mining has destroyed in Ghana. A study revealed that surface mining resulted in about 58% deforestation and a substantial 45% loss of farmland within the western region of Ghana. In some places, riverbanks are mined to a depth of 35 m expanding to about 60 m wide. Land degradation results in a threat to biodiversity conservation, with devastating effects on soil ecosystem leading to increase in soil temperature, loss and depletion of soil nutrients, changes in topography, erosion, destruction of its unique habitats and fauna, and making land less productive. The majority of the mining concessions are found in forests, agricultural lands, and human settlements resulting in competition for land and depriving farmers access to farming land.

Mercury Pollution: Mercury is one of the priority toxic elements and is regarded as an important environmental problem in Ghana. Mercury contamination can be either through natural means such as volcanic eruption and weathering or through a variety of anthropogenic sources such as the burning of fossil fuel or mining.

The use of liquid mercury in small-scale mining is a serious threat to water quality in many parts of the world. Mercury, when used in the gold extraction process, forms an amalgam and turns into a stable methyl-mercury compound, which when ingested, inhaled, or absorbed by fauna and flora becomes toxic to man and the environment. In Ghana small-scale miners use mercury in the processing of their ore. In most cases, the waste products are dumped into water bodies, which cause bio-accumulation in the bodies of aquatic animals and enter the food chain of human beings. Exposure to mercury can cause kidney problems; disorders of respiratory, central nervous, and cardiovascular systems; loss of memory; psychosis; reproductive problems; and, in some cases, severe complications in children resulting in death. In Ghana, contaminations of surface and groundwater bodies have particularly been experienced in gold mining communities. When chemicals such as the cyanide used in processing the ore leak, spill, or leach from the processing site into nearby water bodies, those can be harmful to humans, aquatic organisms, and wildlife, as a whole. The majority of small-scale surface miners in Ghana wash the waste products from the ore processing into the rivers and other water bodies that serve as sources of clean portable drinking water to the mining communities. Others release mine tailings directly into aquatic habitats. Mine tailings are often toxic and pose serious health threats to plant life, humans, and animals. Research revealed that many mining communities in the western region of Ghana are at risk of health issues from heavily polluted water bodies by small-scale mining activities.

9.3   GOLD MINING IN ECUADOR

The mining company Aurelin discovered a large deposit, estimated to be 10 million ounces of gold and 14 million ounces of silver, described by some as one of the most exciting gold discoveries in the past 15 years. It has estimated reserves of 6.8 million ounces of gold and 9.1 million ounces of silver.

Gold mining in the Portovelo-Zaruma district in southern Ecuador is causing considerable impacts; the most important ones are related to the discharge of cyanide, mercury, and metal-rich tailings into the Puyango catchment area. Cyanide and metals in rivers regularly exceed environmental quality criteria. The contamination impacts biodiversity, with cyanide causing a direct lethal effect on biodata further downstream. It is shown that the prevailing neutral or slightly alkaline conditions of the river water further downstream ensure that metals are primarily sediments. However, elevated metal levels in bottom-living larvae collected from contaminated sites suggest that these sediment-bound metals are readily available. Leaching experiments indicate that the relative ease by which metals are taken up by larvae is related to the specification of the sediment under ambient pH conditions. They enter in the dissolved and directly bioavailable state in more acidic conditions. Metal levels in carnivorous fish were found to be modestly elevated, with the exception of mercury. Mercury levels exceeded 10 mg/kg in fish from both contaminated and uncontaminated sites.

9.4   GOLD MINING IN ROMANIA AND ENVIRONMENTAL DAMAGE

On January 30, 2000, the dam containing toxic waste material from the Bala Mare Aurrul gold mine in North-Western Romania burst and released 100,000 m³ of wastewater, heavily contaminated with cyanide, into the Lapus and Somes tributary of River Tisza, one of the biggest in Hungary. The cyanide-contaminated water was carried to River Danube, which flows through Serbia, Bulgaria, and Romania. Besides the ecological damage, the cyanide pollution in the river meant a significant threat to the human health, because in the upper part of the Tisza, the cyanide concentration was 100 times more than the limit value of the drinking water. It was estimated that between Tiszafuered and Szolnok, 80–90% of fish stock would be killed. Other wildlife has been affected, including mute swans, black cormorants, foxes, and other carnivores. The dam was built in 1998.

Since the 1960s, when cyanide heap leaching was introduced, the impact of gold mining has rocketed. There have been tailings dams where the contaminated wastewater from the mining process is stored, which is a frequent cause of environmental disasters. In late December 1999, several thousand cubic metres of cyanide-contaminated effluent was released from the Bala De Aries mine into the Aries river, which subsequently flowed into the Mures, a tributary of the Tisza. The Industria Samuei metal processing combine in Campia Turzil discharges all of its effluents around Hunearound into the Aries. A further cyanide disaster may well be caused by the tailings ponds of the gold mines in Brad, Abrud, and Zlanta. The latest accident that happened in Brad was in May 1998, when the Tisza tributary Crisul Alb was contaminated severely with cyanide and heavy metals. The uranium mines near Brad are a further threat to this river. In Zlanta, the Ampelum noble metal processing combine released sulphur oxides in February 1998, devastating 47,000 ha of farmland and 193 km of the river landscape. Metal smelters around Hunedoara pollute the Mures on a regular basis. In Tiravenin, an accident in the Bicapa combine in December 1999 caused chromium levels to rise to 20 times the permissible values. This poison reached the Mures. In Slovakia, too, magnesite and other ore mines operate on the Tisza tributary Hornad near Koisice; here again, there are reports of heavy metal levels in excess of permissible values in the river water. A cyanide pollution incident was reported in February 2000 from the Slovakian part of the Bodrog plain.

BIBLIOGRAPHY

1. The environmental disaster that is gold industry, Smithsonian.com, 1809497620.
2. Bland Alastair, The mining industry has had a devastating impact on ecosystem worldwide. Is there hope in sight? Smithsonianmag.com (February 14, 2014).
3. Tarras-Wahlberg et.al. Environmental impacts and metal exposure of aquatic ecosystems in rivers contaminated by small scale gold mining: The Puyango river basin, Southern Ecuador, *Science of the Total Environment*, 278: 239–261 (2001).
4. Kanthak, Judith, The Baia Mare Gold Mine Cyanide Spill: Causes, Impacts and Liability, Greenpeace (April 12, 2000).
5. Arah, Isaac Kojo, The impact of small scale gold mining on mining communities in Ghana, in African Studies Association of Australia and the Pacific 37th Annual Conference, Dunedin, New Zealand (January 2015).
6. Hilson, Gavin, The environmental impact of small scale gold mining in Ghana, Identifying problems and possible solutions, *Geog. J.* 168(1): 57–57 (2002). https://doi.org/10.1111/1475-4959.00038.

10 Blasting

10.1 INTRODUCTION

The goal of blasting is to obtain maximum fragmentation of the consolidated material in the overburden with optimum drilling and blasting cost. The amount of fragmentation required is determined by the stripping unit to be used in overburden removal. Many coal seams by surface coal mines must also be broken by blasting; this is conducted before coal removal. Environmental factors, as well as due regard for public safety, health, and welfare, must be considered in choosing the blasting plan.[1]

The blasting plan should be made during preplanning and is based on data from the overburden cores. The analysis of the data will help determine the kind of drilling equipment and bit types that will be needed for overburden preparation.

A variety of complaints has always been received by the industry pertaining to blasting. The population explosion and urban sprawl have acted in concert to bring the industry and the public into closer physical contact. In many cases, structures were built on property adjacent to surface mining operations. As a result, the number of complaints increased drastically and presently constitutes a major problem.

Some complaints registered are legitimate claims of damage from blasting vibrations. The advances in blasting technology and a more knowledgeable explosives profession have minimized real structural damage. However, vibration levels that are completely safe for structures may be annoying and unpleasant for humans. Though no actual damage is done, air blast pressures may cause windows to rattle and the loud noise may be intolerable. Repeated vibrations, such as those from a nearby quarry, may eventually cause damage.[2]

Control of vibrational damage to natural scenic formations is a very important environmental consideration in surface mining. The wind-eroded formations are very fragile, and damage as far as one-fourth of a mile (P:402 km) from the operation has been noted.

Where a conventional detonating cord is used to link blastholes, most airborne noise results from the connecting trunk lines. A new, low-energy detonating cord has been developed that can be substituted for the conventional cord. A 150-ft (45.6 m) length of this cord makes about as much noise as one electric blasting cap or 2 in. (50.8 mm) of the conventional cord.

If a detonating cord is used on the surface, noise can be reduced by covering the trunk lines with up to 10 in. (254 mm) of dirt. When the detonating cord is used only in the holes to fire the primers, a shovelful of dirt at each hole will effectively cover the exposed cord and cap.

Millisecond delays can be used to decrease the vibration level from blasting because it is the maximum charge weight per delay internal rather than the total charge that determines the resulting amplitude. Also, many mines limit the number of holes per shot, using millisecond delays in series to minimize concussion and noise, especially near population centres, natural scenic formations, wells, water impoundments, and stream channels.[3-5]

Weather conditions can increase airborne noise. When temperature inversions prevail, blasting should be avoided. This condition frequently exists in the early dawn and after sundown. Foggy, hazy, or smoky days are unfavourable for blasting. When the wind is directed towards residential areas, blasting should be postponed.

When blasting is performed in congested areas or close to a structure, stream, highway, or other installation, the blast should be covered with a mat to prevent fragments from being thrown by the blast. The possibility of dust problems resulting from blasting is very remote.

The possibility does exist, however, and precautions must be taken to control dust pollution if the operation is close to high-use areas. During periods of dry weather, dust from explosions has been carried by air currents for many miles, and in certain isolated instances, it has been a public nuisance.

Several states, including West Virginia, Tennessee, Ohio, Montana, and Kentucky, have established guidelines for preventing or holding vibrational damage to a minimum. Most state laws concerning blasting pertain only to safety, storage, handling, and transportation of explosives.

When a blast is detonated, the bulk of the energy is consumed by fragmentation and some permanent displacement of the rock close to the location of the drilled holes containing the explosive. This activity normally occurs within a few tens of feet (metres) of the blast hole. The leftover energy is dissipated in the form of waves travelling outward from the blast, either through the ground or through the atmosphere. The ground waves produce oscillations in the soil or rock through which they pass, with the intensity of these oscillations decreasing as distance from the blast increases.

One measurable quantity of interest that is caused by seismic waves or oscillations is particle velocity. This quantity defines how fast a particle (or structure) is moved by passing seismic waves, measured in inches (millimetres) per second. The results of a 10-year study programme in blasting seismology by the U.S. Bureau of Mines concluded that particle velocity is more directly related to structural damage than particle displacement or particle acceleration. It is not how much but how fast the ground under a structure is moved by the passing seismic waves that determines the likelihood of damage. Particle velocity, therefore, becomes the vibration quantity of the greatest concern to those engaged in blasting activities. They also concluded that a safe blasting limit of 2.0 in./sec (50.8 mm) peak particle velocity as measured from any of three mutually perpendicular directions in the ground adjacent to a structure should not be exceeded if the probability of damage to the structure is to be small (<5%). Kentucky is the only coal-producing state that has passed a law based on seismographic measurements. They limit vibrations adjacent to any structure to levels producing a particle velocity of 2.0 in./sec (50.8 mm) or less.[6,7]

Where instrumentation is not used or is unavailable, the U.S. Bureau of Mines found that a scaled distance of 50 feet per square root of pounds (22.62 m per square root of kilograms) can be used as a control limit with a reasonable margin of safety, and the probability is small or finding a site that produces a vibrational level that exceeds the safe blasting limit of 2.0 in./sec. For cases in which a scaled distance of 50 feet per square root of pounds (22.62 m per square root of kilograms) appears to be too restrictive, a controlled experiment with instrumentation should be conducted to determine what scaled distance can be used to ensure that vibrational levels do not exceed the particle velocity of 2.0 in./sec (50.8 mm).[8]

West Virginia uses the scaled distance formula, $W = (D/50)^2$, for the control of vibrational damage. W equals the weight in pounds (kilograms) of explosives detonated at any one instant and D equals the distance in feet (metres) from the nearest structure, provided that explosive charges are considered to be detonated at one time if their detonation occurs within 8 msec or less of each other (Table 10.1) for maximum explosive charges. A blasting plan for each method for a typical blast must be submitted with the permit application.

Citizen complaints concerning blasting on surface mining operations have been drastically reduced since the West Virginia law became effective. This success can be attributed to the conscientious efforts by the operators in using the scaled distance formula and guidelines for blasting issued by the State of West Virginia Department of Natural Resources.

Ammonium nitrate-fuel oil (AN/FO) blasting agents and slurries, used as breaking mediums for overburden, have greatly improved the efficiency of surface mine blasting operations and have reduced the cost of explosives considerably. AN/FO is an excellent heterogeneous fertilizer because it contains readily available ammonia nitrogen and nitrate nitrogen and does not leave unfavourable residues in the soil. As a constituent of various types of explosives, it functions as an oxidizer and an explosive modifier. AN/FO mixes lead all other types of explosives in bank preparation. Several types are available and can be obtained in prilled, granular, crystalline, or grained forms.

TABLE 10.1

Maximum Explosive Charges[a] Using Scaled Distance Formula, $W = (D/50)^2$

Distance to Nearest Residence Building or Other Structure (ft)	Max. Explosive Charge (to be detonated) (lb)	Distance to Nearest Residence Building or Other Structure (ft)	Max. Explosive Charge (to be detonated) (lb)
100	4	2,100	1,764
150	9	2,200	1,936
200	16	2,300	2,116
250	25	2,400	2,304
300	36	2,500	2,500
350	49	2,600	2,704
400	64	2,700	2,916
450	81	2,800	3,136
500	100	2,900	3,364
550	121	3,000	3,600
600	144		
650	169	3,100	3,844
700	196	3,200	4,096
750	225	3,300	4,356
800	256	3,400	4,624
850	289	3,500	4,900
900	324	3,600	5,184
950	361	3,700	5,476
1,000	400	3,800	5,776
		3,900	6,084
1,050	441	4,000	6,400
1,100	484		
1,150	529	4,100	6,724
1,200	576	4,200	7,056
1,250	625	4,300	7,396
1,300	676	4,400	7,744
1,350	729	4,500	8,100
1,400	784	4,600	8,464
1,450	841	4,700	8,836
1,500	900	4,800	9,216
1,550	961	4,900	9,604
1,600	1,024	5,000	10,000
1,700	1,156		
1,750	1,225		
1,800	1,296		
1,850	1,369		
1,900	1,444		
1,950	1,521		
2,000	1,600		

Metric unit conversion: foot = 0.304 m; pound = 0.453 kg.

[a] Where blast sizes would exceed the limits of the scaled distance formula, blasts shall be denoted by the use of delay detonators (either electric or non-electric) to provide detonation times separated by 9 msec or more for each section of the blast complying with the scaled distance of the formula. Explosive charges shall be considered to be detonated at one time if their detonation occurs within 8 msec or less of each other.

A new line of metalized blasting agents has become commercially available. These products are reported to be three times more powerful by weight and five times more powerful by volume than AN/FO combinations. Based on ammonium nitrate in combination with aluminium chips, the blasting agents vary in aluminium content between a low of 5% and a high of 30%. They are soft, silvery gels and maintain that softness even at 0°F (–18°C).

A trend is developing for casting overburden with explosives. The goal is to cast as much overburden as possible into the parallel cut with blasting techniques. With proper loading, spacing, and detonation delays, a good portion of overburden can be moved, thus reducing backfilling costs. This method also minimizes the need for recasting. Some mines report that 30–50% of their overburden is moved with explosives. This method works very well in deep, narrow pits by casting overburden into the pit away from the highwall and up on the spoil pile on the low wall side.

BIBLIOGRAPHY

1. Nicholls, H. R., C. F. Johnson and W. J. Duvall, *Blasting Vibrations and Their Effects on Structures* (Washington, DC: U.S. Department of Interior, Bureau of Mines, 1971) Bull. 656.
2. Perkins, B., Jr. and W. F. Jackson, *Handbook for Prediction of Air Blast Focusing*, Ballistic Research Laboratories, Report No. 1240, U.S. Army, Aberdeen, MD (1964).
3. Perkins, B., Jr., P. H. Lorrain and W. H. Townsend, *Forecasting the Focus of Air Blast Due to Meteorological Conditions in the Lower Atmosphere*, Ballistic Research Laboratories Report No. 1118, U.S. Army, Aberdeen, MD (1960).
4. Berger, P. R., Blasting controls and regulations, *Min. Congr. J. 59*(II):21–25 (1973).
5. West Virginia Code, Chapter 20: Surface Mining and Reclamation: Section 6–1 la, Blasting Restriction; Formula; Filing Preplan; Penalties; Notice, 1971.
6. Siskind, D. E., Damage to residential structures from surface mine blasting, in *SME-AIME Preprint 80-336* (Denver, CO: Society of Mining Engineers, 1980).
7. Linehan, P. and F. J. Wiss, Vibration and air blast noise from surface coal mine blasting, *SME-AIME Preprint 80-336* (Denver, CO: Society of Mining Engineers, 1980).
8. Siskind, D. E. and M. S. Stagg, Structure Response and Damage Produced by Ground Vibration from Surface Mine Blasting, U.S. Bureau of Mines Report of Investigations, RI 8507 (1980).

11 Mining Subsidence

11.1 INTRODUCTION

Underground mining creates a void system that disturbs the existing stress field established by natural processes. The overburden responds with the first cut into the coal and continues until the applied stress reaches a new equilibrium within the overburden. These stress changes result in deformation and displacement of the surrounding overburden material. The magnitude of the change that takes place is controlled by the size of the cavity and the strength characteristics of the affected strata. Figure 11.1 is a simplified illustration of the forces acting on a stratigraphic section as a result of a mine void. The force vectors above the void are deflected towards the void proportionate to the distance from the centre of the void. Below the coal seam, an upward component of force also acts towards the centre of the void. As the cavity increases in size, fractures develop in the overburden, which may result in subsidence of the surface and sometimes heaving of the mine floor.

In general, mine subsidence problems develop where postmining pillar support systems and coal barriers ultimately fail. Many interrelated factors control when, where, and how failure will occur, including:

- Thickness of coal mined
- Size, shape, and distribution of pillars and rooms
- Percentage of extraction of coal
- Thickness and physical characteristics of the overburden
- Method of mining (e.g., longwall, short wall, room and pillar, room and pillar with full or partial retreat)
- Dry or flooded conditions in the mine
- Actual or potential level and degree of fracturing in the overburden
- Mineralogy of overburden (e.g., clay minerals that swell when water is added, sulphide minerals that chemically and physically change in the presence of oxygen and moisture, and minerals that react with water to form new minerals)

Three main types of subsidence are seen: pit, trough, and shaft. Pit subsidence is responsible for more of the extreme danger problems than trough or shaft subsidence. Each of these three types of subsidence has specific movements that separate them from each other. Eventual erosion is a type of movement that accentuates each of these subsidence problems.

Typically, pit subsidence occurs from the failure of the roof (Figure 11.2). The result is an opening from 2 to 2.5 m deep and 0.6 to 12 m in diameter. The main movement, at least initially, is vertical. In the eastern United States, pit subsidence usually occurs where mines are <30 m deep; in Pennsylvania, most pit subsidence is found where the mine overburden is <15 m. Areas in which the mine is deeper than 45–50 m experience decreased frequencies of subsidence occurrence. Pit subsidence usually occurs over rooms and entries, resulting in a surface reflection of the mining pattern.

The second type of subsidence is a trough, which is usually a gradual disintegration of the pillars or squeezing of floor materials into the mine voids. The disintegration and squeezing may also occur simultaneously. Troughs (Figure 11.3) may occur in areas of mining at any depth, but most frequently associated with deeper mines than those in which pit subsidence occurs. Troughs are usually 0.6–1.3 m deep and cover larger areas than pit subsidence. The movements are both vertical and horizontal as the whole area is depressed. The horizontal areas may extend over a broader area

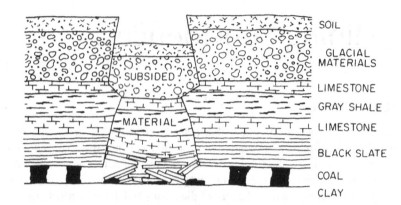

FIGURE 11.1 Pit subsidence. (From Elbert, J. and L. Guernsey. An Analysis of Extreme Danger Problems Associated with Subsidence of Abandoned Coal Mine Lands in Southwestern Indiana, U.S. Bureau of Mines, IC 9184 (1988), pp. 383–389.)

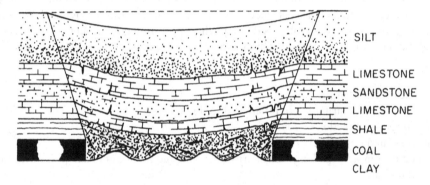

FIGURE 11.2 Trough subsidence. (From Elbert, J. and L. Guernsey. An Analysis of Extreme Danger Problems Associated with Subsidence of Abandoned Coal Mine Lands in Southwestern Indiana, U.S. Bureau of Mines, IC 9184 (1988), pp. 383–389.)

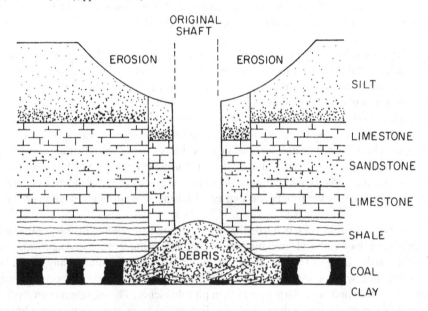

FIGURE 11.3 Shaft subsidence. (From Elbert, J. and L. Guernsey. An Analysis of Extreme Danger Problems Associated with Subsidence of Abandoned Coal Mine Lands in Southwestern Indiana, U.S. Bureau of Mines, IC 9184 (1988), pp. 383–389.)

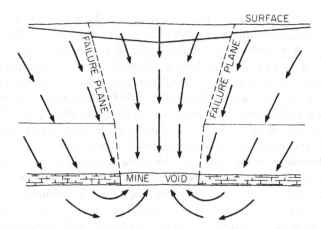

FIGURE 11.4 Ground movement due to subsidence. (From Elbert, J. and L Guernsey. An Analysis of Extreme Danger Problems Associated with Subsidence of Abandoned Coal Mine Lands in Southwestern Indiana, U.S. Bureau of Mines, IC 9184 (1988), pp. 383–389.)

than is undermined. For a deep mine, the area affected may be 20 ft beyond the mined-out areas. This is an important point as many individual homeowners feel that they and their property are safe if the mine does not extend beneath their house.

The third type of subsidence is associated with the shaft itself. This is commonly associated with older mines. The older shafts were not well supported when active mining was ongoing. Frequently the added weight of the buildings and equipment near the shaft was not considered in the required support. Decades later, the supports fail and erosion widens the original shaft opening (Figure 11.4). New subsidence sites pose continuous problems because they are a result of collapsing shafts and tunnels. In the western United States, subsidence occurs around mine shafts. This subsidence is similar in appearance to pit subsidence, except that the pits are much deeper.

11.2   SUBSIDENCE INVESTIGATIONS

The approach and procedures for the study of subsidence over abandoned mine lands have been established. Time is essential because the damaged conditions may be disturbed, and the exposed features may disappear. Therefore, a site investigation is always the first step for subsidence research because most of the information must be collected from the subsidence site, especially during the first-time reconnaissance. For most cases the causes, damage pattern, and the status of ground movement can be determined by the obtained information. The information mainly consists of six categories such as (1) general information, (2) structure information, (3) surface condition, (4) geological information, (5) geometry and characteristics of the subsidence, and (6) mining activity. Each category consists of several subcategories. In order to ensure that complete information can be collected a site investigation checklist has been developed by this research group.

With the data collected from a site investigation, a hypothesis should be established to clear up the logic and narrow down the number of possible causes. In order to determine the cause of the damage, the cause identification system with an identification list has also been developed by this group to evaluate the most possible cause. If the cause cannot be determined due to insufficient data, an in-depth exploration should be adopted.

An in-depth exploration mainly consists of surface and subsurface measurements and instrumentations, including surface survey, borehole TV camera examination, crackmeter, inclinometer, and Sondex settlement measurements. The results can be used to determine the cause(s) of the damage and the direction and magnitude of movements and can assist in the design of a better remedial programme.

When subsidence occurs, several types of ground damage will be induced. The most common damage is depressions, cave-in pits, ground cracks, and compression ridges. The ground depression is associated with trough or sag subsidence. Its mining-related causes are pillar and floor failures; non-mining-related causes are differential settlement on the backfilled area, fluid withdrawal, and organic soil drainage. A ground depression is a gentle settlement over a broad area, which may range from a few tens to hundreds of centimetres in diameter, while the depth of depression ranges from 0.1 to 1 m for mine subsidence. The depth of the depressions due to non-mining-related causes is relatively unknown, but a 4 m depression has been reported to be due to fluid withdrawal.

The cave-in pits are most likely associated with potholes, sinkholes, or chimney subsidence. Their mining-related cause is roof failure, while the non-mining-related causes are limestone solution and soil piping. The newly developed cave-in pits are likely to be steep sided with straight or bell-shaped walls depending on the soil type. The diameter of cave-in pits ranges from 0.6 to 12 m, but most of them are <16 ft in mine subsidence. The depth of cave-in pits is about 2–2.5 m for mining-related cases, and the coal seams are <55 m deep. For the non-mining-related cave-in pits, the diameter ranges from a few metres to 3,000 m, and the depth ranges from a few metres to 20 m. All the pothole subsidence cases investigated in this research showed that the diameter of the holes ranged from 0.0 to 6 m, the depth was about 0.6–5 m, and the coal seam was about 6–18 m deep.

The ground cracks that occur at the margin of the depression and cave-in pits are the result of convex bending and are associated with the stretching of the ground surface. Generally, the ground cracks at the margin of the depression, and cave-in pits indicate the location of the maximum tensile strain. Beyond this maximum tensile strain zone, structure and ground damage is usually minor or non-existent. Therefore, a subsidence damaged area can be determined by locating the ground cracks. In general, ground cracks are wider, more abundant, and more extensive near cliffs and steep terrain than they are in flat or gently rolling topography. Moreover, ground cracks are more often located at the asphaltic pavement than that at soil zone because asphalt is more brittle than soil and can absorb less stress.

Compression ridges occur in the depression where the ground surface is subjected to concave bending and associated with shortening of the ground surface. Generally, compression ridges are located in the central part of the depression. The ground cracks and compression ridges sometimes are not visible when they are located in the soil zone, especially when the surface is covered by vegetation. The crack does not always reach the surface. The stress at the tip of the crack in propagation can be less than the soil cohesion and root binding forces. Crack propagation is either stopped or changes its direction to the weakened part of the soil. In addition, a ground crack may have been developed in the subsurface and cannot reach the ground surface. Therefore, if the ground crack cannot be seen in the area, it does not necessarily mean that a ground crack has not developed.

11.3 STRUCTURAL DAMAGE

Structural damage due to subsidence ranges from slight to very severe. Some structures may have a few hairline cracks while others may have cracks a few inches wide. The damage severity is dependent upon (1) the types of ground surface deformation, (2) the type of stresses, (3) the type and size of structures, (4) the location of the structure with reference to the mine openings, (5) the material of the structures, and (6) the type of foundation of structure attached to the ground. For instance, structures attached firmly into the ground are likely to be affected more by ground movement.

Damage to buildings can be divided into functional (structural) and cosmetic (architecture) damage. The major functional damage consists of misalignment of windows, doors, and walls; slanting floor; leakage in the roof; and breakage of utility lines. This type of damage may impair the function and usage of the structure. The major cosmetic damage consists of cracks along the mortar lines on the exterior and interior walls, cracks on the stone facing and plaster, separation of siding and frames, and cracks on the floor. This type of damage is more annoying than dangerous. Both functional and cosmetic damage can occur separately or simultaneously, based on the level of the damage source. For instance, a low level of vibration due to traffic or industry may cause cosmetic damage; however, a high level of vibration due to blasting or earthquake will cause not only cosmetic damage but also functional damage and sometimes even cause structural collapse. If a building has a basement, the deformation of the building due to subsidence most likely will be initiated from the foundation or basement and propagate upward to the upper structure. The foundation or the basement is likely to suffer more severe damage than the rest of the building because the basement is directly subjected to the stresses. The amount of damage is referred to as the degree of structural deformation, not the repair cost.

Generally, structural damage is associated with ground movements; the structures are assumed to move with the ground. However, resistance or friction exists, which is due to the weight of the structure, between the structure and the ground. Therefore, the ground movement cannot be entirely transmitted to the surface structures. Accordingly, several types of stresses, such as axial, shear, and bending, are induced against the foundation of the structures or the structures directly. The axial stresses, either extension or compression, are caused by horizontal movement. This type of stress will induce tensile and compressive damage to the structures. The shear stresses are caused by the friction between the structure and the ground. The bending stresses are caused by the vertical movement of the ground and will induce tilt and curvature damage to the structures.

As discussed earlier, linear deformation is due mainly to nonuniform horizontal movement, which will induce tensile and compressive damage to the structures. If the structure is subjected to tension, the structure will be lengthened and open cracks will be induced along the joints and weak points, such as the contact line between the windows, door frames, and the walls. All the tensile cracks are uniformly developed along the brick faces and mortar lines because these are the weakest places on the brick wall. If a structure is subjected to compression, the foundation walls buckle inward and induce horizontal cracks perpendicular to the direction of the compression along the mortar on the brick faces. Tensile cracks and buckled structures are the specific characteristics of tensile and compressive damage, respectively. Those characteristics can be used to identify the types of stresses that cause damage to the structure.

When the ground slope changes, tilting is induced on the structures. Tall structures with a small base, e.g., towers and chimneys, are the most sensitive to tilting. The natural ground slope will also affect the stability of the ground surface and structures when subsidence occurs. If the dip of a slope is consistent with the ground movement direction, ground movement is enhanced. On the other hand, when the dip of a slope is in the opposite direction from that of the ground movement, ground movement is reduced. In addition, a slope area may be a location prone to instability or landslide.

Depending on the location of the structure in the depression, the structure may be subjected to sagging or hogging damage. Sagging and hogging are two different types of ground curvatures which are caused by the differential settlement. Sagging is a negative or concave curvature. For structures located on the concave area, the lower portions of the structures are subjected to tension and the upper portions are subjected to compression. As a result, vertical cracks with wider gaps at the bottom occur at the window and door frames, on the walls, and at the wall joints.

Hogging is a positive or convex curvature. For structures located in the convex area, the upper portions of structures are subjected to more severe stretching than the lower portions. As a result, vertical cracks with wider gaps are found at the top of the window and door frame, on the wall, and at the wall joints. This crack pattern indicates that the structure has been subjected to hogging. In general, structures located at the edges or the inflection point of the depression will be subjected to hogging damage. The most specific damage characteristics on the structure due to hogging are the convex bending and v-shaped crack.

11.4 DAMAGE CRITERIA

There are two types of damage criteria for buildings affected by mining subsidence. One type is based on the appearance of the damage and the other on the magnitude of the ground movement parameters. The damage criteria discussed here belong to the second category, and their development involves three steps. The first step serves to classify buildings based on their structural

---

TABLE 11.1

Building Characteristics and Their Ratings

Characteristics	Rating
Foundation	
Isolated footing	1
Continuous footing	4
Raft foundation	8
Buoyancy foundation	16
Superstructure materials	
Brick, stone, and concrete	2
Reinforced concrete	4
Timber	6
Steel	8
L:R ratio	
<0.1	4
0.10–0.25	3
0.26–0.15	2
>0.50	1
H:L ratio	
<1.0	4
1.0–2.5	3
2.6–5.0	2
>5.0	1

---

characteristics, the second to specify the damage levels, and the third to determine the critical subsidence indices for each building category (Table 11.1).

The building classification method used in this section is similar to that used in Poland. Several building characteristics are taken into account, and each of these characteristics is rated into different classes, according to its response to subsidence. The final rating is the sum of all the ratings of the various building characteristics, which in this case include building foundation, superstructure material and L:R and H:L ratios, where L is the length of the building, H the height, and R the radius of influence.

Several methods are available to specify damage levels. For example, under the system proposed by the British National Coal Board, the change in structure length combined with a physical description of damage is used to categorize damage. In a classification proposed seven parameters were considered and rated; damage level was determined by adding the ratings of the individual parameters.

Several subsidence indices have been used when developing building damage criteria. The most commonly used indices are horizontal strain, deflection ratio, curvature, tilt, and angular distortion. It is not necessary to use all these indices in establishing damage criteria as some of them are interrelated.

11.5 REMEDIAL MEASURES

Currently, four types of remedial measures are available for abandoned mine subsidence: point support, local backfilling, areal backfilling, and strata consolidation. In the point support method, a large number of grouting holes are drilled and a relatively small quantity of grouting materials are injected to form the grouting piles in the mine voids or openings and to achieve firm contact

TABLE 11.2
Building Classes

Total Rating	Class
4–10	I
11–17	II
18–25	III
26–32	IV

with the main roof. By doing so, the amount of subsidence potential and subsidence severity can be reduced. The advantages of these methods are that they are simple and use a small amount of grouting materials. The disadvantages are that a large number of grouting holes are needed to be drilled, the distribution of grouting materials cannot be controlled, and its support capacity is uncertain. Based on different procedures and approaches the point support methods can be subdivided into gravel column method, grout column method, fly ash grout injection method, and fabric formed concrete method (Table 11.2).

The local backfilling method is designed to fill the small and shallow potholes or surface cracks without drilling the grouting holes. The backfill materials include gravel, refuse, and dirt. Generally, the backfill materials are directly dumped to fill the cracks and holes.

Areal backfilling methods are designed to fill a large area of underground openings in order to provide general protection to urban areas that are measured in terms of hundreds of acres. These methods involve a large quantity of grouting materials such as coal refuse, fly ash, and gravel which are injected into underground openings under high pressure.

Strata consolidation methods are designed to bind or grout the shallow strata beneath the damaged structure into a single rigid unit. If subsidence continues, the unit will be moved as a rigid body without being damaged. Based on different grouting materials and approaches, the strata consolidation methods can be divided into the polyurethane binder method, the cement grout pad method, and the rock anchor method.

Stress redistribution due to abandoned mine subsidence was initially conducted on a model that represented a typical retreated abandoned room and pillar mine. At the first stage, the pillars at the central area were subjected to higher stress than the surrounding area. Multiple pillar failure and subsidence were initiated. At the second stage, more pillars failed and subsidence increased. Simultaneously, destressing occurred at the failure zone while stress concentration occurred in the vicinity. At the third stage, pillar failure and subsidence were continuously developed. Meanwhile, because of the compaction of gob materials, stress again built up at the failure zone, but stress concentration still remained at the outward area. In summary, the magnitude of subsidence and surface affected area increase as stress concentration continues to cause pillar failure.

BIBLIOGRAPHY

1. Craft, J. L. and T. M. Crandall, *Mine Configuration and Its Relationship to Surface Subsidence* (Washington, DC, U.S. Bureau of Mines, 1988), IC 9184, pp. 373–381.
2. Elbert, J. and L. Guernsey, *An Analysis of Extreme Danger Problems Associated with Subsidence of Abandoned Coal Mine Lands in Southwestern Indiana* (Washington, DC, U.S. Bureau of Mines, 1988), IC 9184, pp. 383–389.
3. Hsiung, S. M., P. M. Lin and S. S. Peng, *Structures and Ground Damages over Abandoned Mine Lands* (Washington, DC, U.S. Bureau of Mines, 1988), IC 9184, pp. 362–368.
4. https://en.wikipedia.org/wiki/Aberfan_disaster.

12 Postmining Land Use

12.1 INTRODUCTION

During surface coal mining activities, all other uses of land are precluded. This results in a public perception that a conflict exists between surface mining and other established land use demands. However, surface mining is only a temporary use of land that can be accommodated in a sequence of future land uses. This can be achieved by successful reclamation efforts in restoring the land to its premining use capabilities.

Land use planning is conducted at two levels: macro- and microscale. Macroscale planning, sometimes called comprehensive planning, is directed towards guiding growth, development, and land use for an entire region through the formulation of a comprehensive plan. Microscale planning, generally referred to as site planning, is concerned with obtaining a site plan that satisfies given performance standards and an overall comprehensive plan for the area. Macroscale planning is generally conducted by public planners at the county, state, and federal levels. Site planning is generally carried out by the private sector, by land developers, or mine planners.

The general process of land use planning is shown in Figure 12.1. The mined land use planning process is illustrated in Figures 12.2 and 12.3.

The processes used for both types of land use planning are essentially the same, but the level of effort may vary significantly. Comprehensive planning is conducted before site planning. However, it is not uncommon that site planning be conducted in an area in which no comprehensive plan exists. This places the burden on the site planner to evaluate regional objectives. Once a mining company has defined the scope of its reclamation and land use planning, this scope is unlikely to change for different mining operations. Scope includes specifying the type of output required from the planning process, such as final design, conceptual plan, and organizational structure of the planning process.

The local goals and objectives are determined by public planners. Mine planners interact with local and regional planners so that the postmining land use plan is compatible with the overall plan for the area. Besides satisfying local goals and objectives, the mining company may wish to establish other goals and objectives, such as improving the value of the land, promoting good public relations, or developing other company goals for postmining land use.

The environmental differences between the various regions impact the postmining land use potential of surface-mined lands. The socioeconomic conditions, such as population trends, employment trends, land values, the influence of public planners, regulatory constraints, and the availability of cultural resources, may have an impact on the actual use of reclaimed land.

Two case studies of postmining land use are discussed as follows.

12.2 APPALACHIAN REGION CASE STUDY2

The Spingola No. 1 mine is located in Clearfield County, Pennsylvania, and is worked by a small operator, Simca Mining Inc. This mine is typical of the region in that the most recent operation has had a relatively short life (about 3 years), but a large portion of the site has been disturbed by past surface and underground operations. All of the land is privately owned, and the parcel has been assembled through lease agreements with six different landowners. Land uses that existed before recent mining includes forest land, open fields, and disturbed land that has either been reclaimed to forests, permanent grasses, or left unreclaimed.

Soils in this area are relatively thin, particularly on the steeper slopes. In many instances, both at this site and over the entire region, previous mining operations have resulted in a complete loss

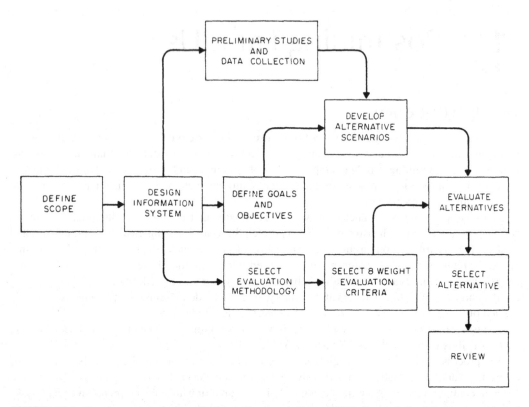

FIGURE 12.1 Land use planning model. (From Sweigard, R.J. and R.V. Ramani. Regional comparison of post-mining land use practices, in *SME Preprint 83-105* (Littleton, CO: Society of Mining, Metallurgy and Exploration), 1983.)

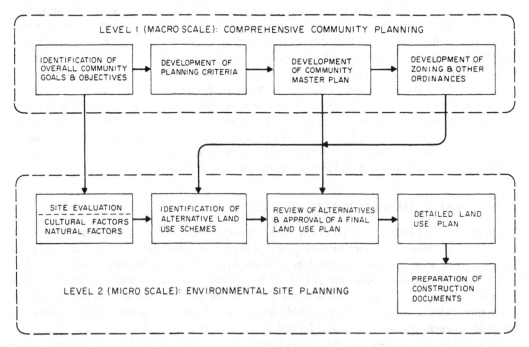

FIGURE 12.2 Macro- and micro-level land use planning. (From Sweigard, R.J. and R.V. Ramani. Regional comparison of post-mining land use practices, in *SME Preprint 83-105* (Littleton, CO: Society of Mining, Metallurgy and Exploration), 1983.)

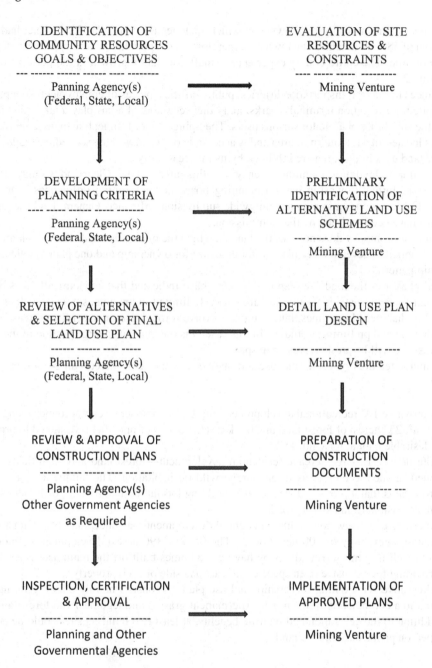

FIGURE 12.3   Mine-land use planning. (From Ramani, R.V. and R.J. Sweigard. Impacts of land use planning on mineral resources, in *SME Preprint 82-418* (Littleton, CO: Society of Mining, Metallurgy and Exploration), 1982.)

of topsoil. Two soil types at this mine have been identified as prime farmland. However, they have been exempted from prime farmland standards due to previous stripping, forest cover, or a lack of agricultural activity. Agriculture does not play a significant role in this area, with only about 8% of the land in Clearfield County devoted to crops or pasture. In addition to the soil limitations, this area has a fairly short average growing season.

It was estimated that 29% of the population lived in urban areas, while the remaining 71% lived in rural areas of the county. Based on a countywide average, the population density was about 70 persons per square mile. Jordan Township, which includes the Spingola No. 1 mine, had a 1980 population of 580 inhabitants. The township, therefore, had a population density of only 25 persons per square mile. The entire county experienced small population growth in the period from 1970 to 1980 of 12%.

Distances to major transportation arteries, public utilities, places of employment, shopping centres, public transportation terminals, parks, and other recreational areas play a large part in determining the suitability of a site for various users. The Spingola No. 1 mine, like many eastern surface mines, is located in a remote rural area and is accessible only by two-lane secondary roads. Higher-intensity land uses in this area are inhibited by its inaccessibility.

The primary land use planning agency for this area is the Clearfield County Planning Commission. Jordan Township has no planning board and no zoning ordinances. The County Planning Commission is adopting a countywide subdivision ordinance. General land use goals and objectives have been outlined by the Commission.

The mining company does not own the land on which the mine is located. The site planning process was applied to develop three plans without regard for ownership and one plan to reflect current ownership patterns.

Initial phases of the land use planning process have indicated that economically feasible post-mining land use alternatives for this site are basically limited to low intensity uses such as pasture and wildlife habitat. Many times, land use plans evolve through common sense and reason during data collection and preliminary studies. In this case, it is obvious that a large portion of the 1.7 km^2 (430 acres) site will be preserved as open space.

The major features of each land use alternative that was considered can be summarized as follows:

- Alternative 1A: recreational development—an 0.2-km^2 (56 acres) campgrounds along with 1 km^2 (253 acres) of forest land and 0.5 km^2 (121 acres) of open fields designed to provide a desirable habitat for wildlife.
- Alternative 1B: large acreage residential development—forest land and open fields combined in such a manner as to encourage wildlife habitation. The unique aspect of this proposal is that it provides for nine large building lots (at least 40,470 m^2, or 10 acres) that are accessible from existing roads.
- Alternative 1C: new community residential development—a conceptual design for a residential community of 100 new homes. The 0.4-km^2 (90 acres) development is intended to fit well into the surroundings by having the homes built off the main access road and providing for extensive open space on at least one side of each property.
- Alternative ID: existing ownership land use plan — consists mainly of revegetating the land to a combination of open fields (permanent grasses and legumes) and forest land. In addition to the vegetation and wildlife benefits, at least two building sites could be developed on prime developable land.

12.3 EVALUATION OF ALTERNATIVES

Each of the postmining land use alternatives was evaluated from an economic, environmental, and social point of view. Quantitative analyses were performed for the economic and environmental considerations, and the social implications for each plan were discussed.

12.4 ECONOMIC EVALUATION

Estimates of the total reclaimed land values based on the various land use plans were made. The unit prices for each land use were determined with the aid of experienced realtors operating in Clearfield County.

Alternatives 1A and 1C are presumed to be infeasible due to market conditions. Alternative 1B would be selected based on economic analysis and the assumption that present ownership presents no barrier to such development. Alternative ID, based on the existing ownership pattern, demonstrates that initial ownership can act to constrain higher land use realization.

12.5   ENVIRONMENTAL EVALUATION

Alternative 1D is the most acceptable plan environmentally because it involves the least amount of change from premining conditions. Alternative 1B also represents a very small change from premining conditions, and the only foreseen difficulty may be some stability problems with building foundations. The negative impacts of Alternative 1A are estimated to be small, but some decrease in infiltration and a slight increase in erosion may occur. Alternative 1C would have the greatest negative impact due to the high level of development. Many of the negative impacts would be mitigated through construction measures engineered to recharge or improve the properties of the soil.

12.6   SOCIAL IMPACT EVALUATION

Alternative 1A is consistent with the county's objective of providing outdoor recreational facilities. Possible negative social impacts could be experienced by adjacent landowners and the residents in the nearby village of McCartney. These inconveniences may be offset by local economic benefits. Aesthetically, a campground could be designed in such a way so as not to detract from the rural atmosphere.

Alternative 1B is consistent with the local desire for rural living. This plan could be aesthetically pleasing if the homes are properly designed and oriented on the lots. No negative social impact should occur to the adjacent landowners or the village of McCartney.

Alternative 1C would result in the greatest social impact due to the addition of 250 to 300 new residents. This increase would place a strain on community facilities such as schools, medical centres, and police and fire protection.

Alternative 1D is viewed as having a little social impact because it calls for no change in the existing social conditions.

12.7   SELECTED ALTERNATIVES

Based on the economic analysis, Alternative 1B is considered to be the most feasible of the three alternatives that are not influenced by the existing land ownership pattern. Although the calculated land value is not as high as those estimated for Alternatives 1A and 1C, it is believed that these values are unattainable in light of the local real estate market. If Alternative 1D is disregarded, as stated at the outset, the environmental analysis also indicates that Alternative 1B is the most desirable. Finally, the social impact analysis does not indicate that Alternative 1B poses any threat to social conditions.

12.8   MIDWEST CASE STUDY

The Chinook Mine of the AMAX Coal Co. is located in west central Indiana, about 9 km (6 mi) east of Terre Haute. Although the area immediately adjacent to the mine is rural in nature, the mine is situated in the heart of the midwest industrial belt. This mine has been in operation since 1928 and has the distinction of being the oldest continually operating surface coal mine in Indiana.

Meadowlark Farms Inc., a subsidiary of AMAX Inc., performs landholding and land management services for AMAX Coal. Meadowlark Farms carries out all revegetation operations at AMAX Coal Co. mines, operates four corporate farms, and conducts a crop-share lease programme on company land. One of these corporate farms is located at Chinook Mine and uses reclaimed land for agricultural purposes. In 1980, Meadowlark Farms produced from all its Indiana operations a total of 656,258 bushels of corn, 196,226 bushels of soybeans, 26,177 bushels of wheat, and 309,502 pounds of marketed livestock. Approximately 62% of the 12-km^2 (2,976 acres) West Field is owned by Meadowlark Farms. The remainder is leased from private landowners.

Locally, the overburden consists mainly of dark grey to black shales and grey sandstones. A brown, oxidized sandstone is also present in some locations. Overburden analyses have shown the consolidated overburden to have high total sulfur values. This is particularly true for the deeper overburden and the interburden between some of the coal seams. Much of this material has the potential for acid production.

High-capacity soils are abundant in the area, as evidenced by the percentage of agricultural land. The climate is also conducive to agricultural production with an average growing season of 207 days and average annual precipitation of about 40 in. Besides cultivated fields, plant communities characteristic of fallow fields, old fields (not cultivated for 3–10 years), and woodlands are found in the vicinity of the mine.

With a few local exceptions, this portion of Indiana has lagged behind the remainder of the state in population growth and economic development. Over a period of 50 years, from 1930 to 1980, the population of Indiana has increased by 69.53%. For the same period, the population of the two counties surrounding the mine increased by only 9.5%.

Although this region has been one of high unemployment in the past, changes made during the 1980s have helped to reverse this trend. One of these changes has been the success of Terre Haute in attracting new industries and encouraging the expansion of existing industries. A second factor has been a diversification of industry in the area. Vigo County has experienced relatively low unemployment during that period. However, the most drastic improvement was seen in Clay County, which went from nearly 4% above the state average in 1971 to 1.4% below the state average in 1975.

Local planning in the project area is conducted at the county level. Of the two counties affected by Chinook Mine, only Vigo County presently has a planning commission. A visit was made to the Area Planning Department of Vigo County during the "site visit" stage of the project. At that time, the Area Planning Department was in the process of formulating a comprehensive county plan, nonexistent prior to this period. Vigo County has no zoning ordinances, but a subdivision ordinance is in effect.

The West Central Indiana Economic Development District Inc. (WCIEDD) serves a six-county region that includes the case study site. Although the major objective of this organization is to promote the economic growth of the region, the WCIEDD has prepared a District Land Use Element. The agency, however, has no authority to implement any of its recommendations. Therefore, the land use plans serve mainly as a source document for future planning activities.

A need has been identified in this area for improvements to the existing county and municipal parks and the acquisition of additional parkland, particularly in and around Terre Haute. The per capita acres of outdoor recreational facilities for Clay County and Vigo County, 0.097 and 0.040, respectively, are considerably lower than the state average of 0.168 acres per capita.

Row crops accounted for 54% of the premining land use in the West Field. Pastureland and forests also occupied significant portions of the premining landscape. About 20% of the field has been disturbed by earlier mining operations and had been either reclaimed to pastureland or was still unreclaimed due to ongoing operations. Minor amounts of land were used for water impoundments, roads, and utilities.

12.9 EVALUATION OF ALTERNATIVES

The three alternatives developed for this case study were

- Alternative 2A: agricultural use
- Alternative 2B: low-density residential development
- Alternative 2C: residential development with integrated open space and recreational uses

These alternatives were subjected to the same evaluation procedure that was developed for the Appalachian case study.

12.10 ECONOMIC EVALUATION

The economic evaluation attempts to estimate the resale value of the reclaimed land for a variety of alternative uses. For each alternative, the largest percentage of the total land value is derived from the land's usefulness in agricultural production. Alternative 2C has a slight advantage over both 2A and 2B because it makes better use of land having marginal agricultural activities. By using some of the residential land for multifamily dwellings, its value is increased. Also, private recreational and commercial uses occupy more land in this alternative than in the other two. Based on this evaluation procedure, it is estimated that Alternative 2C produces the highest potential total land value. However, the increase over the other two alternatives is rather small.

12.11 ENVIRONMENTAL EVALUATION

The environmental impact assessment matrix method was used to rank the alternatives according to the seriousness of the environmental impact that would be carried by each alternative. In reviewing the total environmental impact of the three land use plans, it can be concluded that Alternative 2A has the least impact because it differs the least from the premining uses. Although the impact of Alternatives 2B and 2C varies slightly on certain points, the overall impact points of the plans are essentially identical. It should be noted that while Alternatives 2B and 2C do represent larger negative environmental impacts than Alternative 2 A, the impact is in no way disproportionate than any other development project and can be mitigated, in many instances, through proper design and construction methods.

12.12 SOCIAL IMPACT EVALUATION

Alternative 2A would cause the least social impact for the same reason that it would cause the least environmental impact: the great similarity between this land use and the premining land uses. Alternative 2C would result in much of the same social impact as Alternative 2B. The magnitude of the impact would be increased slightly, however. Because this alternative would provide more housing units, it would more adequately meet the need for geographically suitable housing. It would also help alleviate the recognized shortage of rental housing. The premining study phase of the investigation has pointed to a deficiency in outdoor recreation facilities. Alternative 2C would provide a variety of recreational opportunities. Aesthetically, a properly designed residential community is often much more satisfactory than the haphazard development that is likely to occur without any planning. Alternative 2C is essentially the same as Alternative 2B in preserving prime agricultural land and in providing buffers between developed and rural areas. As Alternative 2B would have some negative impact on utilities, public facilities, and highways, Alternative 2C would have a slightly larger negative impact. It appears that the magnitude and importance of the positive social impact generated by Alternative 2C would outweigh any negative impact.

12.13 SELECTED ALTERNATIVES

Upon reviewing the three different analyses, it becomes apparent that Alternative 2B is the least desirable. The choice, then, is between Alternative 2A, which is similar to the premining land use, and Alternative 2C. Based strictly on the maximization of land value, Alternative 2C would be selected. Although this alternative could result in a slightly larger negative environmental impact than Alternative 2A, it would also provide the largest and most diverse social benefits by helping to meet several regional objectives. For these reasons, Alternative 2C is the recommended alternative.

It must be re-emphasized that this decision was reached strictly from a land use potential perspective and was not influenced by regulatory requirements. Another point that requires clarification is the time frame for implementing the selected alternative. Because the field will be actively mined for more than 10 years, it is necessary to have an interim land use plan that productively utilizes the land until the entire area is mined. Such an interim plan would likely be similar to Alternative 2A, as the land could be used for agricultural production without preempting or diminishing its potential for later development.

BIBLIOGRAPHY

1. Ramani, R. V. and R. J. Sweigard, *Development of a Procedure for Land Use Potential Evaluation for Surface Mined Land*, (National Technical Information Service, 1983), PB-83-253401.
2. Sweigard, R. J. and R. V. Ramani, Regional comparison of post mining land use practices, in *SME Preprint 83-105* (Littleton, CO: Society of Mining, Metallurgy and Exploration, 1983).
3. Ramani, R. V. and R. J. Sweigard, Impacts of land use planning on mineral resources, in *SME Preprint 82-418* (Littleton, CO: Society of Mining, Metallurgy and Exploration, 1982).
4. Kivinen, S., Sustainable post mining land use: Are closed metal mine3s abandoned or reused space? *Sustainability*, 9(10): 1705 (2017).
5. Mbborah, C., Banssah, K. J., Boateng, M. K., Evaluating alternate post mining land uses: A review, *Environ. Pollut.*, 5(1): 14–22 (2016).
6. Bangian, A. H., Ataei, M., Sayadi, A., Optimising post mining land use for pit area in open pit area in open pit mining using fuzzy decision-making method, *Int. J. Environ. Sci. Technol.*, 9(4): 613–628 (2012).
7. Limipitlaw, D., Briel A., Post mining land use opportunities in developing countries a review, *J. South. Afr. Inst. Min. Metall.*, 114(11): 899–903 (2014).

13 Bioremediation

13.1 INTRODUCTION

Microbial processes have a major role in the global cycling of metals. Once in the environment, the metals frequently undergo transformation into various mobile or immobile forms in the environmental sink. Biological activity accounts for a large number of the environmental sinks for toxic metals. Biotransformation and immobilization sometimes produce spectacular effects such as gold precipitation and sedimentation of radionuclides by marine plankton. Metal deposition by microorganism is of great importance in biogeochemical cycles, e.g., microfossil and mineral formation, iron and manganese deposition, and uranium and silver mineralization.

Bioremediation for the toxic metal pollution consists of selectively using and enhancing these natural processes to treat particular wastes. The processes by which microorganisms interact with toxic metals can be very diverse and complex. Microbial transformations can include both redox conversions of inorganic and organic forms, increasing and decreasing the solubility through various complex reactions and altering the pH of the environment, or through adsorption and uptake.[2]

Metals and their compounds can inhibit biological activity. Metals such as copper, manganese cobalt, molybdenum, and zinc are nutrients to microorganisms; they can also become toxic at higher concentrations. Some microorganisms can develop adsorption, oxidation-reduction, or methylation mechanism to protect themselves from the toxic effects of metals. These mechanisms can be utilized in the abatement processes against metal contamination.

Biological activities can be utilized in several ways to remedy metal contaminations, such as alteration of the chemical state, and form or distribution of metals. The common biological mechanisms exploited include biosorption, bioaccumulation, oxidation-reduction, methylation-demethylation, precipitation, magnetic separation, metal transformation, metal binding, and degradation.

13.2 BIOSORPTION

The term "biosorption" includes uptake by whole biomass (living or dead) via physico-chemical mechanisms such as adsorption or ion exchange. Where living biomass is utilized, metabolic uptake mechanisms may also contribute to the process. Bioaccumulation involves the transfer of a metal from a contaminated matrix to a biomass.[2]

Certain plant species and microorganisms can actively accumulate metals. Living cells can also absorb metals and concentrate in organic materials within the cells. Heavy metals may not be essential for the metabolic process; they can be taken up by the biomass as a side effect of the normal metabolic activity of the cells. Active metabolization by biomass can include ion exchange at the cell walls, complexation reactions at the cell walls, intra- and extracellular precipitation, and intra- and extracellular complexation reactions.

Metals are removed primarily through adsorption of metals to the ionic groups either on the cell surface or in the polysaccharide coating found on most forms of bacteria. The metals are bound by exchange of functional groups or by sorption on polymers.

Where living biomass is used, metabolic uptake mechanisms may also participate in the process. As a result of metal toxicity, living cells may be inactivated; therefore, living cells have been used to decontaminate effluent-containing metals at subtoxic concentrations. Such systems may employ a mixture of microorganisms as well as higher plants. Algae and cyanobacterial booms, encouraged by the addition of sewage effluent, reduced the levels of copper, cadmium, zinc, mercury, and iron in the mining effluent. The meander system used in the Homestake Lead mine passes the effluent-containing lead, copper, zinc, manganese, nickel, iron, and cadmium through channels containing cyanobacteria, algae, and higher plants. Metals are removed from the water column with an efficiency higher than 99%. Such complex systems utilize other mechanisms, like precipitation and particulate entrapment, in addition to biosorption, all of which concentrate the metals in sediment forms that have significantly reduced environmental mobility and biological availability.

Fungi, including yeasts, have been utilized in connection with metal absorption, particularly because industrial wastes contain fungal biomass as a by-product from industrial fermentations. Actinide accumulation by intact biomass usually comprises metabolism-independent biosorption, using the cell wall as the main site. Increasing cell permeability with carbonates or detergent can increase uptake, and metals may accumulate in the cells as granules or deposits. Metal-containing particulates, e.g., zinc dust, magnetite metals and sulfides, can be removed from a solution by fungal biomass, a combination of biosorption and entrapment.

13.3 OXIDATION-REDUCTION

Many microorganisms transform metals through oxidation or reduction mechanisms. The most well-known reaction is the oxidation of ferrous iron, and mineral sulfides by chemolithotropic *Thiobacillus*. Manganese can be biologically oxidized. Metals can be reduced by actively metabolizing microorganisms. Examples include reduction of Cr^{6+} to Cr^{3+} and the reduction of SeO_4^{2-} to SeO_3^{2-} and Se^{0}.[1]

Passive metal reduction can also occur. The metal is bound to a microbial organic component acting as a reductant. A common example is the reduction of Au^{3+} to Au^{+} and finally Au^{0}.

Microbially mediated redox reactions are reactions through which an exchange of electrons results in a change in the valence of the metal. Oxidation involves the removal of electrons from the metal and results in an increase in the valence states. In reduction reactions electrons are removed from the metal atom and the valence state is reduced.[3]

Some of the common examples of microbially mediated redox reactions are the following:

Oxidation reactions:

ferrous iron	Fe^{2+}	$\rightarrow Fe^{3+} + 2e^{-}$
manganese	Mn^{2+}	$\rightarrow Mn^{7+} + 2e^{-}$
antimony	Sb^{3+}	$\rightarrow Sb^{5+} + 2e^{-}$
iron	Fe	$\rightarrow Fe^{2+} + 2e^{-}$
cadmium	Cd	$\rightarrow Cd^{2+} + 2e^{-}$
mercury	Hg^{0}	$\rightarrow Hg^{2+} + 2e^{-}$
antimony	Sb^{3+}	$\rightarrow Sb^{5+} + 2e^{-}$

Reduction Reactions:

ferric iron	$Fe^{3+} + 1e_{-}$	$\rightarrow Fe^{2+}$
mercury	$Hg^{2+} + 1e^{-}$	$\rightarrow Hg^{+}$
arsenate	$AS^{5+} + 2e^{-}$	$\rightarrow AS^{3+}$

Microbial oxidation or reduction increases mobility which can be utilized in remedial processes. An insoluble sulfide can be oxidized to form a soluble sulfate, which can be collected by subsequent treatment, disposal, or recovery.

Biological oxidation or reduction can be used to reduce mobility as well. For example, a limited number of bacteria can reduce mobility or toxicity by converting hexavalent chromium, Cr(VI), to less a toxic trivalent form, Cr(III).

13.4 PRECIPITATION

Metals can undergo indirect microbial transformation as a result of sulfate reduction when anaerobic bacteria oxidize simple carbon substrates while the sulfate serves as the electron acceptor. Hydrogen sulfate is produced, and alkalinity is increased. Sulfate reduction is strictly an anaerobic process, and proceeds in the absence of oxygen. A source of carbon to support microbial growth, a source of sulfate, and a population of sulfate-reducing bacteria (SRB) such as *Desulfovibrio* and/or *Desulfotomeculum* are required. The H_2S formed in the reaction can react with many types of contaminant metals to precipitate metals as sulfides, which are insoluble, for example, ZnS, CdS, CuS, and FeS. The process can be described as:

$$H_2S + M^{2+} \rightarrow MS(s) + 2H^+ \tag{13.1}$$

The production of alkalinity in sulfate reduction reactions increases the pH level, which can form and precipitate insoluble metal hydroxides or oxides. The process can be described as:

$$M^{3+} + 3H_2O \rightarrow M(OH)_3(s) + 3H^+ \tag{13.2}$$

In situ treatment of metals utilizing sulfate reduction process precipitates the metals and prevents further migration in contaminated soils and groundwater.[3]

Shell Research Ltd. developed a purpose-designed, $9m^3$ stainless-steel, sludge-blanket reactor, using sulfate-reducing bacteria. This plant successfully removed toxic metals and sulfate from contaminated ground water of a long-standing smelter site by precipitating metals as sulfates. The reactor used a selected but undefined mixture of bacteria with ethanol as the growth substrate, which was capable of tolerating a wide range of inflow pH and operating temperature, yielding outflow metal concentrations below the parts per billion ranges. Methanogenic bacteria in the consortium also removed acetate produced by SRBs, leaving an effluent with an acceptably low biological oxygen demand (BoD). Excess H_2S was removed out of waste gases using a $ZnSo_4$ solution.[1]

This process has been expanded to a commercial pilot-scale using an $1800m^3$ concrete reactor. The plant is capable of treating 7000 m^3/day. In another pilot plant study conducted by the U.S Bureau of Mines, a 4500-liter reactor using spent mushroom-compost as substrate and sulfate-reducing bacteria removed 96% of the metals and 20% of sulfate from 19.3 m^3 of coal mine drainage waters.

13.5 BIOSORPTION BY IMMOBILIZED BIOMASS

Disadvantages in using freely suspended microbial biomass include small particle size, low mechanical strength, and difficulty in separating biomass and effluent. The use of immobilized biomass particles in packed- or fluidized-bed reactors reduces these difficulties. The immobilized biomass can be in the form of biofilms on supports prepared from a range of inert materials. Various bioreactor configurations include rotating biological contactors, fixed-bed reactors, trickle filters, fluidized beds, and air-lift bioreactors. Sometimes different reactor types can be combined. Living-cell biofilms may provide additional capacity for the removal of other pollutants including nitrates and hydro-carbons. A large-scale commercial process has been utilized to treat effluent from metal mining and milling using rotating-disc biofilm contacting units for simultaneously degrading cyanide, thiocyanate, and ammonia, and also removing metals by biosorption.

Living or dead biomass of all microbial groups has been immobilized using crosslinking or encapsulation. Agra, cellulose, alginate, cross-linked ethyl acrylate-ethylene glycol dime- thylacrylate, and silica gel can be used as supports. Cross-linking reagents include toluene diisocyanate and glutaraldehyde. The biomass may be used in its natural state or be modified to increase biosorption efficiency.

For low volume, small-scale systems may be sufficient. For larger systems, conventional bio-reactor technology can be used. Immobilized biomass particles should have properties similar to those of other commercial adsorbents; e.g., 0.5–1.5 mm in size, similar in strength and chemical resistance. High porosity, hydrophilicity with maximum amount of biomass, and minimal amounts of binding agent are necessary. Waste *Bacillus* sp. biomass from industrial fermentations can be subjected to alkali treatment to increase metal uptake. Granulated *Bacillus* has been used to remove Cd, Cr, Cu, Hg, Ni, Pb, U, and Zn in nonselective procedures. Removal efficiency up to 99%, metal loadings up to 10% of the dry weight and, effluent with total metal concentrations around 10–50 parts per billion have been noticed.

Immobilized particles containing *Rhizopus arrhizus*, diameter 0.7–1.3 mm with 12 to 23% added polymer improved uranium removed. Complete uranium removal was possible from dilute ura-nium-ore bioleaching solutions (< 300 mg per liter) with eluate concentrations after desorption of >5000 mg per liter. The particles retained full loading capacity (~ 50mg U per gm) over many biosorption-desorption cycles.[4]

Fluidized beds of alginate- and polyacrylamide-immobilized algae have been utilized to remove a variety of metals, including Cu^{2+}, Pb^{2+}, Zn^{2+}, and Au^{3+}. Alginate and polyacrylamide provide good resistance to mechanical deterioration and hydrostatic pressure.

13.6   METHYLATION

Biological methylation occurs when organisms attach a methyl group($-CH_3$) to an inorganic form of a metal. Methylation occurs when organometallic compounds are formed and are more volatile than the elemental form. The organometallic compounds are removed from the contaminated matrix using volatilization and then collected from the gas stream.[5]

Biological methylation is unlikely to be employed at contaminated sites because it results in a more toxic by-product that can be more difficult to control.

Methylation increases the mobility of the metal. Metals like mercury, arsenic, cadmium, and lead can be methylated. The more mobile forms may leach from a site and impact surface or ground water or may pose an air emission problem.

13.7   PHOSPHATE PRECIPATION

Surface-located acid-type phosphatase enzymes can occur in "Resting cells" of a *Citrobacter*. These can release HPO_4^{2-} from a supplied substrate like, glycerol 2-phosphate and precipitates divalent cations (M^{2+}) as $MHPO_4$ at all surfaces. This process has potential applications where phosphate-containing organic substrates are present in metal- or radionuclide-containing effluent.

13.8   MAGNETIC SEPARATION

Compounds such as sulfides or phosphates are of paramagnetic or ferromagnetic elements. Microbial cells carrying deposits of these materials can be separated from other cells and debris by high-gradient magnetic separation.

13.9   METAL-BINDING PROTEINS

All biological materials frequently exhibit affinity for toxic metals and radionuclides. Some bio-molecules function specifically to bind metals and are induced to do so by the presence of metals. Other molecules with significant metal-binding abilities, e.g., fungal melanins, may be overpro-duced because of exposure to low concentrations of metal and interference with normal metabo-lism. These substances have potential in metal pollution control.[5]

13.10   CELL-WALL COMPONENTS AND EXPOLYMERS

The microbial expolymers that exhibit strongest metal binding properties are those that form capsules or slime layers. Many of these expolymers are composed of polysaccharide glycoproteins and lipopolysaccharide, which are frequently associated with proteins. A correlation is often found between high anionic charge and metal-complexing capability. Additionally, chemical deposition may take place in altered forms.

In activated sludge, the bacterium *Zoogloea ramigara* is used as a flocculent for its extensive exopolysaccharide producing properties. This includes metal-binding properties. A continuous process for metal accumulation using pregrown Z.ramigara removed ~3 mM per gram of dry weight of copper at a biomass concentration of <1g dry weight per liter. Other bacterial polymers have received attention, including polysaccharides from capsulate *Klebsiella aerogenes*, *Arthrobacter Viscosus*, and *Pseudomonas* sp. Bacterial polysaccharide-emulsifying agent has been used for uranium removal.

Cell walls of bacteria have several metal-binding properties which assist in the biosorption process. The carboxyl groups of the peptidoglycan are the main metal-binding sites in all walls of gram-positive bacteria. Many fungi have high chitin content in their cell walls, and this polymer is an effective metal and radionuclide biosorbent.

Melanins are fungal pigments that support survival under environmental stress. Fungal melanin contains carbohydrates, peptides, and aliphatic hydrocarbons.. Melanized cell forms can have high metal-uptake capacities, with all metals being located in the cell walls. Melanin from *Aureobasidium pullulans* can bind significant amounts of metals and organometallic compounds.

13.11   APPLICABILITY OF BIOLOGICAL TREATMENT

The biological treatment of metals has limited application to special wastes because the characteristics of biological alternatives for metal treatment are not suitable for broad application to waste treatment. Biological treatment systems usually operate at a lower rate than the analogous chemical treatment. Biological systems which rely on live organisms are less tolerant to high metal concentrations. However, biological systems have several advantages over chemical treatment technologies.

Biological metal absorbents have been used in conventional water treatment processes such as ion exchange or activated carbon. While using inactive biosorbents, the material would appear at the end much like ion exchange resins. The biosorbents offer several advantages including higher loading capacity at low-metal contaminants level. Therefore, the biological materials are more effective for treating metal pollution at lower concentrations. Biosorbents are more selective for heavy metals, so they can be more effective for treating waters with high sodium or magnesium ion concentrations.[2]

Biological treatment does not commonly require the addition of chemicals that often are used in chemical oxidation-reduction or loading processes. A feasible chemical in situ leaching or oxidation-reduction process may be unacceptable due to limitations on the types of injected chemicals. When applicable, biological mechanisms can achieve leaching, oxidation, or reduction without the use of toxic chemicals.

13.12   CURRENT STATUS OF BIOREMEDIATION TECHNOLOGIES

Biotreatment of metals is currently being developed at the commercial or pilot scale for some metals.

13.13   METAL BIOSORBENT MATERIALS

Chemical biosorbent materials consist of inactive biomass. Inactive biomass materials include algae and a natural biopolymer, chitosan, made from shellfish wastes. Some of the commercial biosorbents are summarized in Table 13.1.

TABLE 13.1
Characteristics of Commercial Biosorbents

MRA	AMT-BIOCLAIM	Caustic treated killed bacteria	Polymer	Available in cationic form and anionic form	
Alga-SORB	Bio-Recovery	Silica gel	Cationic metals		Tolerates low pH and high temperatures better than conventional ion exchange materials
BIO-FIX	U.S.Bureau of Mines and Licensees	Peat moss, *Spirulina*, and other	Polysulfone polymer	Cationic metals	

Source: Reference 3, with permission
AMT = Advanced Mineral Technology
BIO-FIX = Biomass Foam Immobilized Extractant
MRA = Metal Removal Agent

Biosorption by living systems is in testing stage and not yet in commercial use. Nickel has been removed from metal-plating wastes by *Enterobactor and Pseudomonas* species. Organisms are being generally developed to sequester metals such as cadmium, cobalt, copper, and mercury.

13.13.1   BIOREMEDIATION OF MERCURY SALTS

Biological activity can be used to recover mercury by reducing mercury salts to metal. Biological processes can reduce Hg(II) to mercury metal and in some instances may hydrolyze organo-mercuricals. Bioreactors have been developed to process feedwater containing 1–2mg/kg of Hg(II) to yield an effluent with 50 mg/kg or lower Hg(II). The metallic mercury formed can be volatilized from water stream through air stripping. Several organisms, namely *Pseudomonas putida* and *Thibacillus ferrooxidans* have been tested in chemical reduction and recovery of mercury from wastewater. Bioreduction has been demonstrated at the bench scale level.[2]

13.14   BIO REDUCTION OF CHROMIUM

Hexavalent chromium is frequently targeted for chemical or electrochemical reduction to reduce its toxicity and mobility. Biological mechanisms can reduce Cr(VI) by direct microbial action or by microbial production of sulfide.

An anaerobic process has been developed using sulfate-reducing bacteria to produce hydrogen sulfide (H_2S) which in turn reduces Cr(VI) to Cr(III). A natural by-product of metabolism of the sulfate-reducing bacteria is H_2S. The pH-adjusted waste material containing Cr(VI) passes through a anaerobic sludge blanket column reactor. The bacteria coats the gravel in the distribution layer, where the Cr (VI) is reduced to Cr(III) by the H_2S. The bacteria is somewhat protected as it does not metabolize the chromium.

13.15   METAL WASTE REMOVAL IN WETLANDS

Wetlands can be artificially constructed to remove metal contaminants. In the wetland environment, natural geochemical and biochemical processes are combined to accumulate and remove metals from influent waters. The wetland treatment system utilizes organic soil, microbial fauna, algae, and vascular plants to reduce the concentration of metals in the water. Microbial activity accounts for most of the metal remediation activity.[5]

The wetland contains an aerobic zone at the surface and an anaerobic zone below the surface. Redox reactions catalyzed by bacteria play a major role in metal remediation in both zones.

13.16   BIOLEACHING

Microbiological activities assist in dissolving metal contaminants from a solid or semisolid matrix. Bioleaching uses organisms to dissolve metals by direct action of active bacteria, by indirect attack by metabolic products, or by a combination of both direct and indirect attacks.

Copper and Uranium have been subjected to bioleaching. Leaching is believed to be accomplished by a complex combination of *Thiobacillus ferrooxidans*, *Thiobacillus Organosparus*, and other gram-negative acidophilic bacteria.

Mercury removal by microbial activity has received recent attention. Mercury salts can be reduced to elemental metals through biological activity. The native mercury metal is easily separated from aqueous streams after bio reduction.

13.17   BIO BENEFICIATION

Biobeneficiation involves separation of the contaminated waste into contaminant-rich and contaminated-poor streams. Bio beneficiation is applied in the froth flotation process. Biological action could be used to replace the chemical treatment needed in the flotation process. For example, cyanide ions are added to pulp containing iron sulfide to form an iron-cyanide complex. This complex prevents the pyrite particles from attaching to the bubbles. Biotreatment with Thiobacillus ferrooxidans has been used to modify surfaces to improve the performance of the flotation process.

Biological process can be applied to improve settling or filtration by controlling the agglomeration of mineral particles. *Mycobacterium pheli* has been used as a good flocculating agent for coal, phosphate slime, and hematite.

13.18   VEGETATIVE UPTAKE

Some types of plants can uptake metals through their root systems and deposit on their leaves. For example, many plants in the *Leguminous* family require specific minerals and thrive in habitats with those minerals. Alfalfa and clover both require manganese and potassium.

13.19   CYANIDE DEGRADATION

Cyanide degradation in spent heaps and trailing impoundments using biotechnological approach is of recent origin. Cyanide degradation in aqueous discharges has been applied sometimes. Most of the work in microbial cyanide degradation has focused on aerobic degradation; some effects are under way for using anaerobic cyanide degradation. This biotechnical process would have applicability in decreasing cyanide concentration in oxygen-impoverished tailings impoundments.[2]

13.20   APPLICABILITY OF BIOLOGICAL TREATMENT

The biotreatment processes for waste treatment are chosen after considering the following conditions:

- Types and level of concentrations of metal's ph
- Matrix
- Alkalinity

- Temperature
- Oxygen concentration
- Substrate concentrations
- Nutrient concentrations
- Presence of indigenous microorganisms
- Population density
- Cell age
- Use of active or inactive biomass
- Contact time
- Volume of contaminated material
- Depth of the contaminated material
- Site controllability

13.21   BIOTREATMENT PROCESS DEVELOPMENT FOR SUMMITVILLE MINE

L.C Thompson et al. developed a biotreatment process for detoxification of spent ore of the Summitville mine in Colorado.[5]

In the Summitville mining operation, approximately 10 million tons of gold ore was mined, crushed, and stacked on a lined leach pad. Heap leach operations experienced serious environmental control difficulties due to unplanned solution discharge. The Environmental Protection Agency took over site operations to prevent catastrophic release of hazardous substances into the environment.[5]

It is a bowl-shaped structure, underneath of which is located a French Drain structure, which is a network of gravel trenches and perforated pipe designed to intercept groundwater and leakage from the heap. The heap contained about 10 million tons of cyanide-leached ore and 90–150 million gallons of process solution. Contaminated dikes and liners control the process solution in the lower portion of the heap.

Tests were conducted to study feasibility of detoxification of the solution and spent ore through establishing a biotreatment process in the heap leach pad by inoculating bacteria and nutrients into the spent ore. The test design had the following objectives:

- Development of site-specific biotreatment process for spent ore and process solution cyanide detoxification
- Provide treatment data for a complete feasibility study on environment control
- Provide data for immobilization of leachable and soluble metals in the heap including zinc, copper, manganese, iron, and arsenic

Bacteria for decomposing cyanide were isolated from ore samples collected during the drilling program. The bacteria were augmented by waste infusion, nutrient stress, and growth tests.

The column tests procedure preparation included the following:

- Bacteria isolation from spent ore and process solutions
- Augmentation of the bacteria population growth
- Demonstration of cyanide detoxification in flask tests
- Production of treatment bacteria for column ore biotreatment.

In the column treatment demonstration process, the following procedure was used:

Spent ore (about 150–170 lbs) was loaded into 6 in. × 10 ft PVC columns, fitted with a perforated screen and a topped end cap to allow the treatment solution to percolate through the ore, and then collected for analysis at the bottom of the column. The following percolation leach biotreatment tests were conducted:

Column #1: Sulfide unrinsed zone ore
Column #2: Oxidized unrinsed ore, 25–90-ft depth
Column #3: Oxidized saturation zone ore 90–130-ft depth
Column #4: Oxidized ore, 0–90-ft depth
Column #5: Oxidized ore, 0–25-ft depth
Column #6: Control Column, oxidized ore, 90–130-ft depth

The treatment bacteria population was developed to working strength ($> 10^{10}$ cells/ml) and transferred to a nutrient solution for use in percolation leach column tests. Thirty gallons of the barren solution were used in the column tests. The treatment solution was applied to each column at a rate of 0.004 gpm/ft^2; the same rate was used in heap leach application. The barren solutions were collected from the columns and analyzed for total and WAD cyanide, gold, and select metals.[5]

13.22   RESULTS AND CONCLUSIONS

The final results and its analysis indicate that biological treatment columns achieved a 99% removal of WAD cyanide. Total cyanide in column-leached solutions at the end of the test was found to be less than 0.5 mg/L which indicated that the biotreatment processes have the potential to function as an effective field treatment process can metabolize strong metal-cyanide compounds.

13.23   HOMESTAKE CYANIDE BIODEGRADATION PROCESS

The Homestake gold mine in South Dakota treats and discharges about 21,000 m^3 (5.5 MGD) of wastewater from its gold mining and milling operations.[11] Wastewater sources include decanted water from Grizzly Gulch tailings impoundment and underground mining operations, which is about 2,560 meters deep. The water is discharged into a cold-water trout fishery. This discharge from the mining operations amount to 56% of total stream flow under low flow conditions. Prior to the constructions of the mill tailings impoundment and operation of the wastewater biotreatment plant, the creek almost contained no life forms. Currently this receiving stream has been turned into an excellent trout stream.[12]

Engineers, Whitlock and Mudder developed and designed a biotreatment system for treating the wastewater discharged from gold mining and milling operations. The treatment process is based upon biodegradation of free cyanide, thiocyanate, metal complexed cyanides and the cyanide degradation by-product, and ammonia through oxidation. Forty-eight rotating biological contactors are used, and they utilized *Pseudomones* sp. as the predominant microorganism.[12]

13.24   MICROBIOLOGY

Researchers from the Homestake Company isolated a specific strain of bacteria indigenous to process waters at the Homestake mine. These microorganisms were gradually acclimated to the greater concentration of cyanide and thiocyanate in the matrix.

The predominant species of bacterium were found unable to grow microaerophilically (in the presence of small amounts of oxygen or in an atmosphere of hydrogen). As per Whitlock, minimal growth requirements are not complex and organic compounds serve as the energy and carbon source.

The predominant bacterium is a gram-negative (0.7–1.4 μm) motila by means of a single polar flagellum, approximately three times the body length. Optimum growth was found at 30^0 C. Metabolism is respiratory and never fermentative. The pH range favorable to the bacterium was between 7.0 and 8.5. Nitrates and nitrites are not reduced and H_2S is not produced.

Free and complexed cyanides are oxidized by bacteria to carbonate and ammonia. Metals are absorbed, ingested, and precipitated as per the below equation:

$$M_xCN_y + 4H_2O + O_2 \underline{\text{Bacteria}}\ M\text{-biofilm} + 2HCO_3 + 2NH_3 \tag{13.3}$$

The order of degradation from most rapid to the slowest is Zn, Ni, Cu, and Fe for metal complexed cyanides. Iron cyanides are degraded as well as adsorped in the process.

Thiocyanate degradation process follows a similar path:

$$SCN + 2.5O_2 + 2H_2O \underline{\text{Bacteria}}\ SO_4^{2-} + HCO_3 + NH_3 \tag{13.4}$$

Ammonia is produced as a by-product from cyanide oxidation, requiring detoxification. The detoxification of ammonia can be achieved by means of a common pair of aerobic, autotrophic bacteria, which use only inorganic material as nutrients and inorganic carbon as the only carbon source.

The degradation and nitrification stages are represented by the below equations:

$$NH_4 + 1.5O_2 \underline{\text{Bacteria}}\ NO_2 + 2H^+ + H_2O$$

$$NO_2 + 0.5_2 \underline{\text{Bacteria}}\ NO_3 \tag{13.5}$$

Ammonia is oxidized to nitrite slowly and then rapidly to nitrate. The microorganisms are not normally noncompetitive. Thiocyanate and cyanide are utilized as food sources in the degradation stage, but they are toxic to the nitrifying bacteria. The nitrifiers use ammonia as a food source.

If the cyanide degradation process is not effective, it adversely affects the nitrification process. The recovery of the nitrifying bacteria is several times slower than the recovery of cyanide degraders.

13.25   PLANT DESIGN

The principle used in designing a biotreatment plant was to establish a system which would mimic natural surface biodegradation processes which were identified at the tailings impoundments. This required designing a "thin film" of 17,000–21,000 m³ of wastewater under aerobic conditions over a short retention time. The thin film of wastewater is necessary to contact sufficient biomass for cyanide degradation and metal adsorption. Total degradation of cyanide would produce ammonia and therefore a nitrification stage was needed. Trickling and rotating-biocontactor (RBC) type filters were evaluated. The RBC was selected as the most applicable technology.

The biotreatment plants incorporated 48 RBC's, 24 of which contain 9,240 m² (100,000 sq.ft) of surface area for cyanide degradation and metals removals. The remaining 24 RBCs with 13,935 m² (150,000 sq.ft) were used for nitrification.

Supplemental air is added to the liquid phase of the RBC basins to maintain dissolved oxygen at 4.0 mg/L or higher. Biomass is attached to plastic media surface of the RBC in films. The biomass weight may approach 18,000 kg per disk. The biomass weight is self-regulating by sloughing of the biomass from the disks. Turbulent aeration and pH shock have been used to reduce biomass weights.[13]

A train of five disks was used as pilot testing indicated. Two of the disks were located in a common basin for cyanide degradation. The other three disks were located in a common basin with partition baffles to accomplish complete nitrification of ammonia.

The 48 RBCs are arranged in a mirror image plant design with 24 discs per side. The flow of wastewater is arranged perpendicular to the discs.

Each disc is filled with aeration headers and electronic load cells for supplemental aeration and biofilm thickness monitoring. The discs are approximately 40% submerged and rotate at variable speeds of 0.5–1.5 RPM by reversible drives.

Each disc is 3.6m in diameter and 7.6m in length. They are built from corrugated plastic with surface areas of 9,290 m² (100,000 sq.ft) or 13,935m² (150,000 sq.ft) depending upon the density of the media. Each RBC is housed in an insulated, fiberglass cover.

13.26   WASTEWATER ANALYSIS

The design and selection of biotreatment methods and equipment substantially depends upon precise and accurate chemical analysis of the wastewater. The EPA's Total Cyanide Method (substituting phosphoric acid for sulfuric acid) and the ASTM Method-C cyanide method were found most reliable. The Total Cyanide Method measures free and metal-complexed cyanides including iron cyanide. Method-C cyanide measures free and metal exception of iron-complexed cyanides. Method-C cyanide values represent the forms of cyanide which are highly toxic. Iron-complexed cyanide toxicity determines the degree to which photodecomposition to form free cyanide occurs. Method-C cyanide was found to be free of much interference involved in the Total Method.

Tailing impoundment decant water and underground mine water are the two major source of wastewater. Contaminant concentration levels in the wastewater vary with seasons, impoundment freeze-over/turn-over, and mine backfill operations. The wastewaters also contain diesel/fuel spills, degreasers, foaming lubricants, dispersants, biocides, and various other chemicals used in mining.

A general wastewater analysis is presented in Table 13.2 which also includes influent blend characteristics. The operators control the blend ratio of decants and mine water to set cyanide loadings for the treatment plant. The blending operation also helps maintain reasonable temperatures. The temperature range of 10–30⁰C has been found suitable for efficient operation of the plant.

A treatment process designed solely based on chemical analysis of the effluent water may result in unexpected toxicity for the following reasons:

- Synergestic effects
- Antagonistic effects
- Toxic parameters overlooked in analysis
- Lethal concentration values not quantified
- Stress on organisms

An effluent test program involving chemical analysis and whole effluent toxicity testing is more desirable.

TABLE 13.2
Wastewater Matrix mg/l

Parameter	Decant water	Mine water	Influent blend
Thiocyanate	110.00-350.00	1.00-33.00	35.00-110.00
Total Cyanide	5.50-65.00	0.30-2.50	0.50-11.50
WAD Cyanide	3.10-38.75	0.50-1.10	0.50-7.15
Copper	0.50-3.10	0.10-2.65	0.15-2.95
Ammonia-N	5.00-10.00	5.00-19.00	6.00-12.00
Phosphorous-P	0.10-0.20	0.10-0.15	0.10-0.15
Alkalinity	50.00-200.00	150.00-250.00	125.00-225.00
pH	7.00-9.00	7.00-9.00	7.50-8.50
Hardness	400.00-500.00	650-1400	500-850
Temperature C⁰	1.0-27.2	24-33	10.0-25

Source: Reference 13, with permission

13.27 TOXICITY AND BIOMASS FACILITY PROGRAM

A biomass laboratory was established for testing whole effluent and individual parameters on trout, daphnia, and macroinvertebrates collected form the actual receiving stream. A four-phase program was established:

- Bioassessment of the receiving stream and background waters
- Design and construction of the bioassay toxicity testing facility
- A protocol development for operation of the testing facility Improvement of the South Dakota Water Quality Standards and incorporation of modern analytical techniques

The testing procedure involves 48-hour static, range finding tests followed by 96-hour, flow through tests. Finally, 45-day-term flow through tests are used to establish a maximum in stream concentration limits and a 30-day average concentration limits, respectively. Tests were conducted on fingerling rainbow trout, and caddisflies, stoneflies, midges, and daphnia, which are part of the food chain for the trout. Toxicity testing enables determination of whether an effluent is compatible with the dynamic aquatic system into which it is discharged.

13.28 LOADING RATES

Loading rates are based on cyanide concentration levels and vary with season.[13]
    The general loading rates kgs/day/m^2 of surface area for the RBCs is as follows:

- Thiocyanate (SCN): 0.47 kg/d/93m^2
  (1.05lbs/d/1000sq.ft)
- Total cyanide (CNT): 0.06kg/d/m^2
  (0.14 lb/d/1000 sq.ft)
- Week and dissociable cyanide(WAD):0.03kg/d/m^2
  (0.07lb/d/1000 sq.ft)
- Ammonia (as N): 0.11 kg/d/m^2 (0.23lb/d/1000sq.ft)
- Heavy metals: (0.02kg/d/93m^2(0.04lb/d/1000sq.ft)
- Hydraulic retention times are:
- Basin 1 (first and second stage RBCs):1.0 hour
- Basin 2 (third, fourth, and fifth stage RBCs): 1.5 hour
- Clarification/filtration step: 3.0 hour

13.29 PLANT PERFORMANCE

Characteristics of wastewater can be described as high volume and low toxic contaminant effluent.
    Wastewater is blended under strict control for ensuring proper loading to the RBCs. Minimum and maximum values are a result of adjustments in blend ratios as well as changes in the chemical composition of wastewaters, particularly tailings impoundment decant water.
    The order of ease of removal through cyanide degradation (from easiest to most difficult) is as follows:

- Free cyanide
- Thiocyanate
- Zinc cyanide
- Nickel cyanide
- Copper cyanide
- Iron cyanide

TABLE 13.3

Influent, Effluent, and Permit Concentration mg/L, 1988

	Thiocyanate	Total Cyanide	WAD Cyanide	Total Copper
Influent	62.0	4.1	2.3	0.56
Effluent	< 0.05	0.07	0.02	0.07
Permit	1.00	0.10	0.13	
	TSS	mmonia -N		
Influent	--	5.60		
Effluent	6.0	< 0.50		
Permit	10.0	1.0-3.9		

Source: Reference 13, with permission

Iron cyanide is only partially degraded. Adsorption mechanism plays a strong role in the removal of iron cyanide. Iron cyanide can be established approximately by substituting for an individual influent or effluent sample (Table 13.3).

Metal adsorption is determined by biomass conditions as well as influent metal concentrations. Metal concentrations vary seasonally, being higher in winter and lower in summer because of cyanide concentrations and other factors like freezeup in the tailings impoundments. In periods with low metal concentration and high biomass weight on the disks, metal removal rates may decrease. In some instances, when adsorption occurs, effluent values may exceed influent values.

Biological degradation and bioadsorption in the biofilm effectively remove 92% of the total cyanide, 99% of the Method-C cyanide (loosely complexed cyanide) and 95% of copper and other toxic heavy metals. Degradations are represented in Figure 5.15.

Copper cyanide comprises almost totally the Method-C (loosely complexed cyanide) cyanide. Subtracting the Method-C cyanide from the total cyanide determines value of iron cyanides, predominantly ferrocyanides. The degree of removal of iron cyanide complexes is dependent upon adsorption as well as biodegradation.

Extensive bioassay evaluations have indicated no trout mortality for as long as 300 days. Examination of whole fish and fish tissues indicated fish are in fine condition with accelerated growth rates because metabolic rates increased in warm effluent waters compared with stream waters (Table 13.4).

13.30 ADVANTAGES AND DISADVANTAGES

The Homestake Mining Co. biotreatment plant indicated that biotreatment process offers the following advantages:

TABLE 13.4

Effluent Total and WAD Cyanide Concentrations mg/l

YEAR	TOTAL CYANIDE	WAD CYANIDE
1984	0.45	0.09
1985	0.37	0.05
1986	0.38	0.05
1987	.22	0.05
1988	0.07	0.02

Source: Reference 12, with permission

- Compatibility with receiving streams
- Dynamic ability to charges in influent flows
- Capability for recovery from upset without human intervention
- Higher efficiencies than chemical treatment processes
- Lower operational costs than chemical processes

13.31   DISADVANTAGES INCLUDE:

- Research and time required for system design and start-up
- Application difficulties with cold water

13.32   BACTERIAL CYANIDE DETOXIFICATION OF GOLD HEAP LEACH OPERATION

The U.S. Bureau of Mines demonstrated the viability of using bacterial cyanide oxidation for decommissioning a heap leach operation. A major environmental problem in the mineral industry involves the cyanide wastes generated by the heap leaching of low-grade gold ores. Low cost-remediation techniques will be necessary both for contaminated solutions and leachate residues (spent heaps) during closure of the operations.[14]

The Bureau conducted research to evaluate the biological oxidation of cyanide tailings water containing 280 ppm CN at pH 10.5. Cyanide-degrading bacteria, *Pseudomonas pseudoalcaligenes*, were isolated from tailings water and were evaluated in batch and column tests. The results indicated that more than 90% of the cyanide in the pond water was oxidized. Subsequently, the Bureau conducted field tests on biological cyanide oxidation during decommissioning of the Green Springs heap leach operations in Nevada.

Initial laboratory studies were conducted using Green spring solutions which had pH 9 to 9.5, WAD cyanide 20 ppm, 80 Ca, 7 Cu, 0.5 Fe, 0.35 Mg, 11 Ni, and 0.2 Zn.

Before field testing was initiated, extensive laboratory research was performed to determine the feasibility of the bio-treatment process and to evaluate the effects of bacterial nutrient, bioreactor configuration, and retention time for cyanide destruction. The tests were conducted in 2.54 cm ID glass columns, packed with either quartz chips or activated carbon as a bacterial growth surface. The columns were seeded by recirculating PGY broth (a commonly used laboratory nutrient) which had been inoculated with *Pseudomonas pseudoalcaligens* for 2–3 days. After the bacteria had built up on the growth surface, Green Springs solution was pumped through the system. Laboratory test results were utilized in designing the field test.[14]

13.33   FEASIBILITY TESTS

A single pass, trickling filter reactor was used to conduct feasibility tests. The test data, presented in Table 13.5, indicated that the Green Springs effluent can be effectively treated using bacterial oxidation. For optimum cyanide removal, the cutoff point for PGY concentration was 3%.

Use of PGY will be very expensive in field tests. A series of experiments were conducted to determine whether phosphate or plus sugar can replace PGY. The results as presented in Table 13.6 indicate that all nutrients tested were quite effective at the 4.5-h retention time.

13.34   BIOREACTOR STUDIES

Tests were conducted in up flow columns packed with activated carbon to determine if the carbon tanks at the Green Springs site could be used as bioreactors. Green Springs process solution containing 55 ppm WAD CN was treated at a 4.5-h retention time with 10% PGY as the nutrient. The results achieved an effluent WAD cyanide concentration of 0.1 ppm. These tests indicated that:

TABLE 13.5
**Biotreating Green Spring Solution; WAD CN
Feed Concentration of 35–45 ppm, 4.5-h
Retention Time, and PGY Nutrient**

PGY concentration, pct	Effluent WAD CN, ppm
10	0.1
5	0.1
3	0.1
2	1.0

Source: Reference 14, with permission

TABLE 13.6
Nutrient Studies Conducted at 4.5-h Retention Time

Nutrient type	WAD CN, ppm Feed solution	WAD CN, ppm effluent solution
3-pct PGY	28	0.1
Medium 1A[2]	21	0.2
Medium 1A(modified)[3]	26	0.7
Medium 1C[4]	20	1.2
26 ppm PO_4 (as H_3PO_4)	17	0.5

Source: Reference 14, with permission

- Activated carbon would provide an adequate surface for bacterial growth.
- Cyanide can be effectively oxidized in an up-flow column.

A decision was made to use the carbon tanks at Green Springs as bioreactors.

Since the carbon tanks present the problem of short retention time, tests were conducted in a downflow system at a 1.5-h retention time to evaluate the relationship between retention time and nutrient concentration. For example, at a 4.5-h retention time, the effluent WAD cyanide concentration was 0.1 ppm, whereas at a 1.5-h retention time, the effluent contained 7.4 ppm WAD cyanide.

The Green Springs operation consisted of two heaps, each containing 600,000 tons of ore, a pregnant solution pond and a processing plant with five carbon adsorption tanks. The closure of the heap leaching operations involved rinsing the heaps with process solution until the cyanide concentration in the rinse solution was below 0.2 ppm WAD CN.

A fine mist of water is sprayed on the top of the heaps. The solution volume is maintained with freshwater makeup. The last portion of the remaining cyanide in the rinse solution is destroyed using a chemical oxidant such as hydrogen peroxide.

The Bureau of Mines approach involves using the carbon columns as bioreactors to destroy a portion of the cyanide in the rinse solution as it is circulated through the system. This procedure spreads up the closure process and may eliminate the need for chemical treatment. Field testing involved the following steps:

- Inoculating the carbon tanks with cyanide oxidizing bacteria
- Destroying cyanide using the carbon tanks as bioreactors

An inoculum of *Pseudomonas pseudoalcaligenes* was developed in full strength PGY to 10^9 cells/m liter. At the site, 50-gallons of process water was added to each of the five 55-gallon drums. Peptone, glycerol, and yeast were added to each barrel to yield full-strength PGY. Five gallons of inoculum were then added to each barrel, and thorough mixing was done.

The bacteria were allowed to grow for a week in the barrel to a concentration of 10^9 cells/m liter. The resulting inoculum was found to be very pure because the cyanide in the process solution kills most undesirable strains.

One barrel of inoculum was added to each of the five carbon tanks. An additional 60–70 gallons of PGY was also added to each tank. During the inoculation period, the tanks were bypassed. Solution was pumped onto the heaps at 500 gallons per minute. No process solution was pumped through the tanks.

The bacteria were allowed to grow in the tank for a week. An additional 55 gallons of PGY, made up in process solution was then added to each tank. After one week, the solution was circulated through the five tanks system for a day at 10gallons/minute. About two weeks were spent in inoculating the tanks.[14]

After inoculation, a process solution flow of 50 gallons, per/minute was passed through the tanks with a retention time of 80 minutes. The flow rate was maintained for 2 weeks. Phosphoric acid was also added as a nutrient. The test results are given in Table 13.8 The results of the experiments indicate that the WAD cyanide was oxidized in the carbon tanks and the system efficiency was highest when the phosphate concentration was maintained at 26 ppm PO_4 (Tables 13.7 and 13.8).

Field tests successfully demonstrated that the bacteria have the ability to consistently destroy cyanide with a very short retention time using a less-than-ideal nutrient. The WAD cyanide was reduced from 20 ppm to 8.5 ppm in 15-week period. The bacteria oxidized 10–15% of the cyanide present in the process solution, including solution tied up in the heaps.

Sample	WAD CN, ppm
pregnant solution (tank 1 feed)	14.3
Tank 1 effluent	14.0
Tank 2 effluent	13.5
Tank 3 effluent	13.0
Tank 4 effluent	12.7
barren solution (tank 5 effluent)	12.4

Source

TABLE 13.7
Treating Green Spring Solution in a Downflow
System at a 1.5-h Retention Time

Nutrient type	WAD CN, ppm feed solution	WAD CN, ppm effluent solution
0.5-pct PGY	14.1	12.8
1-pct PGY	15.0	12.4
5-pct PGY	15.0	7.4
10-pct PGY	14.6	1.7
26-ppm PO4	16.0	11.4

Source: Reference 14, with permission

TABLE 13.8

WAD Cyanide Oxidation at 50 gal/min and the Effect of Phosphate Concentration

Date	Sample	WAD CN, ppm	PO_4, ppm
6/3	preg solution	19.1	42.9
	barren solution	13.9	20.4
6/4	preg solution	19.3	19.9
	barren solution	14.8	16.3
6/9	preg solution	19.1	3.9
	barren solution	18.3	2.4
6/13	preg solution	18.2	12.0
	barren solution	14.1	9.0
6/14	preg solution	17.5	34.0
	barren solution	13.9	9.0
6/15	preg solution	17.3	25.8
	barren solution	13.9	8.5

Source: Reference 14, with permission

BIBLIOGRAPHY

1. Gadd, G. M. and White, C., Microbial treatment of metal pollution: A working biotechnology, *Trends Biotechnol. 11*:353–359 (1993).
2. Brierley, Corale L., Metals bioremediation in *New Remediation Technology in the Changing Environmental Arena* (Littleton, CO: Society for Mining, Metallurgy and Exploration, 1995), pp. 205–210.
3. Smith, L. A., Alleman, B. C. and Coploy-Groves, L., Biological treatment options, in *Emerging Technology for Bioremediation of Metals* (Boca Raton, FL: Lewis Publishers,1993), pp.1–12.
4. Brierley, C.-L., Environmental biotechnology applications in mining, in BIOMINE'93 (Adelaide, Australia: Australian Mineral Foundation, 1993), pp. 4.1–4.5.
5. Brierley, C. L., Bioremediation of metal contaminated surface and groundwater, *Geomicrobiol. J. 8*:201–212, 223 (1990).
6. Howe, R. H. L., Biodestruction of cyanide wastes-advantages and disadvantages, *Int. J. Air Water Pollut. 9*:463–478(1965).
7. Knowles, C. J., Microorganisms and cyanide, *Bacteriol. Rev. 40*(3):652–650 (1976).
8. Boucabeille et al. Microbial degradation of metal complexed cyanides and thiocyanate from mining wastewaters, *Environ. Pollut. 84*:59–67 (1994).
9. Thompson, L.C., Developments in mine waste biotreatment processes, in *Emerging Process Technologies for a Cleaner Environment* (Littleton, CO: Society for Mining, Metallurgy and Exploration, 1992), pp.197–204.
10. Thompson, L. C. et al., Summitville mine: Development of biotreatment processes for spent ore detoxification test results from focused feasibility studies, in *New Remediation Technologies in the Changing Environmental Arena* (Littleton, CO: Society for mining, Metallurgy and Exploration, 1995), pp. 53–59.
11. Thompson, L.C. et al., Biomineralization of Summitville spent ore, in *New Remediation Technologies in the Changing Environmental Arena* (Littleton, CO: Society for Mining, Metallurgy and Exploration, 1995), pp. 61–70.
12. Wong-Chong, G. M., Biological degradation of cyanide in complex industrial wastewaters, in *Biohydrometallurgy* (Ottawa: CANMET, Canada Center for Mineral and Energy Technology, 1989), pp. 289–300.

13. Fallan, R. D. et al., Anaerobic biodegradation of cyanide under methanogenic conditions, *Appl. Environ. Microbiol. 57*(6):1656–1662 (1991).

14. Noel, D. M. et al., Degradation of cyanide utilizing facultative anaerobic bacteria, in *Mineral Bioprocessing* (Warrendale, PA: The Minerals, Metals and Materials society, 1991), pp. 355–365.

15. Garcia, H. J. et al., Fate of cyanide in anaerobic microbial systems, in *New Remediation Technology in the Changing Environmental Arena* (Littleton, CO: Society of Mining, Metallurgy and Exploration, 1995), pp. 229–234.

16. Whitlock, Jim., The advantages of biodegradation of cyanide, *J. Met. 41*(12):46–48 (1989).

17. Whitlock, J. L. and Mudder, T., The Homestake wastewater treatment process biological removal of toxic parameters from cyanidation wastewaters and bioassay effluent evaluation, in *Mineral Bioprocessing* (Warrendale, PA: The Mineral, Metals and Materials Society, 1991), pp. 327–339.

18. Whitlock, J. L. and Smith, G. R., Operation of Homestake cyanide biodegradation wastewater system based on multivariable trend analysis, in *Biohydrometallurgy* (Ottawa: CANMET, Canada Center for Mineral and Energy Technology, 1989), pp. 613–625.

19. Lien, R. H. and Altringer, P. B., Case study: Bacterial cyanide detoxification during closure of the Green spring gold heap leach operation in *Biohydrometallurgycal Technologies* (Warrendale, Pennsylvanian: The Minerals, Metals and Metallurgical Society, 1993), pp. 219–227.

20. Wildeman, T. R. Pinto, A.P., Tondo, L.A. and Alves, L.A., Passive treatment of a cyanide and arsenic laden process water at the RPM gold mine, Minas Gerais, Brazil, in *Proceedings of the International Conference on Acid Rock Drainage*, March 26, St. Lois, MO, Published by the American Society of Mining and Reclamation, Lexington, Ky.

14 Vegetative Uptake and Use of Metal Tolerant Plants to Reclaim Mining Wastes

14.1 INTRODUCTION

The elemental composition of plants is commonly very different from that of the soil in which they grow. These differences are caused by the plants' ability to fix carbon from the air and to absorb essential nutrients from the soil. The essential elements for plant nutrition are not always established. However, it is well known that plants are not only built of molecules and ions that have functional or structural roles in their development. This is particularly true for many heavy metals, which can readily be detected in field-grown plants.[1]

The term heavy metal is broadly defined as any element that has metallic properties (e.g., ductility, conductivity, density, stability as cations, ligand specificity, etc.) and an atomic number > 20. Heavy metals required by plants include Mn, Fe, Cu, Zn, Mo, and Ni. The phytotoxicity of such relatively common heavy metals as Cd, Cu, Hg, and Ni is significantly higher than that of Pb and Zn. Hexavalent Cr is much more toxic to plants than trivalent Cr.

Heavy metals are present in soils as natural compounds or as a result of human activity. Metal-rich mine tailings, metal smelting, electroplating, gas exhausts, energy and fuel production, down-wash from power lines, intensive agriculture, and sludge dumping are some examples of man-made sources that produce contamination of soils with heavy metals. Natural mineral deposits containing large concentrations of heavy metals often support very characteristic plant assemblages and species. Examples of such distinct plant communities include serpentine (i.e., growing on Ni-, Cr-, Mn-, Mg-, and Co-rich soils), selexiferous (i.e., growing on Se-rich soils), calamint (i.e., growing on Zn- and Cd-rich soils), uraniferous (i.e., growing on U-rich soils) and Cr/Co floras.

Plants utilize three strategies for growing on metalliferous soils. Metal excluders prevent heavy metals from entering their aerial parts; however, they can contain large amounts of metal in their roots. Metal nonexcluders actively accumulate metals in their tissues above the ground. They can be roughly divided into two groups—indicators and hyperaccumulators. In indicator species, metal levels in the tissues indicate metal levels in the soil. Hyperaccumulators can concentrate on metals in the tissues above the ground to levels much above those prevailing in the soil. A plant containing more than 0.1% of Ni, Co, Cu, Cr, and Pb or 1% Zn in its leaves on a dry weight basis is called a hyperaccumulator. Almost all metal-hyperaccumulating plant species were discovered on metalliferous soils, restricted to a few specific geographical areas. Nickel hyperaccumulators were found in New Caledonia, the Philippines, Brazil, and Cuba. Nickel and zinc accumulators were found in Southern and Central Europe. Chromium and Cobalt accumulators were found in Central Africa.[1]

All hyperaccumulator plants can be divided into three major groups, on the basis of their tendency to accumulate different metals: Cu/Co accumulators, Zn/Cd/Pb accumulators, and Ni accumulators.

So far very little is known about the biological and evolutionary significance of metal hyperaccumulators.

The following hypothesis has been put forward:

- Tolerance or disposal of metal from plants
- Drought resistance
- Inadvertent uptake
- Defence against herbivores or pathogens

14.2   MECHANISMS OF METAL ACCUMULATION

Metals in soils are bound to organic constituents and inorganic (clay) constituents in the soil that are present as insoluble precipitates. For accumulation in plants, the soil-bound metals must be mobilized into the soil solution. This mobilization can be achieved through several processes:

- Metal-chelating molecules can be secreted into the rhizosphere to chelate and solubilize soil-bound metal.
- Roots can reduce soil-bound metal ions species' plasma membrane-bound metal reductases.
- Plant roots can solubilize heavy metals by acidifying their soil environment with protons extruded from the roots.

The plant roots assimilate the solubilized metal ions either via the extracellular or via intracellular pathways. After entry into the root, they can either be stored or exported to the shoot. In hyperaccumulating plants, heavy metals accumulate both in the shoot and the root.

14.3   USES OF METAL-ACCUMULATING PLANTS

Recently the value of metal-accumulating plants in environmental remediation processes has been recognized. The application of plants in environmental clean-up has been called phytoremediation, which is classified as follows:

- Phytoextraction utilizes metal-accumulating plants to transport and concentrate metals from the soil into harvestable parts of the roots and above the ground shoots.
- Rhizofiltration uses plant roots to absorb, concentrate, and precipitate toxic metals from polluted effluents.
- Phytostabilization uses plants to eliminate the bioavailability of toxic metals in soils.

In the phytoextraction process, hyperaccumulating plants are utilized to reduce soil concentrations of heavy metals to environmentally acceptable levels. Preliminary trials with Ni and Zn hyperaccumulator plants were successful in partially removing heavy metals from soils contaminated by long-term application of heavy metals containing sludges. Plant residues, highly enriched in heavy metals may be dried, ashed, or composed and isolated as hazardous waste. Plants that accumulate toxic metals can be grown and harvested, leaving the water or soil with a greatly reduced level of toxic metal concentration. There has been growing interest in using metal-accumulating roots and rhizomes of aquatic or semi-aquatic vascular plants like water hyacinth, duckweed, water velvet etc., for the removal of heavy metals from contaminated aquarius streams.

Significant research has been conducted on the physiology and ecology of plants that can grow on soils containing elevated concentrations of heavy metals. Investigations have been concentrated on the mechanisms that allow selected varieties of plant species to survive in phytotoxic environments, and on the potential applications of metal tolerant varieties of plant species to reclaim mining wastes and soils containing heavy metals.

14.4 MINING WASTE CHARACTERIZATIONS

Metalliferous mining wastes present unique environmental restoration problems because of their toxicity. Toxicity is caused by the presence of residual metals produced from weathering of low-grade ore or the waste products of mineral processing.

Where pyrite is present in mining waste, acid generation by oxidation aggravates the toxicity problem by promoting heavy metal mobilization in aqueous solution.

In acid soils, a high concentration of hydrogen ions causes direct damage to vegetation root systems and adversely affects plant nutrition. Reclaiming such wastes by establishing a permanent vegetation cover may not be feasible.

Table 14.1 presents chemical characteristics of some metalliferous mining wastes. The sample sites represented a range of abandoned mine sites in northern England. The sites exhibited a level of variability in chemical levels within and between them. However, each site was characterized by high levels of lead, zinc, and fluoride. The wastes were generally deficient in nitrogen and phosphorus, the essential nutrients for plant life. The mine wastes mainly consist of gravel and coarse sand from old mines as well as fine tailings from modern processing (Table 14.2).

TABLE 14.1

Summary of Chemical Characteristics of Mine Wastes for 24 Mine Sites (mg/kg Air-Dry Oil, Except pH)

Variable	Range	Mean	Standard Deviation	Range in Norm al Soils
pH	3.6–7.4	5.75	1.1	–
Lead (Pb)	288–127,500	36,077	32,427	2–200
Zinc (Zn)	340–51,200	7789	10,657	10–300
Copper (Cu)	27–192	91	50	2–100
Caesium (Cs)	7–44	24.2	9.6	<10
Nickel (Ni)	12–82	37	17.7	5–500
Potassium (K)	1,000–20,000	9,418	6,576	500–3,500
Phosphorous (P)	15–870	382	193	200–5,000
Nitrogen (N)	70–670	285	152	200–2,000
Fluorine (F)	3410–196,329	49,536	46,333	20–500
Barium (Ba)	12,100–16,100	13,730	1,134	100–300

Source: Reference 2, with permission.

TABLE 14.2

Summary of Particle Size Analysis on Mine Wastes from 24 Mine Sites

Fraction	Size Limits (Particles Diameter, mm)	Composition (% Air Dry Soil ± SD)
Gravel	>2.0	67.7 ± 11.5
Coarse sand	2.0–0.2	20.9 ± 7.0
Fine sand	0.2–0.02	5.9 ± 3.6
Silt	0.02–0.002	3.5 ± 2.4
Clay	<0.002	2.1 ± 1.9

Source: Reference 1, with permission.

Both types of mining wastes show evidence of adverse water balance caused by excessive infil-tration and waterlogging in fine tailings. Both these conditions restrict plant growth and its estab-lishment. The range of environmental factors restricting plant growth on these mining wastes is given in Table 14.1.

14.5 PLANT ADOPTIONS TO CONTAMINATED SOILS AND WASTES

The performance of plant establishment and growth depends on soil toxicity and toxicity tolerance of the plants. Soil chemical factors limit normal plant growth.

All plants respond to increases in heavy metal concentration in the soil. The magnitude of the response depends on the plant sensitivity, the concentration level, and the form of metal present. In most cases, the plant growth is depressed (Table 14.3).

14.5.1 OTHER SOIL CHARACTERISTICS

Contaminated soils, besides differing in their metal content, differ in other factors which affect the number and types of plants growing on those soils. Those factors include the nutrient status of the soil, the organic matter content, and the soil texture. Their effects may be direct, or the factors may act indirectly by influencing the toxicity of the metal in the soil.

In general, mine-tip soils are extremely low in the essential nutrients: nitrogen, phosphorous, and potassium. Lead and copper-tolerant plant populations are able to grow in culture solutions containing low phosphate. Studies have shown that the addition of complete fertilizer significantly improves the growth of plants, both native and foreign, to mine soils.

14.5.2 CLIMATIC FACTORS

Contaminated areas are frequently sparsely colonized. The plants on the mine soils are smaller, prostate, and adapted to exposed conditions. Some species found in copper soils in the neighbouring area of Katanga and Rhodesia were different because of the greater aridity of Rhodesia.

The effects and symptoms of metal toxicity differ depending upon plant species and form of metal. The conditions which are most commonly encountered under phytotoxicity are the following:

- Subcellular effects: inhibition of enzyme activity, iron transport, chlorophyll synthesis, photosynthesis, respiration, and mineral uptake and transport.
- Cellular effects: inhibition of root and stem elongation, leaf expansion, and damage to cell membranes.

Visible symptoms indicate poor root development, stunting, and leaf discolouration. These symptoms usually result in tissue necrosis and death.

TABLE 14.3

Environmental Factors Frequently Found to Limit Vegetation Establishment on Mining Wastes

Chemical	Physical	Biological
Low pH	Compaction and runoff	Sterility
Salinity	Excessive infiltration	Lack of nutrient cycling
Toxicity	High surface temperatures	
Nutrient deficiencies	Surface mobility	

Source: Reference 2, with permission.

14.6 TOXICITY TOLERANCE

Mining wastes frequently contain extremely high level of metal concentrations. Despite that, populations of some species are known as natural colonists of those materials. Investigations on these species have indicated that in contaminated mining wastes, colonists of metal-enriched soils are favoured because of their physiological resistance to the effects of metal toxicity, which prevents the normal nonresistant species from surviving.[2]

Metal resistant plants have been classified as "metallophytes" that grow and survive on metalliferous soils. Some species whose growth is restricted on metal-enriched soils are referred to as "obligate metallophytes." Those that can grow on normal and metal-enriched soils have been classified as "facultative metallophytes."

Researchers have identified two general strategies that confer resistance to heavy metal toxicity:

- Tolerance, where the plant can survive the effects of internal stress avoidance and is protected externally from the effects of stress.

  Tolerance arises from specific physiological mechanisms through which the plant can function normally in highly contaminated environments. The genetic mechanisms involve metal accumulation in detoxified forms. The detoxification processes include:
  - Cell wall binding of metals
  - Organic acid complexity
  - Protein metal binding
  - Enzyme modification
  - Modified membrane permeability

Avoidance arises from exclusion mechanisms, for example, root infections by mycorrhizal fungi. The fungal mass acts as a biological barrier against the metal uptake by most plant roots.

14.7 MEASUREMENT AND DETECTION OF METAL TOLERANCE

Several test procedures are available to detect and measure metal tolerance in plants. The root elongation test is commonly used because non-tolerant plants show inhibited root growth in metal-contaminated soils.

The root elongation test compares root lengths measured in plants when grown in liquid growth solutions containing the metal of interest relative to root growth in a normal, uncontaminated solution. Experiments were conducted to determine the existence of tolerance to lead, zinc, and copper in a range of commercially available grass species. Root length and total plant biomass measurements were used to evaluate relative tolerance. The results presented in Table 14.4 indicate that levels of tolerance differ among species and among metals. This indicates that metal tolerance is of a specific nature. Also, the tolerance of metals is polygenetic. However, in root elongation tests, several higher plant species seem to have developed multiple metal tolerances.

14.7.1 USE OF METAL TOLERANT PLANTS IN MINE WASTE RECLAMATION

The establishment of toxicity tolerance vegetation directly over mining wastes without the use of topsoil cover has been widely used. The advantages lie in cost and convenience. Metal tolerant varieties of grass have been adapted to toxic and infertile soil or waste conditions, with little substrate amelioration requirements. Metallophyte species are usually slow in growing and low yielding, with low fertilizer requirements.

However, there are limitations to this approach. Metal tolerant species are very limited, and only a few are commercially available. The available species are restricted to growth in relatively moist, cool, temperate environments. Such species include *Festuca rubra* (lead, zinc tolerant), *Agrostis Capillaries* (copper, acid tolerant) and *Agrostis Stolonifera* (lead, zinc tolerant).

TABLE 14.4

Summary of Toxicity Tolerance across 12 Grass Species

Species	Lead	Zinc	Copper	pH
Alpine bluegrass	**	*	*	**
Creeping bluegrass	**	–	*	**
Chewing fescue	*	**	–	**
Foxtail	**	*	*	**
Hard fescue	**	*	*	**
Hairgrass	*	–	–	–
Meadow fescue	*	**	–	*
Red fescue	–	**	–	**
Red top	–	–	–	**
Sheeps fescue	*	**	–	**
Timothy	–	–	–	**
Tall fescue	*	*	–	**

Source: Reference 2, with permission.
* Moderate tolerance
** High tolerance
– Intolerance

For several years, research work has been conducted to identify and select various metal and acid-tolerant varieties of grasses for use over a broader range of geoclimatic areas. Research data indicate that normal varieties of commercial grasses can exhibit metal tolerance. Reclamation programmes to establish vegetation over mining wastes at Leadville, CO and Cataldo, ID have been completed with the use of metal tolerant commercial species.

Active mines in most states, as per environmental legislation, require topsoil conservation and replacement. So adaptive reclamation approach is restricted only to abandoned mined land reclamation programmes and superfund applications.

14.8   METAL TOLERANCE IN PLANTS

Plants have been found growing in toxic soils produced from mining activity. The following is a brief description of some of these plants.

Vascular plants including ferns have been found in nickel- and copper-bearing mining soils in Rhodesia and Katanga. Plants that are largely abundant on metal-contaminated soils have been used as indicators of heavy metal ores and so have been used in metal prospecting.

Plants regarded as metal indicators are either metallophytes or pseudo metallophytes. A classification of plants found to grow on metal-contaminated soils is given in Table 14.5.

Several species of mosses have been found on copper-contaminated soil. Hence, they are called copper mosses. They are restricted to copper soils by virtue of their being able to use hydrogen sulphide into photosynthesis. Copper mosses have been observed in copper-rich areas in Scandinavia and Australia.

Lichens are found both as colonizers of bare rock and as general components of the established vegetation on contaminated soils. Large concentrations of iron and copper are found in the cell walls of the species.

Field observations can be summarized. The species growing in toxic soils are very varied and differ according to the local ecological conditions and geographical area. There appears to be a close affinity between the flora of toxic and serpentine soils. Serpentine soils are often high in nickel and chromium, and these metals appear in part at least to determine the serpentine flora.

TABLE 14.5

Classification of Plants on Metal-Contaminated Soils with Examples from Metal-Contaminated Soils of Europe

1. *Metallophytes*—taxa found only on metal-contaminated soils
   (a) *Absolute metallophytes*—found only on metal-contaminated soil overall their distribution.
   (b) *Local metallophytes*—found only on metal-contaminated soil within a given region but occurring also in a phyto-geographically distinct non-contaminated area.
2. *Pseudometallophytes*—taxa occurring both on contaminated soils and on normal soils in the same region.
   (a) *Elective pseudometallophytes*—abundant and often more vigorous on contaminated soils.
   (b) *Indifferent pseudometallophytes*—live on contaminated soil regularly but show neither abundance nor particular vitality.
   (c) *Accidental metallophytes*—usually weeds and rederals appearing sporadically and showing reduced vigour on metal-contaminated soils.

Source: Reference 3, with permission.

It is clear that the plants growing on metal-contaminated soils are often characteristic of such soils. Toxic areas are usually colonized by some plants, and although the colonization of many areas is sporadic, other areas carry regular communities. The diversity of the flora on a contaminated site may be influenced by the diversity of the surrounding flora, the extent of the site, its antiquity, and the concentration of toxic metal present.[3]

14.9 FACTORS DETERMINING PLANT DISTRIBUTION ON CONTAMINATED SOILS METAL CONCENTRATIONS AND TYPES

The main characteristic of contaminated soils is high metal concentration. The contaminated area is usually recognizable by a unique flora different in the surrounding areas. But observation information is lacking as to what extent metal concentration and type are important in determining the distribution of mine plants.

The metal is usually present in the soil in several forms: water soluble, exchangeable, and bound to inorganic or organic substrates, or unavailable as insoluble compounds or minerals. The simplest measure to express metal concentration has been the total metal content.

Another complicating factor has been the vertical distribution of metals in the soil. Leaching accumulation of metals in humus layers, localization of ore bodies at certain depths, and the spread of toxic debris in normal soils cause nonuniform metal distribution in soils.

Soils with extremely high concentrations are usually barren, whereas those with only slightly enhanced concentration may not have any visible effect on the vegetation. But the plants growing there may have increased metal content. Different plants can withstand different amounts of metal.

Three types of "cuprophyte" (plants restricted to copper soils) have been observed in copper-rich soils in copper mining districts in Katanga in Rhodesia. Plants classified as "polycuprophytes" grew in the areas of 5,000 to 10,000 ppm copper, whereas the "eurycuprophytes" could colonize a whole range of copper concentrations.

Three initial phases have been identified in the colonization of zinc-contaminated areas. The last of these phases contained a considerable number of species on soils with generally lower concentration than the first two.

Sharp changes in vegetation coincided with changes in total contamination. Changes in relative amounts of lead, copper, and zinc also had significant effects on vegetation characteristics. *Eriachne mucronate* seemed to tolerate high concentration of all three metals; *Bulbostylis barbata* and *Polycarpea glabra* were found in areas of high copper, whereas *Tephrosia* sp. nov. was found where copper concentration was low. In general, changes in lead and copper were more determinant than changes in zinc.

14.9.1   EVOLUTION OF METAL TOLERANCE

In Katanga, 26 species have been recognized as endemic to the copper outcrops. Endemism of this kind is not observable in other copper-bearing areas in Africa, but over four hundred species have been identified as copper tolerant.

This above fact suggests a common phenomenon. However, some of the species which abound in the region have not given rise to Cu-tolerant forms. For example, dominant flora in the region includes deciduous trees of leguminous genera as *Isoberlinia* and *Brachystegia*. But these species have not developed any high degree of Cu tolerance. The grasses on the heavily contaminated soils include *Loudetia*, *Eragrostis*, and *Danthoniopsis*. But the most frequent genus of grasses in the region, *Hyparrhenia*, are virtually absent in Cu outcrops.

The restricted propensity for evolving metal tolerance is more evident in other geographical regions. In Britain, restricted propensity for evolving metal tolerance has been observed. But the Cu-contaminated soils are of relatively recent origin. Only eight species frequently occurring on Cu-contaminated soils have been observed. However, some species showing copper tolerance showed lead-zinc tolerance. The relative abundance of species on Pb-Zn-contaminated soils suggests that genes for lead-zinc tolerance may be more widespread than those of the Cu tolerance.

14.10   GENERIC STUDIES ON METAL TOLERANCE

Cu tolerance in plants is usually supposed to be genetically determined. In the genetics of Cu tolerance in *Mimulus guttatus* from copper mine sites in California in which the index of tolerance of the plants broadly correlated with the total Cu content of the soils, a high degree of variability was demonstrated. When F_1 progeny was tested from crosses involving Cu tolerance and a non-tolerant parent, Cu tolerance was dominant at low Cu concentrations, intermediate at intermediate concentrations and recessive at high Cu concentrations.

Tolerance to high levels of zinc in the soil has been found in some species. The specie *Silene inflata* has indicated inheritance of Zn tolerance.

A similar study involved Pb tolerance in *Festuca ovina* using a series of crosses between parents of known tolerance and analyses of half-diallel crosses. The crosses between selected parents did not yield clear results because both the degree and direction of dominance varied.

With respect to Al tolerance, the vegetation of acid soils such as the heathlands of the temperate zones and the evergreen forests of the humid tropics are adapted to growth in the presence of concentrations of Al. In the temperate zones' certain families of plants, notably *Ericaceae* and *Epacridaceae* are associated with the soils containing high levels of Al. In humid tropics, Al-tolerant species are distributed among a large number of plant families. Al tolerance in certain lines of barley has been attributed to a single major dominant gene. Similar single gene control of Al tolerance has been reported in a line of wheat. In other lines, two of the genes plus modifiers have been observed. As with other forms of metal tolerance in plants, genetic evidence of controlling Al tolerance has been disappointing.

Summarizing the research on the genetics of metal tolerance in plants, a few general observations can be made. Most manifestations of metal tolerance have been found to be genetically determined. In some species, populations containing a low frequency of genes for metal tolerance exist. The genetic work in Cu tolerance and Pb tolerance shows that the manner of inheritance is probably Mendelian but involves quantitative effects in which dominant relationships are rarely clear or confusing.

14.11   PHYSIOLOGICAL AND BIOCHEMICAL MECHANISMS OF METAL TOLERANCE

From detailed research on tolerance of different plant species to particular metals, it has become evident that many mechanisms of tolerance may have evolved even in respect of a single metal.

TABLE 14.6

Species Showing Traces Tolerant to Heavy Metals

Species	Metal
Melandrium silverstre	Cu
Tarazacum officinale	Cu
Tussilago farfara	Cu
Minulus guttatus	Cu
Agrostis tenuis	Cu, Pb, Ni, Zn, Cu + Ni
	Pb + Cu, Zn + Pb
	Cu
	Pb
	Pb, Zn
Rumex acetosa	Cu
	Zn
Festuca stolonifera	Pb
	Zn
Agrostis conina	Pb
	Zn
Viola lutea	Zn
Alsine (=Minuartia)	
Verna	
Silene vulgaris (=inflata=cucubalus)	Zn
	Cu, Zn
Plantago lanceolate	Zn
Linum catharticum	Zn
Campanula rotundifolia	Zn
Agrostis tenuis X stolonifera	Zn
Festuca rubra	Zn
Holcus lanatus	Zn
Anthoxanthum odoratum	Zn
Thlaspi alpestre	Zn
Armeria maritime	Zn

Source: Reference 3, with permission.

More than one mechanism has been identified to evolve Cu tolerance. As shown in Table 14.6, leaves of species growing on extremely contaminated soils exhibited great differences in the amount of Cu present in the leaves. Some species accumulate the metal to very high concentrations; some species avoid Cu toxicity by excluding the metal while others occupy an intermediate position. The results also indicate that both exclusion and accumulation

mechanisms evolve in different species of the same genes. Research on the physiology of metal tolerance has revealed the following:

- In some species, the tolerance to Cu can be selected and inherited independently of tolerance to other metals.
- In some species the selection for Cu tolerance alone gives rise to increased tolerance to other metals.
- In some species genetically, determined resistance may be found to certain metals but not to Cu.

14.12   METALS OF CLASS A

This class contains all metals which usually function as macronutrients for plants, e.g., Mg^{2+}, Ca^{2+}, and K^+. The sole member of this group involved in metal pollution problem is aluminium. Aluminium toxicity occurs in soils when the pH level is below 5, which often happens in heavily leached soils such as laterites of the humid tropics. Aluminium shows a preference for ligand containing oxygen. In Al-tolerant plants, three patterns of response may be distinguished.

- Species or genotypes which accumulate Al in the shoots and in the roots
- Species in which tolerant forms accumulate Al in roots but do not transport it to the shoot.
- Species in which Al tolerance is associated with the exclusion of the metal

In species accumulating Al in shoots, soluble complexes are involved in mediating the transport of the metal from root to shoot; and these complexes also store the metal in all compartments.

The species which tolerate Al through exclusion mechanism possess special means for protecting the biosynthesis of their cell walls through precipitation of Al at the surface. In some species, Al is precipitated because of elevated pH at the root surface.

14.13   METALS OF CLASS BORDERLINE

Manganese, zinc, chromium, nickel, iron, and cobalt are included in the borderline class of metals. Manganese is regarded as an important nutrient for plant growth. In normal soils Mn occurs predominantly in oxidized form from normal soils, it must be reduced at the root surface to the manganous Mn^{2+} form. Toxicity of Mn^{2+} occurs on extremely acidic or water-logged soils where reduction of Mn^{4+} to Mn^{2+} takes place.

The visible symptoms of Mn toxicity include marginal chloroses, necrotic legions, and distorted development of the leaves. But a wide range of variation in tolerance to Mn^{2+} between and within plant species has been observed. There seem to be several mechanisms of tolerance. Species tolerant to waterlogging are also tolerant of Mn^{2+} toxicity.

The level of Mn in plant tissues at which symptoms are observed varies widely. This very wide range in Mn contents of leaf tissues reflects differences in the capacities of the species to immobilize the Mn.

Zinc is another borderline class metal. Zinc is essential for plant growth. High levels of zinc usually occur in soils derived from mine wastes, slags from smelter, and sites where there are applications of Zn-containing sludges. It is generally observed that Zn toxicity is highest in acidic mineralized soils, low in organic matter, in Ca^{2+} and in available phosphate.

Zinc toxicity inhibits root growth in many species. In leaves zinc toxicity produces chlorosis. However, both with respect to root growth and leaf chlorosis, the effects of zinc are less severe than those of many other metals of borderline class, specifically copper.

The Zn tolerance in plants has evolved more readily than Cu tolerance. Research on zinc tolerance has identified four types of mechanisms: the immobilization of Zn in cell walls, the compartmentation of Zn in soluble complexes, the compartmentation of zinc in non-diffusible complexes, and the development of enzymes systems resistant to Zn^{2+} ions.

Chromium is usually considered a phytotoxic agent when present in the soil. Chromium is not generally considered an essential element for the growth of plants. Serpentine soils derived from ultrabasic serpentine rocks usually contain substantial levels of CrO_4^{2-}. Serpentine soils are toxic to many plants. However, plants adapted to serpentine soils probably require some degree of adaptation to Cr toxicity. Most species which occur in Cr-rich soils do not contain significant quantities of metal. Communities of Cr-adapted plants have been described from Southern Africa. In these plant communities, seven species were recognized accumulating Cr to very high levels.[2]

Nickel is not generally regarded as essential for plants. Apparently, Ni becomes essential for plants if urea forms a significant part of the nitrogen supply. Nickel is frequently found at higher concentrations in serpentine soils. High levels of Ni are also common, in association with Cu and Zn, on mine spoils and smelter wastes, where oxidation of sulphides gives rise to acid soils containing toxic concentrations of Ni^{2+}.

Ni toxicity produces chlorosis in many plant species. Ni^{2+} shows infractions with Fe^{2+} and Zn^{2+}. In some cases, the presence of toxic levels of Ni in soils may eliminate the more susceptible species, leading to the development of distinctive plant communities. Ni^{2+} has been found to be strongly inhibitory to root growth. Ni-tolerance studies indicate that the tolerance involves some kind of exclusion mechanism involving modification of carrier systems. Some species have been found to contain extraordinarily high concentrations of Ni, described as hyperaccumulators.

Iron is classified as a borderline metal because of the wide range of ligand environments in which it occurs in biochemical molecules. Iron is abundant in most soils, but predominantly in the form of the highly insoluble $Fe(OH)_3$. Under many circumstances, the plants face the problem of obtaining enough iron rather than cope with the excess. There are two approaches to the problems: to create acidity conditions close to the root surface so as to increase the availability of Fe^{3+}, and to create a reducing system at the root surface in which Fe^{3+} is converted to Fe^{2+} which is more readily taken up. So, the availability of iron depends upon the prevailing pH value and redox potential of the soil and upon the ability of the plants to modify these factors at the root surface.

Iron toxicity to plants is caused by conditions in which the acidity and reducing conditions become extreme.

Physiological effects of Fe toxicity are difficult to identify because they are often combined with the effects of Mn^{2+} and Al. Tolerance to Fe toxicity differs between one plant species and another. The basis of this difference in tolerance arises primarily from waterlogging, low pH, and the generation of a reducing system at the root surface.

The cobalt content of most soils is low. Weathering under poor drainage conditions can lead to the release of Co; for example, soils of the copper outcrops of Haut Shava province of Zaire may contain high concentrations of Co^{2+} in addition to Cu^{2+}. Cobalt toxicity has been found to cause diminished root growth, accompanied by the accumulation of carbohydrates and diminished export of assimilates from the leaves. Co^{2+} can act to inhibit the biosynthesis of ethylene, which can lead to widespread disruption of plant metabolism.

Species from a mine waste in Ontario, Canada, indicate Co tolerance to Co^{2+} and Cu^{2+}. The vegetation of the Co-rich copper clearings in Southern Africa also indicates some degree of Co as well as Cu tolerance. In these plant species, only a trace amount of Co was detected in the foliage, which leads to the inference that Co tolerance involves an exclusion mechanism. The vegetation of copper clearings includes a number of hyperaccumulators of Co. The nature of the Co complexes, their manner of formation and compartmentation, and the role these processes play in developing tolerance are not yet understood.

14.14 METALS OF CLASS B

Cadmium can be classified as a Class B metal in terms of its attributes. cadmium is not normally abundant in soils derived from metal outcrops. High levels of cadmium in soils can be produced because of Zinc smelting. Uptake of Cd from the soil by plant roots depends upon the pH value and available P.

At low concentration, Cd is not toxic to plants. In some plant species, Cd accumulation to levels can be toxic to animals. At higher concentration, Cd^{2+} is toxic to plants, and often causes leaf chlorosis and lowering of photosynthesis rate.

Tolerance to Cd^{2+} has been reported on the deposits which were also contaminated with Zn and Pb, and Cd tolerance was accompanied with Co tolerance to these metals.

Copper in the form Cu⁺ is included in Class B, but Cu^{2+} is classified in the borderline group. Much of the Cu present in the soils is in forms which the plants cannot make available for uptake by roots. Acid conditions tend to increase the availability of Cu^{2+}. Under such conditions, through oxidation of sulphides, Cu^{2+} reaches toxic concentrations in many copper mining wastes. Copper can be found in high concentrations in soils near outcropping ore bodies.

The most extensive occurrences are found in the copper belt of Central Africa, where copper content in soils can be as high as 10%.

Copper, even though an essential element for all organisms, becomes toxic at high concentrations. A common symptom of Cu toxicity is a reduction in chlorophyll content and yellowing over the surface of the leaf. In some instances, Cu-induced chlorosis is caused by reduced uptake of Fe. *Becium homblei*, a hyperaccumulator of Cu in Central Africa, develops chlorosis through Cu accumulation in leaves, but the chlorosis is not associated with lower iron content. Copper ions may mediate in the acceleration of per oxidative degradations of the lipids of the chloroplast membranes.

Measurements of root elongation are the used method for assessing degrees of tolerance to toxic metals. Cu^{2+} has been found to be an inhibitor to root elongation.

The primary toxic action of metal ions takes place at the cell surfaces where alterations in cell membrane properties can cause leakage of K⁺ and other ions and solutes. Cu-tolerant species are less sensitive to Cu-induced damage. They exhibit increased resistance of root cell membranes to Cu-induced leakiness.

Tolerance to Cu toxicity has been observed in many plant species. The facts of Cu tolerance include mechanism for the exclusion or diminution of Cu^{2+} uptake, immobilization of Cu in cell walls, compartmentation of Cu in insoluble complexes, and enzyme adaptations.

Lead can be classified as a Class B metal. Lead is not known to be essential for plant growth. The amount of Pb in the soil is not the only determinant of toxicity. When organic matter and other mineral nutrients are in abundant supply, Pb toxicity does not occur. Pb toxicity occurs most commonly on waste heaps from mining operations where organic matter and nutrient content of the soil are low. Phosphate can be a major factor in detoxification of Pb in soils, but other factors such as level of the surface, pH, and organic matter content may also influence the level of available Pb in the soil.

Lead toxicity causes a wide range of symptoms in plants: root elongation is inhibited; roots form on stems; stems elongation and leaf expansion are inhibited. The inhibiting effects of Pb arise from the influence of Pb with auxin-regulated cell elongation. A primary cause of inhibition of cell growth is caused by Pb-induced stimulation.[4] Where Pb is present in mine soils of low fertility, high levels of Pb may be taken up in the leaves of susceptible species causing stunted growth, chlorosis, and purple discolouration of foliage.

Lead tolerance can be an inherited characteristic as in *Festuca ovina*. In mine soils, *Agrostis tenuis* and *Minuartia verna* have been observed to develop Pb-tolerant ecotypes. Plants which succeed in colonizing Pb soil heaps do not contain an abundance of Pb as tolerance involves some kind of exclusion mechanism.[4]

Mercury is regarded as a Class B metal. These metals tend to bind preferentially to nitrogen and sulphur centres in living organisms; binding to such centres in proteins is usually irreversible and causes loss of biological activity in the protein.

Mercury is not essential for plant growth and is highly toxic to living organisms. Soils containing high levels of Hg are rare, but significant amounts of Hg are sometimes found in soils derived by drainage of ponds, lakes, and marshes. In such formations, Hg remains in the soils with very little (less than 0.2%) passing to fish, plant, and other organisms.

There are no direct observations of Hg tolerance in higher plants. In species growing in sediments containing Hg residues, the Hg is usually confined to the root system.

Silver is the most toxic of all metals to higher plants, followed by Hg. Plant leaves show necrosis caused by water shortage as the roots are killed by Ag+ ions. Ag+ ions can cause widespread damage to living cells.

TABLE 14.7

Species Showing an Index of Tolerance Indicative of Root Growth Being Stimulated by Presence of Metal Ions in the Testing Solution

Species	Metal
Agrostis tenuis	Cu
	Pb
	Pb, Cu, Zn
Mimulus guttatus	Cu
Anthoxanthum odoratum	Zn
Holcus lanatus	Zn
Armeria maritime	Zn

Source: Reference 3, with permission.

14.15 THE NEED FOR METALS BY TOLERANT PLANTS

Tolerant plants may have a positive need for metals, and for this reason they are confined to contaminated areas. However, tolerant plants also grow well in normal soil, but the growth of those tolerant plants is stimulated by levels of metals considerably above the normal micro-nutrient levels. Indices of tolerance greater than 100% (more root growth in metal than without metal) have been found and are listed in Table 14.7.

The apparent confinement of metal tolerance species, despite their normal growth on non-contaminated soil, has led to the belief that these plants are largely restricted to mine soils because they are competitively inferior to normal plants.

Serpentine species did not survive when grown in competition with their non-serpentine counterparts. Zinc tolerant *Anthoxanthum odoratum* is competitively inferior to non-tolerant plants when growing on non-contaminated soil.

14.16 EXTERNAL TOLERANCE MECHANISMS

The external mechanisms of tolerance represent those conditions which prevent the entry of the metal ions. These conditions are not under the control of the organisms but are important ecological factors Table 14.8.

The form of metal may not be readily soluble in water. If dissolved, surrounding water causes rapid dilution. For example, manganese can be released from root nodules by microbial action, but

---

TABLE 14.8

Possible Mechanisms of Metal Tolerance

External

  (i) Form of metal is not directly soluble in water and/or if dissolved then rapidly diluted by surrounding water.

  (ii) Actual amount of freely diffusible metal ions is small compared to the total amount present.

  (iii) Lack of permeability to heavy metals under specific conditions.

  (iv) Metal ion antagonisms.

Source: Reference 2, with permission.

---

the effective manganese concentration is low due to high water content. It has been observed that vegetation growth increases considerably in wet places on toxic soils.

The majority of metallic ores which occur naturally are in the form of sulphides, which are insoluble. So the amounts of metal available to plant life are very small. In a peat swamp over bedrock rich in copper, it was found that deeply circulating waters dissolve the copper and it percolates up to the surface. The peaty humus on the surface acts as an efficient natural chelating system. The high toxic quantities of copper are immobilized through absorption. This type of situation can occur in microorganisms.[5]

Another external mechanism of heavy metal tolerance is the lack of permeability to heavy metals. Changes in the environment can produce alterations in the permeability of the cells and in their metal uptake characteristics. *Escherichia coli* has been found to be copper tolerant only under aerobic conditions. *Chorella vulgaris* is tolerant of barium, manganese, nickel, lead, and copper under anaerobic conditions. Cells under anaerobic conditions absorbed less copper than under aerobic conditions. Copper-tolerant fungi increase their tolerance under conditions of low pH. The ability to develop tolerance under low pH depends on the impermeability of the cells to copper. Lowering of pH produces a net positive charge at the cell surface, which would reduce uptake of the metal cations Table 14.8.

Another external mechanism involves the reduction in the effect of toxic ions in the presence of other ions. Additional ions may reduce the net concentration of toxic ions, or they may interfere in uptake mechanisms by competing for entry sites. It has been found that non-tolerant plants of *Agrostis tenuis* could grow on normally toxic mine soil if a full nutrient culture was provided daily, and toxicity could be markedly reduced if only potassium nitrate is added. Copper toxicity of mine water was reduced by the presence of other ions. Calcium is important in cell permeability and influence metal tolerance.

14.16.1 METAL UPTAKE STUDIES

Investigations on plants growing on metal-contaminated soils have attempted to measure the levels of the various metals in the plants and in the soil on which they grow. These studies help in understanding the tolerance mechanism since they indicate whether metal tolerance involves some exclusion mechanism, or whether the mechanism is internal.

14.16.2 ZINC

Almost all researchers found extremely high values of zinc in plants on zinc soil. There have been extensive observations on the characteristics of zinc uptake:

- Different species from the same contaminated area differ in the degree to which they take up zinc.
- Different plant organs accumulate different quantities of zinc. Roots and leaves take up most zinc, stems and inflorescence the least. The metal distribution pattern depends on both the species and the metal concentration.
- The quantity of zinc in a plant changes with the growing season.
- The cellular resistance of plants to solutions of different concentrations is related to the amount of zinc they absorb.

14.16.3 COPPER

Investigations on plants on metal outcrops in Australia have revealed patterns common to several species. Copper uptake in the parts above the ground was found to be low and constant at low levels of soil, but at certain higher copper concentrations in soil, this resistance to uptake breaks down.

TABLE 14.9

Copper Content of the Leaves of Plant Species from Katanga, Growing on Soils Containing >5 × 10^{-4} g Cu g^{-1} Dry Soil

Group 1	Species containing <3.8 × 10^{-5} g Cu g^{-1} dry wt
	Olax obtusifolia
	Protea hirta
	Cryptosepalum maraviense
	Andropogon filifolius
	Triumfetta cupricola
Group 2	Species containing between 5 × 10^{-5} and 1 × 10^{-4} g Cu g^{-1} dry wt
	Becium homblei
	Becium aureoviride
	Gladiolus tshombeanus
	Dissotis derriksiana
Group 3	Species containing > 1 × 10^{-4} g Cu g^{-1} dry wt
	Triumfetta dikuluwensis
	Becium aureoviride ssp., lupotoense
	Pandiaka metallorum

Source: Reference 3, with permission.

Above this threshold level, the quantity in plant tops increases abruptly. At only slightly higher levels in the soil no plants were found. However, each species differed in its response to the increase in copper content in the soil.

The differing response to copper content levels in soil indicates that the copper did not just become available to the plant at a given total soil level, but that a genuine exclusion mechanism was in action at low copper levels Table 14.9.

Therefore, evidence suggests that the mechanism of copper uptake is different from the uptake of zinc. Researchers observed that most species growing on soils containing 1,000 ppm copper only showed enhanced copper contents. Roots consistently contained more copper than shoots Table 14.10.

14.16.4 LEAD

Investigations indicate that the pattern of lead uptake resembles that of copper rather than zinc. The uptake is constant with increasing levels of lead in soil until a certain point is reached when uptake becomes unrestricted and rises abruptly. The species rarely survive when the soil lead content is above a certain level.[3]

14.17 VEGETATIVE STABILIZATION OF MINING WASTES

Metalliferous waste, produced by mining and heavy smelting metals, such as lead, zinc, or copper present potential environmental problems, is apparent from the chemical characteristics of the wastes. Chemical compositions of some metalliferous waste soils in England are given in Table 14.11. Fine-grained spoils from some old derelict mines can contain levels of heavy metals as high as 13%. In modern tailings, the levels are much lower due to the improvement in mineral processing technology. Associated metals, including cadmium, arsenic, antimony, and silver may be present.[6]

Stabilization of the metalliferous waste spoils can be achieved through physical and chemical stabilization techniques. Vegetation stabilization offers several advantages, through reducing

TABLE 14.10

Examples of Cu Tolerance in Some Major Groups of Organisms

Groups	Organisms
Flowering plants	*Melandrium silvertre*
	Silene inflata
	Mimulus guttatus
Bryophytes	Several species from a copper-contaminated area in Austria
Liverworts	*Cephaloziella* spp.
Algae	*Ectocarpus* spp.
	Scenedesmus acutiformis
	Chlorella fusca
Lichens	Genera: *Acarospora* and
	Lecanora
Fungi	Brown rot fungi (many spp.)
Bacteria	*Thiobacillus ferrooxidans*
	Thiobacillus thiooxidans
	Ferrobacillus ferrooxidans
Annelids (Polychaetes)	*Nereis diversicolor*

Source: Reference 4, with permission

erosion due to the wind and surface metallic salts in percolating water. A fully developed ground vegetative cover can transpire water to a considerable extent. This assists the dewatering and consolidation of inactive tailings lagoons and regulates the rate of infiltration of precipitation water.[7]

The physical characteristics of recent tailings are in contrast with those of old mines. Tailings from new mines contain a high proportion of fines which do not exhibit excessive permeability. The old tailings possess excess permeability which severely limits germination and seedling establishment on coarse-grained spoil. Other physical limitations include surface compaction, lack of normal profile development, and unfavourable structure and texture.

Chemical composition of the soil matrix is important for plant growth. A continuous vegetation cover rarely exists over spoil heaps, mainly due to several factors: residual phytotoxic heavy metals, low levels of plant nutrients, and acidity and salinity.

Despite all limitations, sparse vegetation cover over metalliferous spoils exists. The vegetation consists mostly of grasses which have evolved metal tolerant populations to colonize toxic spoil. There is no threshold value of soil metal content over which only tolerant species can survive, as other factors, including particle size distribution, phosphate status, and the calcium content of the substrate influence toxicity of heavy metals.[6]

The mine spoils are deficient in nitrogen, phosphorous, potassium, and other essential nutrients. One of the most important determinants of nutrients status is the gangue minerals. Quartz (silica) and calcite (crystalline $CaCo_3$) are the commonest associated minerals. Spoil heaps rich in calcite are calcareous with a pH of 6.5–8.0 and are relatively hospitable substrates for plant growth because calcium is an essential plant nutrient. Liming reduces the toxicity of zinc, lead, and copper because calcium forms sparingly soluble metal carbonates and metallic complexes.[7]

Acidic mine spoils are more toxic because of the higher solubility of heavy metals in an acidic environment, and accessory metals, such as manganese and aluminium have increased solubility at low pH(<5). Spoils containing high levels of pyrite (>5%) are inhospitable because acidity problems are more active due to sulphuric acid formed through the weathering of pyrite. This toxicity can be reduced by the repeated application of lime or limestone.

Excessive salinity is a limitation to plant growth. This problem is created where strong acids or alkalis are used during the processing of crude mineral ores. Tailings containing high levels of

TABLE 14.11

Non-Ferrous Metals in Spoil from Abandoned Mines in the United Kingdom

Mining Region	Countries	Number of Sites Surveyed	Principal Base Metals			Associated Metals
			Cu	Pb	Zn	
Southwest England	Devon and Cornwell	16	65–6,140	48–2,070	26–1,090	Ag: <5–350 As: 68–7,080 Cd: 5–145 Sn: 80–6,200
West and northwest England	Salop and Cheshire	12	15–7,260	840–26,000	980–21,000	As: 93–1,970 Ba: <100–14,000 Co: <5–95 Ni: 17–720
North Pennines	North Yorkshire and Durham	8	—	605–13,000	470–28,000	Ba: 400–62,000 Sr: 125–2,530
South Pennines	Derbyshire	17	—	10,800–76,500	12,700–42,000	Ba: <100–104,000 Cd: <2–195 Sr: <50–8,790
Lake District	Cumbria	7	77–3,800	2,070–6,370	4,690–7,370	Ba: <100–3,800
Central Wales	Powys and Dyfed	10	—	1,670–54,000	475–8,000	Ag: 8–100 Cd: 15–445 Cu: 77–560
North Wales	Clwyd and Gwynedd	19	30–5,750	6,400–76,000	11,300–127,000	Ag: 18–95 Ba: <100–7,500 Cu: 10–680 Ni: <5–665
South Scotland	Dumfries and Galloway	6	—	4,730–28,300	1,600–31,400	Ag <1, As:1–50 Ba: 100–3,000
			2–100	2–200	10–300	

Source: Reference 7, with permission

pyrite with dolomite can also cause this problem. Secondary reactions between dolomite and the oxidation products of pyrite produce magnesium sulphate, which is readily soluble. Salinity problems can be alleviated prior to the establishment of vegetation by allowing the percolating rainwater to leach out soluble salt.

Several approaches have been used for the revegetation of metal-contaminated land mine spoils. The differences in these procedures are due to the variation in the physical and chemical properties of mine spoil. The highly toxic waste heaps can be vegetated by metal tolerant varieties. On recent tailings which are less phytotoxic, it may be possible to establish normal commercial plant material directly. In both situations, it is important to isolate the surface vegetation from the metal-contaminated material. The spoil can be covered with a layer of an innocuous amendment.

14.18   METAL TOLERANCE

Metal tolerant grasses have been established on a wide range of metal-contaminated soils. Varieties of *Agrostis tenius* tolerant to copper, lead, and zinc are used on acidic spoils. Similarly, lead-zinc tolerant *Festuca rubra* are commercially available for calcareous spoils. Certain limitations are faced in their use which should be compared with costs incurred in alternative restoration procedures.

Regular applications of fertilizer may be needed to prevent deterioration of the established vegetation cover caused by nitrogen deficiency. Heavy metals produce adverse effects on the process of nutrient cycling through the decomposition of organic plant residues by microorganisms. Phosphate deficiency may also occur on established vegetation unless fertilizers are regularly applied. This effect is induced through the formation of insoluble heavy metal-phosphate complexes.

The major limitation is the heavy metal content of the vegetation, which prevents their agricultural use because levels of toxic metals in the herbage exceed allowable standards for grazing. Revegetated areas become low graded ungrazed grassland. In the absence of livestock, the breakdown of organic matter and nutrient recycling is restricted.

Metal tolerance is a useful characteristic in certain plant species but rarely does any population show tolerance to more than two metals. Mine spoils which contain elevated levels of several metals may not be successfully vegetated by those tolerant varieties.

However, tolerant plant populations provide the most economical and practical means of vegetation of abandoned spoil heaps. If agricultural seeding methods are impractical, hydraulic seeding can be used.[8,9]

14.19   ORGANIC AND TOPSOIL AMENDMENTS

A top layer, consisting of about 100–500 mm thick non-toxic material, such as topsoil, sewage sludge, domestic refuse, or peat is applied to the surface of the toxic mineral waste. Experience with non-ferrous smelter wastes has indicated that roots can penetrate through the boundary between the surface layer and the underlying toxic material. Treatment of toxic spoil with sewage sludge or screened pulverized domestic refuge has also proved unsatisfactory because root growth is restricted to surface amendment. The benefits obtained from any of those techniques are a temporary visual improvement and stabilization of the reinstated area. The vegetation gradually declines because of the decline in nutrient levels and a gradual lowering of the organic matter. Regular maintenance is required. The annual application of inorganic fertilizer and 20–100 mm of organic matter might be beneficial, but costs are high.

Organic matter lowers the phytotoxicity of heavy metals, but metals are re-mobilized when organic matter decomposes. A sufficiently thick layer may reduce this problem, but at a high cost. Furthermore, regardless of the depth of the surface layer, upward movement of soluble metal salts may cause toxicity, yet water balance between precipitation evaporation and drainage is an important factor.

The surface amendment used in restoration schemes should have high nutrient status, which will encourage the development of productive, deep-rooting vegetation. If the restoration is for recreational purposes, the development of continuous ground cover with a low maintenance requirement is feasible, and naturally infertile amendments are more suitable for surface treatment with seed mixtures containing less productive grass species.

14.20 INERT MINERAL WASTES

Experiments have been conducted using mineral wastes produced from mining and quarrying, such as limestone chipping, slate quarry waste, and coal mine shales for surface treatment of metal mine wastes. This type of materials has several important advantages. They are available at low cost and in the vicinity of waste piles. Their course, granular in nature deters upward movement of metal salts by capillary action. A physical discontinuity is introduced at the amendment. The rooting characteristics of developing vegetation develop according to the species sown and the quantity and type of fertilizer applied. The required application depth of the amendment material may be higher than other commercial materials. Advantages include low cost and widespread local availability.

If the objective is to develop an ecosystem which would not require regular application of fertilizer, legum, particularly *Trifolium repens*, are essential. A balanced combination of grasses and legum in the vegetation can be achieved with the appropriate combination of inorganic fertilizers during the early stages of vegetation development. After vegetation develops equilibrium, nitrogen fixation and nutrient cycling maintain substrate fertility.[10]

14.21 DIRECT SEEDING

Because of advances in modern mineral processing technology, the recently produced mineral tailings are less toxic than the old mine spoils. The toxicity level of the individual metals is usually less than 0.5%. Revegetation can be achieved without the use of metal tolerant varieties or mineral-based surface amendments. Normal agricultural grasses and legumes can be sown directly with the appropriate nutrient application.[11]

Surface treatment with suitable organic material is often advantageous. The organic layer serves as a source of slow release of nutrients during the initial stages of vegetation development. Regular legum application to supply nitrogen requirements is necessary.

Seed mixtures and organic materials may be applied with modifications of the hydroseeding method without the use of heavy equipment, as they cannot be used over unstable tailings with low load-bearing capacity. In some situations, seeding with agricultural equipment may be feasible.

14.21.1 TREES AND SHRUBS

In some situations, trees and shrubs instead of grasses and legumes can be desirable. Experimental results indicate that the use of commercial conifer species like *Pinus sylvestris* and *Pseudotsuge douglasil*, planted in lead-zinc-contaminated soils containing more than 1% heavy metals, is often unsuccessful. The rate of development is totally unsatisfactory for commercial timber production. However, a suitable combination of species, planting treatments, and management techniques can reduce losses and improve growth rates.

The planting stock should have a well-developed root system before transplanting. Poorly developed roots, for their reduced resistance cause establishment failure. Two-or three-year old stocks are most suitable for transplanting. Larger trees are difficult to establish.

Without surface amendments, metalliferous waste is unsuitable as a substrate even with fertilizer application. Topsoil, subsoil, peat-sand, domestic refuse, and sewage sludge are suitable. Trees and shrubs must be planted in excavated pits or trenches filled with suitable rooting materials. The selection of species is made after considering chemical composition and physical properties of the planting medium.

Root penetration in highly toxic material may be restricted. Only fibrous roots close to the surface develop. Less toxic spoils allow wider rooting.

For very toxic spoils, a thick (1–2 m) surface amendment is essential. The roots do not penetrate beneath the top layer so as to ensure growth in the long term, especially if shallow rooting trees are planted.

14.22   WASTE ROCK

Surface mining of low-grade orebodies produces large quantities of waste which contain low levels of metals. Vegetation can be established. If the rock is hard and the waste heaps contain voids, some surface amendment is necessary. Often, metal levels are higher in processed tailings, metals tolerant species can be useful. The wastes produced widely vary in their characteristics. The relatively soft and innocuous materials should be identified and used as a covering for hard, toxic wastes.

14.23   MAINTENANCE AND MANAGEMENT

Vegetation cover will require maintenance. Costs can sometimes be recovered through livestock grazing, crop production or foresting. Toxicity of heavy metals to livestock is well established. This implies that continuous grazing by livestock is often not possible. Short-term rotational grazing may be feasible if the vegetation is more or less isolated from the toxic substrate or if the metal level is low on modern mill tailings. Regular sampling and analysis of the vegetation are essential.

Where restoration is for amenity or recreational purposes, management will be required during the initial stages even if self-perpetuating, maintenance-free vegetation cover is the ultimate objective. Maintenance would be necessary as a temporary measure (for 2–3 years) provided that toxicity of the soil is effectively countered and a self-supporting ecosystem with its own nutrients is established.

Metal tolerant vegetation can withstand substrate toxicity, but initial maintenance for several years is needed because of reduced mineral cycling and poor retention of applied nutrients.

For maintaining substrate fertility, an annual application of about 400–800 kg balanced inorganic NPK fertilizer is likely to be required. Organic surface amendment usually needs less fertilizer. If commercial, non-tolerant species are used, organic or inert mineral amendments are essential. Clay minerals and organic matter, if present in the amendment, because of their cation exchange capabilities, contribute to the retention properties and long-term nutrient supply. Inert mineral amendments when used as a thick surface layer can support a wide range of species. Seed mixtures should be adopted according to site-specific problems, e.g., drought, exposure, trampling pressures. The choice of appropriate fertilizer application is important.

14.24   REVEGETATION OF TACONITE TAILINGS IN MINNESOTA

The current taconite mining industry in Minnesota produces about 36Mt./a of taconite and employs about 4,000 employees. At the peak of the operation in the 1980s, eight taconite mines were in operation on the Mesabi range. Mined land reclamation rules adopted in Minnesota in 1980 require that all mined land be left in a stable and safe condition. The vegetation standards require that 90% vegetative cover be present after three growing seasons and require that the vegetation be self-sustaining after 10 years. Establishing vegetation on the soil removed during mining and glacial till overburden portion of the pit walls was not difficult. However, establishing 90% vegetation cover on the tailings required special methods through research.[12]

The typical standard reclamation approach uses 448 kg/hm (400 lbs/acre) of diammonium phosphate, 56 kg/hm (50 lbs/acre) of a cool-season grass mix and 4.6 t/hm(2 st/acre) of hay mulch.

This method was successful on the medium- to fine-grained tailings, it did not prove to be effective on the coarse tailing material used to build the retaining dams. After 5 years, the vegetative cover rarely exceeded 60%. Considerable research demonstrated that applying organic amendments improved vegetative success on coarse tailings.[13]

14.24.1 SITE DESCRIPTION

EVTAC taconite processing plant and its tailing basin cover about 300 hm (750 acres) and is made from coarse tailings materials from the processing of the ore and truck hauled to construct the dams. The tailings dams are about 51-m (167 ft) high and contain three 17-m (56ft) high lifts, a final slope of 3:1, and an 8-m bench (26 ft) bench located at the top of each lift. Coarse tailings are low in nitrogen, phosphorous, water holding capacity and organic matter. The coarse tailings are alkaline with a pH value of 8 and contained about 33% coarse sand, 40% medium sand, 22% fine sand and 5% silt and clay.

The fine tailings were also low in nutrients and organic matter, and range in size from fine sand to clay. Fifty per cent of the material is silt and clay.

14.25 APPROACH

14.25.1 COARSE TAILINGS

In the early 1990s, pilot studies were conducted at EVTAC to examine the feasibility of using organic amendments for growing vegetation on coarse tailings. Tests indicated that the use of peat, yard waste compost, municipal solid waste component, and paper mill residue increased the percent cover than in plots where standard mined land reclamation practices were adopted. The per cent cover generally increased with increasing amounts of organic material, 44.8 t/hm (20 st/acre) being selected as the cost-effective rate.

In 1996, tests were conducted to investigate the use of inorganic materials to improve vegetation on coarse tailings. The other four plots contained 50 mm (2 in) of fine tailings from two different sources, 50 mm (2 inch) of silty material, and 50 mm (2 inch) of peat. The materials were applied to the surface and mixed into underlying tailings.

In 1997, six 2-hm (5 acre) plots were established on the tailing's dams. These were treated with municipal solid waste (MSW) compost at 44.8 t/hm (20 st/acre), paper mill residue at 44.8 t/hm (6 st/acre), Class B material from the nearby city of Virginia, MN and 44.8 t/hm (10 st/acre) biosolids. Standard mined land reclamation was used on one plot as a control. Producer of each material determined the application rate and by the results of previous research. The biosolids were applied at a rate that would provide the same amount of nitrogen as the inorganic fertilizer in the standard reclamation practice—100 kg/hm (89 lbs/acre). Because the carbon to nitrogen ratio in the paper residue far exceeded the desirable ratio of 30:1, extra nitrogen was added to these plots.

In the spring of 2000, EVTAC covered about of 18 hm (45 acres) of the coarse tailings dam with 100 to 150 mm (4 to 6 in.) of peat. EVTAC then applied seed, fertilizer, and mulch. In order to improve vegetation, biosolids from the Western Lake Superior Sanitary District were applied in the fall of 2000 as a top dressing on the peat areas. Application rates ranged from 100 kg/hm (89 lb/acre) of nitrogen to 448 kg/hm (400 lb/acre) nitrogen.

14.25.2 FINE TAILINGS

EVTAC was interested in developing lands with a potentially higher economic value the grass meadows. Timber production was a major industry in northern Minnesota, with most of the harvest being used for paper and wood products. EVTAC found that it would be desirable to create tree plantations on reclaimed mined lands for economic benefits. The hybrid poplar has been used extensively in throughout northern Minnesota. Hybrid poplar can grow rapidly and, on natural forest soils, can be harvested in about 12 years, producing about 89 to 109 cords/hm (36 to 44 cords/acre). EVTAC investigated the potential of creating a hybrid poplar plantation on the fine tailings. Test plots were established in 2001 to examine the survival and growth of hybrid poplars in the fine tailings with and without organic amendments (paper mill residue and biosolids).

14.25.3 Wetland Creation

EVTAC was originally built on wetlands. The US Army Corps of Engineers required that EVTAC conduct a study on the feasibility of creating wetlands on the tailings basins when they were permanently closed. In 1992, the State of Minnesota passed the Wetland Conservation Act, the idea of reclaiming tailings basin as wetlands received attention. The EVTAC task force-initiated studies to:

- Examine wetlands that had developed naturally, "incidental wetlands"
- Examine vegetative growth with various soil covers and seed mixes
- Study hydrology in the tailings basin and its effect on wetland establishment

In 1998, the State of Minnesota initiated a study to examine the use of dredged material from Lake Superior as a substrate for creation of wetlands. Two small areas of about 0.1 hm (0.25 acre) were established to examine the suitability of dredged material as a substrate for wetland vegetation and to investigate the impact on water quality. After 2 years, the per cent cover on the dredged material ranged from 80% to more than 90%. No impact on the water was measured.

The success of this study led to a large-scale demonstration project with the participation of the US Environmental Protection Agency. In the project 2 hm (5 acres) of land would be covered with a 100 mm (4 in) of dredge material and a wetland seed mix would be applied. In addition, EVTAC would cover about 400 hm (100 acres) with 50 mm (2 in) of the wetland soil obtained from road building projects.

14.26 RESULTS

14.26.1 Coarse Tailings

In 1996 plot tests, the addition of fine-grained inorganic material was not successful in producing vegetation that met mine land reclamation standards. After 3 years, the per cent cover ranged from 29.8% to 47.1%. The application of fine-grained material changed the infiltration characteristics of the coarse tailing slopes, and major erosion gullies formed during the first year following applications.

The peat plot had the highest per cent cover (71.9%) which did not meet the reclamation standard, but the slopes were stable, and erosion was minimal.

In 1997 test plots, after 3 years, the per cent cover on all amended plots exceeded the cover produced by the standard mine land reclamation practice. The cover ranged from 34% on the standard mine land reclamation plots to 86% on the MSW Table 14.12.

Although none of the covers met the legal requirements of 90% coverage, the cover was judged to be acceptable. Detailed cost information collected during the study indicated that since no inorganic fertilizer was required on the biosolids plot, the company saved $125/hm ($50/acre).

2000 reclamation: Reclamation was performed on the slopes and on the coarse tailings' road. In the first year, the slopes which were covered with peat and top-dressed with biosolids were covered with a dense cover of grasses and legumes.

The first-year cover on the coarse tailings' road grew a thick cover annual rye cover crop and the annual weed, Russian Thistle. The cover ranged from 14% on the standard mine land reclamation

TABLE 14.12

Per cent Cover, 1997 Coarse Tailings Demonstration Plots

Treatment	% cover	% cover	
Third Year	without Biosolids	with Biosolids	Fourth Year
Standard mine land Reclamation	34	42.1	51.0
Blandin paper mill Residue	57	63.9	79.8
Consolidated paper Mill residue	53	74.8	58.1
Biosolids	68	71.5	90.8
Blandin paper mill Residue and biosolids	83	89.9	97.4
Municipal solid waste Compost	86	92.8	95.1

control to 45% on the plot with 448 kg N/hm (400 lb N/acre). No nitrogen was added to the inorganic biosolids plots, but some plots received phosphorous.

14.26.2 FINE TAILINGS

Tree plots: Visual observation indicated that the survival was good, and the first-year growth ranged from 0.3 to 1.2 m (1 to 4 feet). Wetland creation: Dredge and peat material would be spread in the fall of 2001. About of 100 mm (4 in) of dredge material will be placed on 2 hm (5 acres), while about 50 mm (2 in) of wetland soil will be placed on 40 hm (100 acres)

BIBLIOGRAPHY

1. Raskin, I., PBA Nanda Kumar, Slavik Dushenkar and David E. Salt, Bioconcentration of heavy metals of plants, *Curr. Opin. Biotechnol. 5*:285–290 (1994).
2. Morrey, David R., Using metal tolerant plants to reclaim mining wastes, *Mining Engineering 47*(3):247–249 (1995).
3. Antonovics, J., A. D. Bradshaw and R. G. Turner, Heavy metal tolerance in plants, in *Advances in Ecological Research* (London, UK: Academic Press, 1971), p. 73.
4. Morrey, D. R. and M. S. Johnson, A Comparison of metal tolerant and non-tolerant varieties of *Fastuca rubra* for use in the direct hydraulic seeding of metalliferous fluorspar mine tailings, *J. Environ. Manage. 19*:99–105 (1984).
5. Woolhouse, H. W., Toxicity and tolerance in the responses of plants to metals, in *Physiological Plant Ecology* III (Berlin: Springer, 1983), pp. 245–287.
6. Smith, R. A and A. D. Bradshaw, The use of metal tolerant plant populations for the reclamation of metalliferous wastes, *J. Appl. Ecol. 16*:595–612 (1979).
7. Johnson, M. S. and A. D. Bradshaw, Prevention of heavy metal pollution from mine wastes by vegetative stabilization, *Trans. Inst. Min. Metall. 36*(A):47–55 (1977).
8. Antonovics, J., Metal tolerance in plants: Perfecting evolutionary paradigm, in *Proceedings of the International Conference on Heavy Metals in the Environment*, Vol. 2 (Toronto, Canada, 1975), pp. 169–186.
9. Antonovics, J., A. D. Bradshaw and R.G. Turner, Heavy metal tolerance in plants, in *Advances in Ecological Research* (London, UK: Academic Press, 1971), pp. 1–85.

10. Morrey D. R., A. J. Baker and J. A. Cooke, Floristic variation in plant communities on metalliferous mining residues in northern and southern Pennines, England, *Environ. Geochem. Health* *10*(1):11–20 (1988).

11. Morrey, D. R., Reclamation of mining wastes using stress adopted varieties of plant species, in *Integrated Mining and Land Reclamation Planning II* (Reno, Nevada: Morey, 1993).

12. Eger, Paul et al., Thinking outside the box: New ways to close old tailings basins, *Min. Eng. 56*(1): 45–51 (January, 2004).

13. Norland, Michael R. et al., Revegetation of coarse taconite iron ore tailing using municipal solid waste compost, *J. Hazard. Mater. 41*:123–134 (1995).

15 Permit for Mining

15.1 PERMIT FOR MINING

A permit is a permission from the U.S. government and the state government and local agencies for conducting any part or the whole of a mining operation. The U.S. government enacted the Surface Mining and Reclamation Act (SMCRA) in 1977. This Act established guidelines for the regulation of surface mining and the reclamation of mine sites. This Act is enforced under the administration of the Office of Surface and Mining, Reclamation and Enforcement in the Department of the Interior. This law establishes that minimum requirements should be uniform for all surface coal mining on Federal and State lands including exploration activities and the surface effects of underground mining. Mine operators are required to minimize disturbances and the adverse impacts on fish, wildlife, and related environmental elements and maintain the integrity of such resources when possible. Restoration of land and water resources is a priority in reclamation planning.

There are alternative ways that the reclaimed land can be used. One of the main goals of the main goals of the SMCRA is to restore the land as close to its original or better and use it for the good environment. Some popular ways this reclaimed land can be used includes range land, prime farmland, wildlife refuges, wetland, recreation, and military training outposts.

This includes getting the mining plan approved by the federal government under the requirements of the National Environmental Policy Act, or NEPA (http://ceq.doe.gov/).

15.2 PERMITTING PROCESS IN WEST VIRGINIA

In West Virginia, the main uses for reclaimed mining sites are recreational, farming, and military training. For example one such farming use is the growing of lavender. There are many types of lavender, but they all thrive in the dry rocky soil that surface mining leaves behind. For example, the Green River Mining Project in West Virginia took full advantage of this and planned for many lavender fields on reclaimed fields in West Virginia.

State of West Virginia enacted its surface coal mining legislation as follows:

Legislative findings and purpose: Jurisdiction vested in Division of Environmental Protection:

- The Legislature finds that it is essential to the economic and social well-being of the citizens of the State of West Virginia to strike a careful balance between protection of the environment and the economical mining of coal needed to meet the energy requirement.
- Further, the Legislature finds that there is great diversity in terrain, climate, biological, chemicals West Virginia in particular needs an environmentally sound and economically healthy mining industry, and therefore it may be necessary for the Secretary to promulgate rules which vary from federal regulations as provided in the federal Surface Mining Control and Reclamation Act of 1977, as amended, Public Law, "Public Law 95-87".
- Further, the Legislature finds that unregulated surface coal mining operations may result in disturbance of surface and underground areas that burden and adversely affect commerce, public welfare and safety by destroying or diminishing the utility of land for commercial, industrial, residential, agriculture, and forestry purposes, by causing erosion and landslides, by contributing to floods, by polluting the water and river.

The State of West Virginia has established detailed information and requirements for obtaining permits for mining operations: State of West Virginia code chapter 22 Environmental Resources: Articles 1, 2, 3 and Surface Coal Mining and Reclamation Act

15.3   THE MINE PERMITTING PROCESS IN MINNESOTA

The permitting process for metal mining is similar to coal mining, with some differences in different states. For example, in the state of Minnesota, the main metal industry is based on mining of copper, nickel, PGE deposits of the Duluth complex. Due to advances in hydrometallurgical processes and increase in PGE metal prices. In Minnesota, the permitting process for a non-ferrous metal is complicated and involves dealing with numerous different federal, state, and local agencies. This process has not been carried through completion for the establishment of either an open-pit or a underground mine.

The first step in the mine permitting process in Minnesota is the preparation of an Environmental Impact Statement (EIS). Because an EIS is mandatory, it must be preceded by the preparation of a scoping Environmental Assessment Worksheet (EAW). Many of the permits that are mining related require this process to be completed. The scoping EAW is designed to identify potentially significant issues including possible environmental, sociological, economic, health risk impacts that will be associated with a proposed mine and will need further study in an EIS.

The Minnesota Department of Natural Resources (DNR) is the responsible government unit for both the scoping EAW and EIS.

15.4   OREGON MINERAL LAND REGULATION AND RECLAMATION

For surface mining three types of surface approvals are required, including Operating Permits, Exploration Permits, Exploration Permits, and Exclusion Certificates. The type of permit/certificates required for a specific operation depends on quantity, acreage, and/or planned activities.

An Operating Permit is required for a material activity that exceeds one acre of disturbance in any 12-month period and/or 5,000 cubic yards of excavation in any 12-month period. When the total disturbance exceeds 5 acres, an Operating Permit is required unless the activity is exempt. An annual operating permit renewal and reporting are required until mining and reclamation requirements are complete.

For reviewing the Operating Permit application, a preapproval meeting is recommended. The review requires proposed surface mine plans, proof-of-land ownership, permit boundary surface map, site plan map, reclamation plan map, etc. Additional information may be required on sites on flood plains or steep slopes. For evaluating slope stability or proximity to water bodies, the permit is reviewed every year.

15.5   COLORADO MINE PERMITTING

Mine permits are issued by the state Division of Reclamation, Mining, and Safety, Department of Natural Resources. The Minerals Program issues four types of reclamation permits based on the type of operations and characterization of the mineral being mined. Reclamation permits issued under the Hard Rock Act in the minerals Program are issued for the life of the operation. There are provisions for operators to enter into two 5-year periods of temporary cessations, but, however, by the end of the second 5-year period, operators must reinitiate operations or begin reclamation. Once an operator has completed a phase of mining, the operator has 5 years to complete reclamation for that phase. Operators may make revisions to their approved permits at any time through the appropriate revision process.

15.6 COMMONWEALTH OF VIRGINIA

The Division of Mineral Mining (DMM) issues mining permits and licenses for all commercial mineral mining operations in the Commonwealth. The permit application includes suitable operating plans to ensure that mining is conducted in an environmentally sound manner.

Before a permit can be issued, applicants must provide suitable operations drainage, drainage, and reclamation plans, which will be reviewed by the DMM. A reclamation performance bond must be provided to ensure that those are available to contract final reclamation of the mine. Through permitting and bonding of mineral mines, reclamation is essentially guaranteed.

BIBLIOGRAPHY

1. Environment Australia (2002) "Overview of Best Practice Environmental Management in Mining." https://books.google.ie/books/about/Overview_of_Best_Practice_Environmental.html?id=dW5-cgAACAAJ
2. International Institute for Environment and Development (2002) "Breaking New Ground: Mining, Minerals and Sustainable Development: Chapter 9: *Local Communities and Mines. Breaking New Grounds.*" http://www.iied.org/pubs/pdfs/G00901.pdf
3. Amar Nath Singh, Medini Srinivas, and Bikash Narayan Naik, Forecasting the impact of surface mining on surrounding using cloud computing, http://pubs.sciepub.com/jcsa/3/6/1/index.html (2015).
4. World Bank Group, Oil, Gas and Mining Policy Division, Guidance notes for the implementation of financial surety for mine closure, https://www.idl.idaho.gov/wp-content/uploads/sites/2/rulemaking/20.03.02-2019/research/WorldBank_2018.pdf (2008).

16 Monitoring of Mine Environment

16.1 INTRODUCTION

Every Environment Impact Statement and mine permit must include a requirement for mine environment before, during, and after the mining process. The government will monitor the environmental parameters, as stated in the permit. The monitoring program should be a part of the company's overall environmental management system. The monitoring program should be developed using a set of objective commitments of the existing conditions. Monitoring programmes begin with a baseline sampling program performed to characterize the predevelopment environmental conditions. Environmental issues addressed in and management by the plan generally to issues such as land clearing and top soil, water, waste rock, tailings, hazardous wastes, biology (species, health, biodiversity), dust, noise, and transportation.

The Environmental Monitoring Plan (EMP) needs to provide details about where, when, what, and how often a mining company will monitor the quality of water, air, and soil in the vicinity of the mining project, and the quantity of pollutants in effluents and emissions being released. The EMP must specify how this information will be provided to government decision makers and to the general public so as to ascertain that the mining company is complying all of its promises, and environmental standards and regulations.

16.2 WATER QUALITY MONITORING

Monitoring changes in water quality within a mine site is very important for the protection of water quality. A comprehensive water quality monitoring program is essential so as to ensure that the mining company is fulfilling promises in its Environmental Monitoring Plan. According to the Department of Minerals and Energy, Western Australia:

> Monitoring of mine site water quality is an essential part of the environmental management of the mining operation. It enables water quality performance to be assessed. Undesirable impacts can thus be detected at an early stage and remedied.

Surface Water Quality monitoring should be conducted for the following:

1. Discharge or seepage existing mine site sources
2. Discharge or seepage exiting the property boundary
3. On-site water bodies and water bodies downstream from the site
4. Background reference sites

According to the IFC/World Bank Group: "Monitoring frequency should be sufficient to provide representative data for the parameter being monitored."

Groundwater quality monitoring is one of the most important aspects of protecting groundwater resources. This is best achieved by constructing a network of bores. Assessing groundwater quality before an operation commences can set the environmental management needs of a project. Monitoring undertaken during the Environmental Impact Assessment (EIA) process can also establish the baseline data by which the important environmental performance of an operation can be assessed. Undesirable environmental impacts can thus be detected at an early stage and remedied effectively.

Bores are commonly required upstream and downstream, in the direction of groundwater flow to monitor changes in water level and quality across a site and to monitor the performance and stability of tailings facilities. In hard rock areas, bores must be located within geological features that are most likely to transmit groundwater, such as along fault lines, within weathered zones with coarse granular soil or in alluvial sand.

Monitoring bores should be sampled at least once in three months for likely key pollution indicators associated with the project. Surface water chemical monitoring should be conducted for the following.

1. Discharge or seepage exiting on-site sources.
2. Discharge or seepage exiting the property boundary
3. On-site water bodies and water bodies downstream from the sites

Water quality monitoring parameters for mining projects typically include: pH, conductivity, total suspended solids, total dissolved solids, alkalinity, acidity, hardness, cyanide ammonium, sulphates, aluminium arsenic, cadmium, calcium, copper, iron lead, mercury, molybdenum, nickel, and zinc.

A mining company can demonstrate that a particular parameter is not relevant to the mining project. Otherwise, the environmental management plan should require all the above parameters.

16.3   AIR QUALITY MONITORING

A mining operation must have an air quality plan to record the emissions of the most significant air pollutants. The selection and location of monitoring equipment should comply with technical assessment and specifications. Weather conditions, topography, residential areas, and wildlife habitat should be considered in order to determine the best location of air quality monitoring equipment.

Key issues include:

Air quality monitoring plan as detailed in the EIA, equipment and methods to be used, criteria that were to select the location of the monitoring points, frequency of data collections, whether an independent agency will assess the calibration and implementation of the air quality monitoring plan, whether the results would be available to the public.

16.4   VEGETATION AND SOIL QUALITY MONITORING

Key issues include:

The methods would be used to quantify the excavated or disturbed land, erosion and disturbance of surface soils, etc.

16.5   MONITORING IMPACTS ON WILDLIFE AND HABITAT

Key issues include:

- How are primary effects on fauna, flora, and habitats going to be monitored?
- Is any independent agency going to assess the potential effects, including cumulative effects on terrestrial and aquatic wildlife and habitat?

What methods would be used to report and monitoring data?

Monitoring of key species:

Large-scale mining operation activities that could significantly affect the natural functions of terrestrial and aquatic ecosystems. Ideally, an environment monitoring plan for a large-scale mining project would include periodic assessments of impacts on key wildlife species, with support from an independent group of qualified professionals. The baseline section of the EIA should identify wildlife species listed by national or local authorities and/or endemic species.

Key issues include:

- Evaluation of habitat loss, key species as identified in the baseline section, surveys to assess the reduction or alteration of key specific populations, overview of changes in the ecosystem and potential exposure of key species to hazardous pollutants.

Monitoring habitat loss:

An environmental plan must include plans to perform regular surveys to assess the state of the habitat. The key issues include:

- Habitat types should have been adequately identified and mapped previously.
- The persons responsible for habitat monitoring. This activity requires qualified independent experts.
- Surveys must determine habitat density changes in several locations.
- Assessment of the current status of key species based on field work (count and observe species, densities) population.

16.6   MONITORING IMPACTS ON AFFECTED COMMUNITIES

Mineral development can cause serious disruption in local communities related to benefits and costs that may be unevenly shared. The economic gains of a national or a foreign mining corporation do not necessarily contribute to local development. However, environmental degradation affects the livelihood of local people.

Community health:

Key issues include:

- Incidence of pollution-related disease and deaths
- Assessment of water quality and availability for domestic use, agriculture, and other activities
- Results of air quality assessment in populated areas.
- Records of regular or episodes of high air pollution
- Incidence of alcoholism, prostitution, and sexually transmitted diseases related to the presence of mining workers in the area.

16.7   MONITORING OF THREATS TO PUBLIC SAFETY

If a mining project chooses to dispose of its tailings in a wet tailings' impoundment, then the failure of the impoundment would constitute one of the most serious threats to public safety. For example, in 1966 in England, the Aberfan disaster where a collapse of a coal mine spoil tip claimed the lives of 144 people including 116 school children, mostly between the ages of 7 and 10. The failure of an impoundment would constitute one of the most serious threats to public safety. For this kind of risk, the Environmental Monitoring Plan should include details about how the operation and structural integrity of the tailing's impoundment would be monitored to promptly detect possible structural problems and prevent potential disasters.

BIBLIOGRAPHY

1. Meriam A. Bravante and William N. Holden, Going through the motions: The environmental impact assessment of nonferrous metals mining projects in the Philippines, https://www.tandfonline.com/doi/abs/10.1080/09512740903128034 (2009).
2. United States Environmental Protection Agency 40 CFR Part 440, Effluent limitation guidelines for metallic minerals mining, https://www.elaw.org/content/us-epa-effluent-limitation-guidelines-metallic-mineral-mining-40-cfr-part-440.

17 Evaluation of Environmental Impact of Open-Pit Mining, Iran Case study

17.1 INTRODUCTION

The Folchi method of evaluating the environmental impact of open-pit mining was first applied in Italy. It is a numerical method for evaluating impacts of open pits. This method utilizes the following seven steps:

- Evaluation of pre-existing environmental conditions, including geology, geomechanics, hydrology, weather economy, etc.
- Identification of impacting factors, which could alter the pre-existing environmental conditions in the mine site
- Determination of possible ranges of variation in the existing conditions during mine life identifying the important impacting factors
- Singling out the environmental components whose pre-existing conditions could be altered as a result of mining
- Correlating each impacting factor and each environmental component
- Estimating the specific magnitude for each impacting factor, using previously defined ranges
- Calculation of the weighted sum of the environmental impact of each environmental impact on each environmental component

In this method, some parameters, such as general health and safety, social relationship, weather and climatic conditions, vegetations, and animals are defined first. Then consequences of effective mining indexes on each of the environmental component are determined, by applying a rating system for parameters, based on various concerning scenarios. The sum of all the ratings of effective parameters determines the overall effect on each of the environmental indexes. According to this method, impacting parameters are the following:

- Alteration of the area's potential resources
- Exposition, visibility of the pit
- Interference with surface water
- Interference with underground water
- Increase in vehicular traffic
- Atmospheric release of gas and dust
- Fly rock
- Noise
- Ground vibration
- Employment of a local workforce

The possible scenarios for each impacting factor are then considered, and a magnitude is assigned to each of them.

Table 17.1 shows various scenarios and their related magnitude for each impacting factor.

The Folchi method was applied to an important copper open-pit mine in Iran, namely Sarcheshmeh.

This mine is located 50 km south of the city of Rafsanjan in Kerman province and is the largest copper mine in Iran. The geology of Sarcheshmeh porphyry copper mine is very complicated, and various types of rocks can be found. The main minerals are chalcopyrite, chalcocite, covellite, bornite, and molybdenite. Other minerals include molybdenum, gold, and silver. The oxide zone of the deposit consists mainly of cuprite, tenorite, malachite, and azurite. Pyrite, being mineral, which causes acidity in mine sewage. The proven reserve of the deposit is approximately 826Mt with an average grade of 0.7%. The mine worked by open pit. The distance of crusher to mine is 3 km. The annual capacity of the mine mill is 51,000 tons of concentrate with an average grade of 30% and a recovery of 65%. The environmental data related to the Sarcheshmeh mine was evaluated using the Folchi method and magnitude ranges given in Table 15A. Each impacting factor of the proposed mining activities was assessed (Table 17.2). Final scoring for each environmental component can be

TABLE 17.1
Ranges of Magnitude for Impacting Factors (Monjezi, M et. al)

Acting factors	Scenario	Magnitude
Alteration of area's potential resources	Parks, protected areas	8–10
	Urban area	6–8
	Agricultural area, wood	3–6
	Industrial area	1–3
Exposition, visibility of the pit	Can be seen from inhabited areas	6–10
	Can be seen from main roads	2–6
	Not visible	1–2
Interference with above groundwater	Interference with lakes and rivers	6–10
	Interferences with non-relevant water system	3–6
	No interference	1–3
Interference with under groundwater	Water table superficial and permeable grounds	5–10
	Water table deep and permeable grounds	2–5
	Water table deep and un-permeable grounds	1–2
Increase in vehicular traffic	Increase of 200%	6–10
	Increase of 100%	3–6
	No interference	1–3
Atmospheric release of gas and dust	Free emissions in the atmosphere	7–10
	Emission around the given reference values	2–7
	Emission well below the given reference values	1–2
Fly rock	No blast design and no clearance procedures	9–10
	Blast design and no clearance procedures	4–9
	Blast design and clearance procedures	1–4
	Peak air overpressure at 1 km distance	
Noise	<141 db	8–10
	<131 db	4–8
	<121 db	1–4
Ground vibration	Cosmetic damage, above threshold	7–10
	Tolerability of threshold	3–7
	Values under tolerability threshold	1–3
Employment of local workforce	Job opportunities	7–10
	High	3–6
	Medium	1–2
	Low	

TABLE 17.2

Correlation Matrix with Values of the Weighted Influence of Each Impacting Factor on Each Environmental component (Monjezi, M et. al)

Impacting factors	Human Health and Safety	Social Relationship	Water Quality	Air Quality	Use of Territory	Flora and Fauna	Above Ground	Underground	Landscape	Noise	Economy
						Environmental Components					
Alteration of area's potential resources	Med 0.80	Min 0.77	Nil 0	Nil 0	Max 5.71	Min 0.63	Nil 0	Nil 0	Max 2.86	Nil 0	Nil 0
Exposition, visibility of the pit	Nil 0	Min 0.77	Nil 0	Nil 0	Med 2.86	Nil 0	Nil 0	Nil 0	Max 2.86	Min 2.00	Nil 0
Interference with above groundwater	Max 1.60	Nil 0	Max 4.44	Nil 0	Nil 0	Max 2.50	Med 6.67	Nil 0	Max 2.86	Nil 0	Nil 0
Interference with under-groundwater	Min 0.40	Nil 0	Max 4.44	Nil 0	Nil 0	Nil 0	Nil 0	Med 6.67	Nil 0	Nil 0	Ni 01
Increase in vehicular traffic	Max 1.60	Max 3.08	Nil 0	Nil 0	Min 1.43	Max 2.50	Nil 0	Ni 01	Min 0.71	Nil 0	Nil 0
Atmospheric release of gas and dust	Max 1.60	Min 0.77	Min 1.11	Max 10.00	Nil 0	Max 2.50	Min 3.33	Nil 0	Min 0.71	Nil 0	Ni 01
Fly rock	Max 1.60	Nil 0	Nil 0	Nil 0	Nil 0	Med 1.25	Nil 0	Nil 0	Nil 0	Nil 0	Nil 0
Noise	Med 0.80	Max 3.08	Nil 0	Nil 0	Nil 0	Min 0.63	Nil 0	Nil 0	Nil 0	Max 8.0	Nil 0
Ground vibration	Max 1.60	Med 1.54	Nil 0	Nil 0	Nil 0	Nil 0	Nil 0	Min 3.33	Nil 0	Nil 0	Nil 0
Employment of local workforce	Nil 0	Nil 0	Nil 0	Nil 0	Nil 0	Nil 0	Nil 0	Nil 0	Nil 0	Nil 0	Max 10.00
Total	10	10	10	10	10	10	10	10	10	10	10

TABLE 17.3

Rating of Environmental Parameters in the Case Study of Mines (Monjezi, M et. al)

Impacting Factors	Sarcheshmeh
Alteration of area's potential resources	3
Exposition, visibility of the pit	3
Interference with the above groundwater	7
Interference with the under groundwater	7
Increase in vehicular traffic	9
Atmospheric release of gas and dust	10
Fly rock	5
Noise	7
Ground vibration	7
Employment of local work force	7

acquired by multiplying Table 17.2 into Table 15C. The overall effect on each environmental component is calculated by summing the weighted magnitudes of all the impacting factors (Table 17.3).

The Folchi method indicates that specific aspects of environmental impact can be quantified. The most significant impacts in the Sarcheshmeh copper mine are air quality, above ground, and flora and fauna with score values of 100, 80, and 77.6, respectively. The sum of scores for the environmental components can be calculated and then evaluated. The sum of components of scores for the Sarcheshmeh copper mine is 766. The evaluation indicates that some remedial measures must be taken for the affected environmental components, such as air quality and water condition, which are essential for the living creatures. A plant for converting sulphur dioxide to sulphuric acid has been constructed for the Sarcheshmeh mine which has been seriously affecting air quality. Additionally, the Sarcheshmeh mine has seriously affected the water quality in the area as indicated in the Table 17.4 and shows high ground vibration for the Sarcheshmeh mine for which reducing charge per delay is a reasonable solution.

This evaluation indicates that the Folchi method accounted for many environmental parameters that were not recognized by other approaches. This method is the best approach for evaluating mine operations. This method can be potentially used as an environmental regulation tool as needed. There are several advantages to applying this method. The method makes it possible to simplify complex analysis by splitting it in a number of easily quantified components, which can be treated one at a time, being reconstituted in a standardized matrix to give a total magnitude value. This value can then be used to compare mining operations of different types. This is a key requirement for using it as a regulatory tool.

17.2 PEBBLE MINE, ALASKA

The *New York Times* (September 22, 2020) reported that the Pebble project, a major open-pit mine project that would be dug in a remote sparsely populated part of Southwest Alaska, has been fought over a decade and a half by those who say it will provide much-needed economic development and others who argue it will cause environmental harm, especially to salmon that are the basis for subsistence fishing by Alaska natives and the wild salmon fishery in Bristol Bay, one of the world's largest.

The Corp of Engineers issued a final environmental impact statement last month, which found that the project as proposed would not cause long-term changes in the health of the commercial fisheries. The Corp is expected to make a final decision on a permit for the project within weeks.

TABLE 17.4

Final Scorings of Each Environmental Component in Sarcheshmeh Copper Mine (Monjezi, M et. al)

Impacting Factors	Human Health and Safety	Social Relationship	Water Quality	Air Quality	Use of Territory	Flora and Fauna	Above ground	Under-ground	Landscape	Noise	Economy
Alteration of area's potential resources	2.4	2.3	0	0	17.1	1.9	0	0	8.6	0	0
Exposition, visibility of the pit	0	2.3	0	0	8.6	0	0	0	8.6	0	0
Interference with above groundwater	11.2	0	31.3	0	0	17.5	46.7	00	20	0	0
Interference with under-groundwater	2.8	0	31.3	0	0	0	0	46.7	0	0	0
Increase in vehicular traffic	14.4	27.7	0	0	12.9	22.5	0	0	6.4	0	0
Atmospheric release of gas and dust	16	7.7	11.1	100	0	25	33.3	0	7.1	0	0
Fly rock	8	0	0	0	0	6.3	00	0	0	0	0
Noise	5.6	21.6	0	0	0	4.4	00	0	0	56	0
Ground vibration	11.2	10.8	0	0	00	0	0	23.3	0	0	0
Employment of local workforce	0	0	0	0	0	0	0	0	0	0	70
Total	71.6	72.4	73.3	100	38.6	77.6	80	70	50.7	62	70

That would allow construction to proceed after state permits are obtained, which is expected to take three years.

Pebble Mine is a large porphyry copper, gold, and molybdenum mineral deposit in the Bristol Bay region of Southwest Alaska, near Lake Iliamna and Lake Clark. As of August 2020, mine developers are seeking federal permits from the U.S. Army Corps of Engineers, the U.S. Coast Guard, and the Bureau of Safety and Environmental Engineering. State permitting will follow, and developers expect it to take up to three years.

The Pebble prospect is in a remote, wild, and generally uninhabited part of the Bristol Bay watershed. The nearest communities, about 20 miles (32 km) distance, are the villages Nondation, Newhalen, and Iliamnana. The site is 200 miles (320 km) southwest of Anchorage, Alaska.

Pebble is approximately 15 miles (24 km) north of, and upstream of Lake Liliamna. Upper Talarik Creek flows into Lake Iliamna, which flows through the Kvichak River into Bristol Bay. Waters in the Koktuli River drain into the Mulchatna River, a tributary of the Nushagak River, which empties into Bristol Bay at Dillingham. Water from Lake Clark, approximately 20 miles (32 km) east of Pebble, flows down the Newhalen River to Lake Iliamna.

17.2.1 THE ORE BODY

The Pebble Project is a copper gold-gold-molybdenum porphyry deposit in the advanced exploration stage. The project is located on state land in the Bristol Bay region of southwest Alaska, approximately 17 miles northwest of the community of Illamna. Pebble consists of two contiguous deposits. Pebble West is the near-surface resource of approximately 4.1 billion metric tons. Pebble East is significantly deeper than Pebble West and contains an estimated resource at 3.4 billion metric tons.

In 2008, Pebble was estimated to be the second-largest ore deposit of its type in the world. In February 2010, a resource estimate reported that the combined Pebble deposit mineral resources of East and West comprise 5.94 billion tons of ore, grading 0.78% CUEQ, containing 55 billion pounds of copper, 67 million of gold, and 3.3 billion pounds of molybdenum, as well as 4.84 tons of ore, as inferred mineral resources grading 0.53% CuEQ containing 25.6 billion pounds of copper, 40.4 million ounces of gold, and 2.3 billion pounds of molybdenum.

The 2014 resource estimate includes 6.44 billion tons in the measured and indicated categories, containing 57 billion lbs of copper, 70 million oz of gold, 3.4 billion lbs of molybdenum, and 344 million oz of silver, and 4.46 billion tons in the inferred category, containing 24.5 billion lbs. of copper, 37 million oz of gold, 2.2 billion pounds of molybdenum, and 170 million oz silver. Quantities of palladium and rhenium also occur in the deposit.

By dollar value, slightly more than half of the value is from copper, with the remainder split roughly equally between gold and molybdenum. By-products of silver, rhenium, and palladium metals would also be recovered.

17.2.2 FISHERIES

Bristol Bay is home to the world's largest salmon run. All Eastern Pacific species spawn in the bay's freshwater tributaries. Commercial fisheries include the world's largest sockeye salmon fishery. The Kvichhak River has the single largest red salmon run in the world. The Kvichak drains from Lake Iliamna, which is downstream of the deposit. Along with herring and other fisheries, salmon account for nearly 75% of local jobs.

Sport fishing is another important local industry. Many lodges cater to sport fishermen, exploiting the salmon and trout populations in the freshwater tributaries. Freshwater species include whitefish, trout, Arctic grayling, and rainbow trout.

Seasonal subsistence harvesting of salmon and year-round subsistence of fresh water fish is a critical part of life for rural residents, most of whom live downstream of the mine site.

17.2.3 HUMAN POPULATIONS

The Pebble site is within Lake and Peninsula Borough, about 1,600 inhabitants, adjacent to the Bristol Bay Borough, about 1,000 inhabitants, and the Dillingham Area, 4,800 inhabitants. Some 7,500 people live largely rural lifestyles downstream of the pebble site.

The populations in the area rely heavily on wild resources for subsistence living, harvesting moose, caribou, and salmon. Wild resources play an important part in the region's cultural heritage.

There are more than 30 Alaska native tribes in the region who depend on salmon to support their traditional subsistence way of life, in addition to other inhabitants and tourists in the area. If the Pebble mine is developed, the subsistence culture of thousands of people who live in the Bristol Bay region will be threatened.

17.2.4 ECONOMICS

Pebble is the largest-known undeveloped copper ore body in the Northern Dynasty estimated that the Pebble contains over $300.00 billion worth of recoverable metals at 2010 prices.

A report released by Northern Dynasty in 2011 predicted profits for mine owners from a large-scale open-pit mine at Pebble, given the assumption of about $4.7 billion, scale 200,000 tons per day, lifetime 45 years, metal prices over that lifetime (2011), and the mine design plan. The study assumes that a slurry pipeline will deliver ore concentrates from the mine to the new port in the cook inlet and that trucks will haul ore concentrates to Cook inlet.

The plan expected the mine to return the initial capital investment in 32.2 years, employ over a thousand people for the first 25 years, and provide a lifetime 23.2% pretax internal rate of return. The expected pretax cash flow was approximately $2 billion per year. The report states that 58% of the ore resources will remain at year 45.

17.2.5 PERMITS

Pebble Limited Partnership has applied for federal permits to develop a portion of the Pebble deposit as an open-pit mine along with an associated transportation corridor, Amakded port facility, and natural gas pipeline. Pebble Limited Partnership's application has triggered the need for an Environmental Impact Statement (EIS) under the National Policy Act (NEPA). The U.S. Army Corp of Engineers is the lead federal agency for the EIS process. The Alaska Department of Natural Resources is leading the State of Alaska's engagement in the EIS process as a cooperating agency.

Northern Dynasty has applied for a water rights permit to about 35 billion gallons of ground and surface water per year. In April 2017 Northern Dynasty reported that it had received a notice of approval of a Miscellaneous Land Use Permit from the Alaska Department of Natural Resources for ongoing activities at Pebble.

17.2.6 STUDIES

Site-specific baseline data and scientific studies of potential environmental and social effects have been and are being conducted by the project operators and their consultants. These studies address water quality and other concerns. Among these are:

Quantification of acid mine drainage, the chemical stability, and weathering of products of the tailings (waste rock, which would be waste stacked without dewatering) generated by the mine, as well as of the newly exposed and blast-fractured rocks within the proposed mine.

Seismic risks to the impoundment systems (earthen tailings dams) designed to contain the tailings and intended to control their chemical behaviour in perpetuity.

The federal Environmental Protection Agency began conducting a scientific review of Bristol Bay watershed in 2011 focused on the Nushagak and Kvichak river drainages in response to petitions from tribes, commercial fishing organizations, and other organizations that oppose the Pebble project.

The U.S. Army Corps of Engineers released its final environmental impact statement for the Pebble in July 2020, concluding that the project would not lead to any long-term changes in the health of the commercial fisheries in Bristol Bay. The U.S. Fish and Wildlife Services stated that the EIS did not consider the habitat destruction of salmon that could lead to the destruction of salmon stocks in Bristol Bay. The Corp of Engineers did not account for earthquake or volcanic risks, a significant omission due to Alaska's status as the most seismically active state. A 2019 failure of an iron mine in Brazil killed over 250 people under similar conditions.

17.2.7   POSSIBLE MINING PLAN

The mine would be an open-pit mine, a mile square in area and a third of a mile deep. Ponds would be dammed to contain tailings, including some toxic materials. A 165-mile natural gas pipeline would be constructed to provide power, as well as an 80-mile road and pipeline to transport the mined concentrate to Cook inlet. Operating the mine would use and impound large amounts of surface water. The roads would carry fuel, industrial chemicals, and supplies.

Pebble East would most likely be an underground mine.

Pebble would be similar to existing large copper porphyry mines such as the Chuquicamata, Bingham Canyon, and Ok Tedi. The environmental settings and various technical considerations of Pebble distinguish it from these desert and tropical examples. Development and construction would consume years and cost billions of dollars. Required infrastructure includes miles of roads, bridges, and powerlines with pipelines for fuel and rock slurries.

17.3   CONTROVERSY

The controversy over the proposed Pebble mine centres largely on the potential risk to watershed, salmon, and other fisheries. Mining opponents claim that the mine poses a significant and unacceptable risk to downstream fish stocks and could cause an environmental disaster if built. Mining proponents claim that the mine can be developed and operated without significantly harming Bristol Bay area fish.

A steady stream of electoral, legislative, and legal challenges to possible future Pebble mine development are lodged in Alaska. Some of these assert that even the drilling and other scientific investigations conducted to date have caused significant adverse effects to the land and wildlife near the Pebble site.

Pebble has been a major issue in Alaska politics since the mid-2000s; national environmental and sport fishing organizations are involved, while national publications cover the issue.

The *New York Times* reports that, as of 2020, public opinion polls in Alaska indicate more opposition than support for the mine. In 2006 one poll reported 28% of Alaskans in favour of and 53% oppose to Pebble, and another reported 45% of Alaskans in favour of Pebble and 31% in opposition. A poll of Bristol Bay residents reported 20% in favour and 71% opposed.

Organizations including the Resource Development Council, Alaska Mining Association, and the Alaska Chamber of Commerce support the project. The proposal has strong support among statewide elected officials.

Opposition to the proposal was led by organizations including the Renewable Resources Council (formed in 2005 to oppose the Pebble project), local native groups such as Bristol Bay Native Association, commercial and sport fishing organizations, and conservationist groups such as American Rivers and Trout Unlimited. Deceased Senator Ted Stevens, a strong proponent of resource extraction, in 2007 expressed opposition to the Pebble proposal. Multiple UK jewellers pledged not to buy gold from the Pebble mine if it is built, joining several American jewellery retailors and manufacturers who had done the same in 2008.

The Natural Resource Defense Council (NRDC) and other well-known groups and figures (like movie star Robert Redford) have opposed the construction of the mine. In an effort to stop or pause the construction of the mine, the NRDC has created a petition that now has more than one million signatures. The environmental justice nonprofit, Earthworks, has more than 100 gold buyers who have agreed to boycott gold found in the Pebble Bay mine. These retailors include Tiffany & Co., Jostens, and Zales, along with others.

17.3.1 ARGUMENTS AGAINST THE PROPOSAL

Opponents to the mine point out that it is about jobs, current sustainable fishing, world-class fishery with a long history that is perpetually sustainable versus the time-limited mining employment that the mine will generate. All mines have limited lifespans.

The fish in the watershed, and wildlife that depend on them, are too important to risk in exchange for the mine's economic benefits. Bristol Bay is the most valuable Sockeye Salmon fishery in the world, generating $1.5 billion in annual profit.

Accidental discharge of process chemicals and byproducts, heavy metals, and acid mine drainage to the environment is a concern in mine design and operation. Heavy metals are mobilized by acids. Downstream salmons and freshwater fish are vulnerable to mine-generated pollutants. A threat to the fisheries would amount to a threat to the regional subsistence lifestyle.

Hardrock mining already has a notable track record in terms of the permanent and costly legacy of heavy metal-laden acidic leachate that continuously flows from inactive, depleted old mine sites. According to the EPA, mining has contaminated portions of the headwaters of 40% of watersheds in the western continental United States, and the reclamation of 500,000 abandoned mines in 32 states could cost tens of billions of dollars.

A recent study of 25 modern large hard rock metal mines compared water quality outcomes with environmental impact statement (EIS) predictions from the permitting stage. About 76% (19 mines) of the 25 mines violated water quality standards in releases to either surface or groundwater. When the 15 mines with high acid drainage and high contaminant leaching potential and proximity are considered separately, this number is 93% (14 mines). A report commissioned by opponents criticizes for community worker safety, public health, and environmental problems at their mining operations in South Africa, Zimbabwe, Ghana, Mali, Ireland, and Nevada and notes the difference between the previous owner's stated corporate goals and their actual corporate performance. Anglo Americans gave up on the Pebble Project due to environmental concerns. These concerns remain under subsequent owners of the project. Some organizations claimed that earthquake hazards in the area are poorly known.

17.3.2 ARGUMENTS FOR THE PROPOSAL

The mine and supporting activities would provide significant tax revenues to the state. The State of Alaska predicts that direct mining tax revenue will be one of the most important, even without Pebble, sources of non-oil tax revenue (exceeding from fishing).

The mine will create well-paying jobs in an increasingly poverty-stricken region. A 2007 estimate indicated roughly 2,000 jobs for construction, dropping to 1,000 permanent jobs during the 300- to 60-year expected life span of the mine.

The mine would be a domestic resource of raw materials lowering the US reliance on foreign sources. The protection of the environment and fisheries will be ensured by the stringent environmental review and permitting process, including an EIS, that is required before the development is allowed.

Much of the poor environmental track record of mining occurred before current technologies and regulations. Northern Dynasty has no net loss for fisheries.

Opening the Pebble mine could pave the way for another large project, the Donlin gold mine, some 175 miles away across the Tundra. By reducing the cost of building a road to that mine, the Pebble project could help make it economically viable.

President Trump, whose son Donald Jr. is among those opposed to the project, tweeted that "there will be no politics in the decision-making process." The companies said that for several years they had taken the position that there would be potential for subsequent phases for development. However, currently there are no plans for any development beyond the 20 years of mining. However, expanding the mine would not be unusual. The State of Alaska would support the expansion because it would gain revenue from the mine expansion.

Mr. Shiveley, the interim Chief of Northern Dynasty, stated said that our current plan through the regulatory process is that we can prove to the state political leaders, regulatory officials, and all Alaskans that we can meet the very standards expected from us.

The *Washington Post* (dated August 24, 2020) stated that the Trump administration delayed a key permit for the proposed Pebble Mine in Alaska, on Monday, saying the company that wants to build the biggest gold and copper mine in North America needs to take extensive action offset the harm it will cause to the environment. While the Trump administration supports the mining, the Pebble Mine proposal would be too damaging to the Bristol Bay region in southwest Alaska. Therefore, the Corp finds the project, as currently proposed, cannot be permitted under section 404 of the Clean Air Act. Pebble Limited partnership must outline how it will address the damage it will do to nearby wetlands and waterways as it extracts gold, copper, and other minerals from what would be North America's largest mine, Chris Wood, President of Trout unlimited, said it will be difficult for the company to demonstrate how it would make the region's environment whole, given that few fishing camps, the area is pristine. It is hard to mitigate pristine.

Tribes in the region stated: We are thankful the Corps has come to the same conclusion as the rest of the scientific community. It is impossible that Pebble will have such severe impacts that there is no way to mitigate the destruction it will cause. It is impossible for the company to mitigate the devastation the mine will have on the Native cultures, their way of life that has been sustained for thousands of years by the pristine lands, and waters of the Bristol Bay watershed.

BIBLIOGRAPHY

1. Monjezi, M Shahriar, Dehghani, H and Namin Samini, Environmental impact assessment of open pit mining in Iran, *Environ. Geol. 58*:205–216 (2009).
2. Pebble Mine, Wikipedia, the free encyclopedia.
3. Pebble Project, Northern Alaska Environmental Center Archived from the original on 2015-06-22.
4. Pebble project, Alaska Department of Natural Resources.
5. Trump administration says: Alaska's Pebble Mine can't be permitted 'as currently proposed', *The Washington Post*, August 24, 2020.
6. *New York Times*, September 25, 2020.
7. *New York Times National*, Tape casts doubt on Alaska Mine Plan, September 22, 2020.

Index

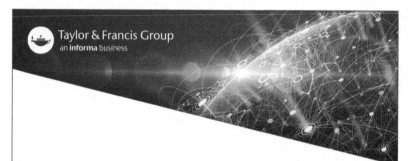

Printed in the United States
by Baker & Taylor Publisher Services